Distributed Multi-Generation Systems: Energy Models and Analyses

Distributed Multi-Generation Systems: Energy Models and Analyses

Pierluigi Mancarella

Gianfranco Chicco

Nova Science Publishers, Inc.
New York

For permission to use material from this book please contact us:
Telephone 631-231-7269; Fax 631-231-8175
Web Site: http://www.novapublishers.com

Library of Congress Cataloging-in-Publication Data

Distributed multi-generation systems: energy models and analyses / Pierluigi Mancarella and Gianfranco Chicco.

p. cm.

ISBN: 978-1-60456-688-8 (hardcover)

1. Energy facilities-Mathematical models. I. Mancarella, Pierluigi. II. Chicco, Gianfranco.

TJ163.2.D573 2008

621.31—dc22 2008015735

Published by Nova Science Publishers, Inc. ✝ New York

To Our Families

PREFACE

This book contains a compendium of various research activities carried out by the authors in the latest years in the field of energy efficiency and territorial sustainability of distributed energy production. These activities have been developed by following the specific conceptual line of characterizing distributed systems for combined energy production in a synthetic fashion and by exploiting simple and intuitive energy-based approaches. In this outlook, the energy system components are described through output-to-input black-box efficiency models. The whole energy system model is built on the basis of the component black-boxes and of suitable characterization of the energy flow interactions within the plant and with external energy networks. We identify the interconnected structures analysed with the label "*Distributed Multi-Generation*" and denote it with the acronym DMG.

The content of this book is relevant to a number of aspects related to *energy efficiency*, with further potential extensions to environmental as well as economic analyses. The main objectives are to provide a general overview on DMG structures, with specific focus on Combined Heat and Power (CHP) and Combined Cooling Heat and Power (CCHP) plants, although other energy carriers can be entailed in the framework developed. The reader is assumed to be familiar with basic thermodynamic concepts taught in undergraduate courses of science and engineering. Some of the key issues addressed are DMG plant structures and equipment, component characteristics and models, assessment framework to compare DMG structures with conventional separate production, and examples of energy assessment of various technologies for planning studies. Various openings to other key issues concerning environmental impact, interactions with renewable sources and external networks, and distributed multi-generation economics, are also included. A relevant outcome of the approach presented is that the structure of the indicators derived is consistent with the representation of the primary energy saving indicators already adopted for regulatory purposes for cogeneration. Indeed, the indicators presented extend the concept of primary energy saving to multi-generation systems, and create a set of further indicators to assess the environmental impact from such systems, covering a missing element in the current regulation. In addition, as also briefly discussed, starting from the energy performance assessment methodology formulated, environmental impact assessment models could be derived as well, leading to a unified framework defining the related indicators with the same formal structure.

Most of the material illustrated in this book has been elaborated starting from research works recently presented by the authors at scientific conferences and taught in doctoral courses and international seminars. The concepts and methods presented are intended to provide a base set of information, from which it is possible to start carrying out more detailed studies in various directions. For instance, these directions include the set up of advanced time-domain simulations and comparisons with actual system operation, economic evaluations based on multi-objective or multi-criteria optimisations, the incorporation of environmental analyses according to life-cycle assessment concepts, and so forth.

The DMG concepts and applications are also addressed in various papers written by the authors, listed in the references and recalled in the text. The material has been rearranged under an original and systematic structure, to reflect the authors' interdisciplinary view concerning the contents presented. This makes this book unique. Several other references to important and recent works appeared in international books and journals are provided to assist the interested reader in gathering the relevant information concerning the state of the art of the various topics addressed in this book. Our intention and hope is that the readers will find in our DMG energy models and analyses useful hints and benefits for their research and relevant activities.

Pierluigi Mancarella and Gianfranco Chicco

September 2008

CONTENTS

NOTATION

ACRONYM LIST

AB	Auxiliary Boiler
AC	Alternating Current
AGP	Additional Generation Plant
AHP	Absorption Heat Pump
ATHR	Absolute Trigeneration Heat Rate
CCGT	Combined Cycle Gas Turbine
CCHP	Combined Cooling Heat and Power
CERG	Compression Electric Refrigerator Group
CFC	Chlorofluorocarbons
CGP	Cooling Generation Plant
CHCP	Combined Heat Cooling and Power
CHG	Combustion Heat Generator
CHP	Combined Heat and Power
CHR	Cooling Heat Rate
CITHR	Cooling-side Incremental Trigeneration Heat Rate
COP	Coefficient Of Performance
DC	Direct Current
DCN	District Cooling Network
DER	Distributed Energy Resources
DFC	Direct-Fired Chiller
DFHP	Direct-Fired Heat Pump
DG	Distributed Generation
DH	District Heating
DMG	Distributed Multi-Generation
DSM	Demand Side Management
DR	Demand Response
DS	Distributed Storage
EDC	Engine-Driven Chiller
EDHP	Engine-Driven Heat Pump
EDS	Electricity Distribution System
EER	Energy Efficiency Ratio
EIHR	Electrical Incremental Heat Rate
EITHR	Electrical-side Incremental Trigeneration Heat Rate
EHP	Electric Heat Pump
EHR	Electrical Heat Rate
ESCO	Energy Service Company
EUF	Energy Utilisation Factor

FC	Fuel Cell
FESR	Fuel Energy Saving Ratio
GAHP	Gas-fired Absorption Heat Pump
GARG	Gas-fired Absorption Refrigerator Group
GAX	Generator Absorber heat eXchange
GDS	Gas Distribution System
GHG	Greenhouse Gas
GT	Gas Turbine
GWP	Global Warming Potential
HC	Hydro-Carbons
HCFC	Hydro-Chlorofluorocarbons
HDS	Hydrogen Distribution System
HHV	Higher Heating Value
HPR	Heat-to-Power Ratio
HR	Heat Rate
HRC	Heat Recovery Condenser
HRSG	Heat Recovery Steam Generator
HVAC	Heating Ventilation and Air Conditioning
ICE	Internal Combustion Engine
ICT	Information and Communication Technologies
IFAC	Indirect-Fired Absorption Chiller
IHR	Incremental Heat Rate
IPLV	Integral Part Load Value
ITHR	Incremental Trigeneration Heat Rate
LCA	Life Cycle Assessment
LHV	Lower Heating Value
MG	Multi-Generation
MPPT	Maximum Power Point Tracking
MT	Microturbine
mu	monetary units
ODP	Ozone Depletion Potential
OTHR	Overall Trigeneration Heat Rate
PCO2ER	Poly-generation CO_2 Emission Reduction
PCU	Power Conditioning Unit
PER	Primary Energy Rate
PES	Primary Energy Saving
PPES	Poly-generation Primary Energy Saving
pu	per unit (referred to a relevant base quantity)
PV	Photovoltaic
PV/T	Photovoltaic/Thermal
QI	Quality Index
RC	Rational Criterion
RES	Renewable Energy Sources
SE	Stirling Engine
SP	Separate Production
TD	Transmission and Distribution
TEUF	Trigeneration Energy Utilization Factor
THR	Thermal Heat Rate
TIHR	Thermal Incremental Heat Rate
TIT	Turbine Inlet Temperature

TITHR	Thermal-side Incremental Trigeneration Heat Rate
TPES	Trigeneration Primary Energy Saving
UHC	Unburned Hydro-Carbons
VPP	Virtual Power Plant
WAHP	Water Absorption Heat Pump
WARG	Water Absorption Refrigerator Group

SYMBOLS

B	hot water thermal energy [kWh_t]
C	cost, in [mu] or [mu/year]
$\mathbf{D}$	set of useful energy demand outputs
E	exergy [kWh]
F	fuel thermal energy [kWh_t]
G	solar irradiance [W/m^2]
H	hydrogen thermal energy [kWh_t]
L	losses [kWh]
Q	heat [kWh_t]
R	cooling [kWh_c]
S	steam thermal energy [kWh_t]
T	temperature [K]
W	electricity [kWh_e]
X	generic energy vector [kWh]
m	mass [g]
$\dot{m}$	mass flow rate [kg/s]
α	dispatch factor
β	capital charge factor
χ	electricity production cost [mu/kWh_e]
ε	effectiveness
η	efficiency
λ	cogeneration ratio
μ	emission factor [g/kWh]
ρ	price [mu] (related to specific quantity units)
ν	generic model parameter
Λ	lambda transform

SUBSCRIPTS AND SUPERSCRIPTS

Subscripts and superscripts are used to represent equipment, energy sources or end use, and to specify the measuring units. The above defined acronyms and uppercase Latin letters can be used as superscripts or subscripts as well. Numbers can also appear as subscripts to represent specific instances of the corresponding variables.

a	artificial
b	hot water
c	cooling

d	demand
e	electricity
f	fuel
h	hydrogen
i	input
o	output
p	poly-generation
rev	reversible
t	thermal
u	useful
vw	value-weighted
x	generic type of energy
y	cogeneration
z	trigeneration
A	ambient
C	Carnot cycle
I	investment
I	first law of thermodynamics
II	second law of thermodynamics
M	operation and maintenance
δ	dispersion
σ	stack

LIST OF FIGURES

LIST OF TABLES

INTRODUCTION

The recent developments in the energy sector are radically changing the scenario of energy production and usage. The diffusion of cost-effective new technologies for local generation is moving the focus of the production of electricity from large centralized power plants to distributed generation systems scattered over the territory. Likewise, the present research scenario emphasizes more and more the role of solutions for combined production of various energy vectors, aimed at improving the energy generation efficiency and thus the sustainability of the overall energy sector. Extending the concept of cogeneration of electricity and heat, multi-generation solutions, in which local cogeneration systems are coupled to various types of chillers and heat pumps for combined production of electricity, heat (at different enthalpy levels), cooling power, and so forth, are gaining more and more interest. The further natural step is the interconnection of the local generation systems to the energy networks, to form a distributed multi-generation system. In such interconnected system, generation of the final demand energy outputs close to the users allows reduction of the losses occurring in the energy conversion and distribution stages, as well as enhancement of the overall generation efficiency. Furthermore, the presence of different types of networks provides as a key benefit the possibility of shifting the consumption from one type of energy to another. This option is significantly promising in the outlook of increasing flexibility in managing the overall energy load, for instance depending on the evolution of the energy prices, and to reduce the energy system vulnerability against possible lack of supply of any individual energy network.

On the above concepts, this book presents a comprehensive view on various small-scale *local* multi-generation solutions interconnected through energy networks. The indicative size of the energy systems addressed is below few megawatts, with most applications below one megawatt. The presentation of the manifold cross-sector aspects and issues is accompanied by a review of the state of the art of the scientific literature in various relevant areas. In this respect, the list of references is mainly selected from the most recent journal papers and high-profile research reports relevant to multi-generation issues. The emergent issues on decentralized energy generation are illustrated under a unified approach that, stepping beyond the already innovative distributed generation approach, introduces the *Distributed Multi-Generation* (DMG) paradigm. Energy planning and performance assessment of energy systems are addressed within the DMG framework. In particular, DMG solutions could be effectively deployed in urban areas, in which interconnected infrastructures are mostly available and there is a significant variety of demand of multiple energy vectors. Extending the classical models based on the analysis of the heat-to-power cogeneration ratio in cogeneration plants, a unified characterization and modelling of the production side, the demand side, their interactions in multi-generation systems, and their connection to energy networks is provided. This characterization resorts to the *transformation* of an array of original energy or cogeneration ratio values into an equivalent set of values. We call this approach *lambda analysis*. Correspondingly, the energy transformations are expressed in mathematical terms by means of appropriately defined *lambda transforms*. The lambda analysis and other mathematical formulations aimed at addressing the effectiveness of exploiting DMG solutions in various scenarios are illustrated and applied to various exemplificative case studies. Typical plant schemes and components are modelled, with special focus on real small-scale applications for trigeneration of electricity, heat and cooling power. A specific example also shows how the lambda analysis concepts can be extended to assess the impact of multi-generation alternatives on energy networks.

Energy performance analyses and evaluations are described by focusing on a system-based point of view. The plant components are typically represented through a *black-box approach*, and various types of plant solutions are in turn reduced to equivalent black-boxes aggregating the component black-boxes. The overall energy system assessment is addressed according to the *energy chain* concept, that is, by tracking the different types of energy needed to supply a given energy load back to the primary thermal energy contained in the relevant fuel. The performance of the energy systems is assessed through a set of indicators of different kind, formulated either in absolute or in incremental terms, to address the energy and environmental benefits brought by exploiting combined local energy production as opposed to separate centralized production with large energy networks.

A unified theoretical framework, based on the energy chain principles and synthesizing different performance assessment techniques, is described in detail. In particular, different indicators are presented for evaluating the potential energy benefits of distributed multi-generation systems with respect to classical case of separate production and centralized energy systems. The same framework can be adopted to deal with CO_2 emission impact assessment of multi-generation systems, resulting in a unified formulation of relevant energy and environmental indicators, as briefly discussed in the final chapter.

The potential increase of economic profitability of DMG solutions within emerging energy-related markets is also discussed and assessed through an explicit economic model formulated taking into account the possibility of trading energy-related commodities such as energy efficiency certificates and emission allowances.

Several case study applications are illustrated to exemplify the models presented and to point out some numerical aspects relevant to equipment available on the market. In particular, schemes with different types of cogeneration prime movers, as well as electric, absorption and engine-driven chillers and heat pumps, are discussed and evaluated. The aspects analysed and the numerical outcomes provided highlight the prominent role of DMG systems towards the development of more sustainable energy scenarios.

Keywords: black-box model, cogeneration, combined cooling heat and power, cooling generation equipment, CO_2 emission reduction, distributed generation, energy chain, energy efficiency, energy networks, energy planning, energy-related markets, energy saving, environmental impact assessment, lambda analysis, multi-generation, performance indicators, poly-generation, power systems, small-scale applications, sustainable energy, trigeneration.

Chapter 1

THE DISTRIBUTED MULTI-GENERATION FRAMEWORK

This chapter discusses the basic concepts leading to formulate the distributed multi-generation framework. The illustration starts from a view on the recent evolutions of the power and energy systems. The reference frameworks, related to distributed energy resources, cogeneration and multi-generation are then recalled, in particular by highlighting the conceptual line followed to extend the classical cogeneration rationale to address more comprehensive multi-generation structures. The distributed multi-generation paradigm emerges from these frameworks to encompass the corresponding ideas and provide further extensions to energy network and environmental impact aspects. Most of the topics introduced in this chapter are extensively detailed and applied in the remainder of this book.

1.1. RECENT ENERGY SYSTEM EVOLUTIONS

The energy sector is undergoing relevant, widespread and fast changes. These changes are a result of several driving forces referred to various technical and economic aspects.

From the technical point of view, in recent years a strong technological evolution has led to the development of efficient solutions for energy production from various types of sources. In particular, major advances have concerned small-scale applications (*i.e.*, with rated power below a few megawatts). Among these solutions, the most interesting ones are aimed at creating local alternatives to the centralised energy production based on fossil fuels, for instance adopting renewable sources, or enhancing the energy efficiency of the overall generation system by exploiting the benefits arising from the combined production of multiple energy vectors. Other relevant technical aspects are linked to the increasing attention towards environmental impact of energy production systems, to cope with the climate change threat and keep the obligations taken up by several countries signing the Kyoto Protocol. The legislation and the technical regulation are more and more inclined to promote sustainable generation solutions capable to reduce greenhouse gas (GHG) emissions and to set up binding local emission constraints for specific types of pollutants affecting air quality. A further key issue refers to the need of enhancing the intrinsic security of vulnerable infrastructures. For this purpose, availability of multiple local generation sources scattered in the territory provides a less vulnerable option, with respect to centralised systems, to guarantee diffuse energy provision and to avoid the effects of extended outages. Possible service outages include shortages in the various energy networks, like interruptions or blackouts in the electrical networks. Furthermore, the development of technologies capable to exploit the energy mix in a more flexible manner paves the way towards the possibility of changing the type of energy supply more easily, reducing the user's dependence on specific types of fuel. All the aspects outlined above correspond to a new and in some cases prominent role of demand side in the energy business. This is due in particular to the key impact of consumers' preferences and to the level of acceptability of the technologies available on the market, which strongly affect the diffusion of these technologies. On the other hand, the connection to energy networks of multiple sources operated without centralized control may heavily affect the uncertainty on the amount of the consumption seen from the energy distributor and supplier side.

From the economic point of view, the new millennium has seen the development of consolidated electricity market structures in most countries worldwide, with the unbundling of electricity generation transmission, distribution, and supply. The evolution towards liberalised markets is being extended beyond the electrical side to encompass the whole energy sector. In addition, the formulation of regulatory incentives to promote energy generation from renewable sources or energy-efficient combined production of multiple energy vectors is creating huge opportunities for different types of stakeholders. The energy system arena is thus becoming densely crowded of a number of subjects such as energy traders, energy service companies, service providers, consultants and various newcomers. These subjects are aimed at scoring consistent profits, taking advantage also of the larger global uncertainty in addressing a variety of energy-related issues under a comprehensive and interdisciplinary view.

Besides market-driven forces, legislative and regulatory actions are required to ensure an acceptable level of quality of products and services. These actions are directed towards setting up suitable prescriptions related to aspects of common interest, such as environmental impact, establishing base options and price or revenue caps within tariff systems, and defining properly selected reference levels driving the technological evolution through specific incentives and penalties.

1.2. Background Frameworks: Distributed Energy Resources

The scientific community is actively participating to the efforts towards energy system evolution by developing widespread approaches and detailed studies to handle manifold technical, environmental, economic, and social issues. Advanced studies are in progress to deal with the various solutions for local production of electricity, taking into account both individual energy vectors and coexisting multi-generation solutions. Part of these studies is carried out at the conceptual stage, with the aim of categorizing the current and incoming solutions and to provide suitable definitions to be used as international references. For instance, from the decentralized electricity production standpoint a commonly used paradigm is the one of *Distributed Energy Resources* (DER) [Ack07]. In more details, the acronym DER encompasses three different aspects, namely:

- *Distributed Generation* (DG) [AcA01][PeD05]: local energy production from various types of electric power sources connected directly to the distribution network or on the customer side of the meter;
- *Demand Response* (DR) [Lev06][VaO07][SeG07]: voluntary customer participation to specific energy saving programmes aimed at improving the effectiveness of the

electricity supply and reducing demand peaks, associated to agreed economic incentives for the consumers;

- *Distributed Storage* (DS) [RiJ01][ClI04]: energy storage from various technologies connected to local plants and supplied with energy produced locally or taken from the network.

The bases of the DER theoretical framework were set up more than two decades ago, with early proposals of the "*dispersed generation and storage*" and "*demand side management*" (DSM) concepts [KiN81][DuT84]. However, notwithstanding huge and positive efforts aimed at formulating specific theories, the expected widespread application of these theories remained very limited. This mainly occurred because of lack of advanced technologies, for instance concerning energy production and conversion equipment, and of associated information and communication systems. The main success on the application side was obtained in the industrial sector, in which typically private investors decided to purchase and operate local power plants, prevailingly dedicated to energy backup and to improve electricity supply availability. Only in few cases there was deployment of local systems for self-production of energy. Indeed, in the vertically integrated industry development of self-production systems was often subject to various obstacles, even at the regulatory level. Lately, the wide development of liberalised electricity markets in several countries has led to a significant diffusion of various types of DG solutions, according to which DG has been defined as "the paradigm of the new millennium" [BoK01]. Yet, this "new" paradigm is already being replaced by the wider concept of DER. This is due to the fact that, in addition to DG, the DR option is now being considered as more and more interesting [ShR05], even though the implementation of the DR concept has not yet reached a mature stage. Similarly, DS solutions are gaining interest, but their application is still relatively limited to classical solutions aimed at improving local continuity of supply in transient of short-duration periods, whereas effective deployment of storage solutions supplied by peaking productions, *e.g.*, from renewable sources, is a promising option.

The most interesting aspect is that the present level of technological evolution, associated to the existence of more open economic frameworks, enables extended diffusion of DER applications. One of the actual key points characterizing the ongoing changes is the *scale* at which these changes are occurring. A number of adapted or innovative technologies have already reached the commercial stage, providing solutions available to *small-scale* consumers (with nominal power below few megawatts). There is a clear trend towards the diffusion of "micro-power" applications (indicatively below 50 kW_e), that could make the user more independent of the connections to the various energy networks, provided that the corresponding technological solutions are cost-effective, easy to use and able to guarantee a sufficient degree of reliability. The key question today is whether or not there will be a massive diffusion of small-scale technologies, with a huge number of different devices bursting into the market. The dimension of such diffusion could heavily impact on the future of energy system applications, in a way similar to what happened in recent years in other technological markets, for instance for mobile phone. In addition to the larger flexibility of usage, small-scale and micro-power applications intrinsically bear further significant value. For instance, the increased availability of local energy sources makes it possible to reduce both energy supply dependency [AsM07] and vulnerability of the electrical system from the effects of grid congestions. Larger exploitation of user-specific solutions is expected to be of major help to keep the lights on in case of service interruptions and blackouts [MaR05][PoK07], and also of lack of supply caused by vandalism or external attacks [AsM07][CoG07]. In the latter cases, DER solutions can play a key role in the formation of self-healing energy areas [HaV03][AmW05] and in driving some phases of the service restoration process [MoR07]. A further key aspect is the role DER can play to provide suitable ancillary services in the electricity markets, or to induce a change in the way the existing ancillary services are procured or deployed [RaR06][ReK07a][ReK07b].

The extended adoption of DER solutions is also changing the views concerning the commissioning and development of large energy plants. On the one hand, small-scale technologies are fast improving their performance, and new solutions can be designed and made available in a relatively limited time. This could be seen as a good reason for *deferring* huge investments in new large generation plants, substations or infrastructures, whose long return time of investment does not fit with the current trend towards setting up rapidly profitable and market-orientated solutions [GiJ06][Ack07]. However, even an explosion of the market of customer-orientated solutions for energy generation could not limit the future development of large-scale infrastructures. These infrastructures, that are often cost-effective because of technological scale reasons, are needed in any case to supply high-power demand and to guarantee service supply with adequate parameters of power quality, reliability and stability. In particular, adequate centralized infrastructure is necessary to cope with the presence of increasing intermittent sources at the various network levels, whose additional cost to the system is a major research topic (see for instance [GrH07]).

The connection of local resources to the electrical grid may also provide different types of benefits to the electrical network [ZoS07]. Some benefits can be found on reliability ground. For instance, it is possible to coordinate the actions performed during the service restoration process after an interruption to involve local resources with suitable characteristics, with the aim of reducing the total energy not supplied. Other benefits may come from the reduction of network branch loading and losses, provided that the diffusion of local resources is not excessive [JeA00]. Further advantages can result in power quality aspects, because of the power conditioning options [ChB05][CaF06] available in electronic converter-based grid interface systems [MaV04][CaF06][PrD07]. In particular, some control systems contained in the power electronics packages tend to limit low-order harmonics (shifting them to high-order harmonics of relatively low amplitude), leading to reduce the waveform distortion at the grid interface for low harmonic orders [ChS09]. Thus, power quality at the connection point with the upstream grid could improve. More generally, the evolution of DER technologies is raising the need for revisiting the standards dedicated to interconnection to electrical distribution networks, in particular concerning the definition of updated requirements for protection devices [DuK06].

Specific projects and relevant energy system models developed worldwide are facing the challenges of improving the operation and management of network-interconnected DER. The *Virtual Power Plants* [AwP97] model is aimed at enhancing the DER visibility at upper grid levels, by developing suitable interfaces among the local components, adopting distributed control strategies through exploitation of Information and Communication Technologies (ICT), and addressing the optimal usage of the available capacity by taking into account the interactions with the energy markets [PuR07]. The *Micro-Grids* [LaA02] model has the objective of studying the characteristics of small local distribution systems containing controllable generation and loads, which could be operated either separated or connected to the main distribution system. In the latter case, the grid interface applies specific power electronics and control system technologies. Micro-grids are intended to be monitored and operated from a control centre, where suitable evaluations are carried out for coordinating and optimising the available DG, DR and DS systems. The present efforts are directed to explore the future of *Intelligent Power Grids* [CoP07b], with approaches like *GridWise*™ (2004) [GrW04] and *Smart Grids* (2005) [EU06][SmGww], characterized by an open view on the architecture of the system infrastructures and addressing in particular the role of security.

From a market standpoint, DER diffusion is increasing the customers' willingness to actively participate to market economics, deciding when and whether to purchase energy from the networks, or to self-produce it from local sources, on the basis of the energy prices. This leads to exploit the concept of price-demand elasticity in electricity markets [Kir03][Ack07][NaS07]. In addition, a multiplicity of local generation sources could orientate the customer preferences towards changing the present balance of usage of the various types of energy, already significantly different from country to country. In perspective, consumers could express a more pronounced interest towards solutions with higher local availability of the corresponding "fuel". This would impact on the acquisition of energy resources and on the development of fuel supply infrastructures [ShF05].

1.3. Background Frameworks: Cogeneration

Focusing our attention on systems for *combined* production of different energy vectors, the major developments are occurring on the wave of the commercial success of cogeneration or Combined Heat and Power (CHP) plants designed for applications in residential districts or other small users' consortia. Cogeneration solutions can be profitably deployed when there is simultaneous demand of the relevant energy vectors in a broad time period throughout the year. In past applications, it was possible to satisfy this condition in industrial and district heating (DH) installations. Today's availability of DG technologies with quite good efficiency has enabled cogeneration to be adopted also on a small-scale and even micro-scale [Peh06b] basis. In particular, suitable applications range from residential houses to schools, restaurants, hotels, and so forth [EDU01][OnU06][DeL07][TaC07].

CHP solutions are typically based on Internal Combustion Engines (ICEs) [CTwww] and more recently on Microturbines (MTs) [CTwww], while Stirling Engines (SEs) [BoK01] targeted for households (with typical capacity range 1-10 kW_e) are playing an increasingly important role, above all in some countries such as Germany or the UK [Peh06b], Furthermore, Fuel Cells (FCs) could represent promising alternatives for the next future, owing above all to very high electrical efficiency, that could also be used in conjunction with some turbine-based technologies for setting up efficient hybrid cycles [ZhT02]. All these applications can be profitably set up today on a scale much smaller than the one of more classical district heating or industrial applications [OnU06]. The main source of interest of adopting CHP plants is due to the high potential capability of these plants of providing significant *energy saving* with respect to the separate production [Hor97][Mar98] of the same cogenerated energy vectors. This results in a higher convenience of adopting small-scale cogeneration technologies with respect to an electricity-only thermal DG system, provided that there is sufficient simultaneous demand of electricity and heat. In addition, the enhanced energy saving may reflect into a high effectiveness of CHP systems to obtain *global* CO_2 (and more generally greenhouse gas) emission reduction [Meu02][ChM08][ChM08b][MaC08] according to the objectives of the Kyoto Protocol.

More generally, the benefits from combined production of different energy vectors on a wide basis (for instance, for a country or region) have to be assessed by taking into account the characteristics of the power system in the specific country or region [Meu02], with particular reference to the nature and share of the existing generation mix. In some countries, the characteristics of the energy production from centralized plants such as hydro or nuclear leave little or no margin of convenience to the development of local cogeneration systems for environmental benefits. For these reasons, the development of cogeneration solutions is mainly driven by the rules set up in national regulations. Indeed, cogeneration has been mainly pushed forward from a regulatory point of view in those countries where the power generation mix is mostly based on thermal plants [CaP05][AEEG][EU04b], in which CHP production can also bring consistent CO_2 emission reduction [ChM08][MaC08].

Besides classical cogeneration system economics issues [BiS06][OhL07], different incentives for promoting cogeneration can be set up depending on its intrinsically enhanced energy efficiency and CO_2 emission characteristics. Some possibilities could be to introduce fiscal deductions on capital expenses or on the energy consumption/generation. In addition, profitability of adopting cogeneration systems can be enhanced by means of the possible access to new *energy-related*, such as the ones based on white certificates, green certificates, or emission allowances [DaL03][BeH06][Bod06][ErV06][DeL07][MaS07] [RoL07][TsH07].

In addition to the global warming issue, the *local* emission problem [AlL02][Gul06] [CaC08][MaC09], which is often not addressed, has to be carefully studied. Indeed, specific technologies could increase the amount of local emissions of certain pollutants such as NO_x, SO_x, as well as partial combustion by-products such as CO, Unburned Hydro-Carbons (UHC), and Particulate Matter of different size [CaC08][MaC09], above all in densely populated *urban areas* with plenty of potential receptors [ExE05]. In urban areas, the contribution of local generation sources to the local pollution added up to the base level of pollution due to other agents existing in the environment (in particular, road traffic and results of human or industrial activities) could lead to exceeding the specific regulatory limits concerning the local emissions. In this respect, the presence of stringent policies regarding emission constraints on specific pollutants could be a limiting factor to the diffusion of cogeneration plants, especially if the benefits deriving from the combined energy production are not adequatey taken into account.

1.4. From Cogeneration to Multi-Generation

Combined production of electricity and other energy vectors can be addressed by following the classical lines drawn by applications of the cogeneration concept. Nowadays, this line has evolved into a more comprehensive perspective of adopting multi-generation solutions [GeK07][ChM09], encompassing an increased number of energy vectors simultaneously produced [Man06][Gei07].

For instance, the presence of a threefold energy demand of electricity heat and cooling leads to the possibility of setting up *trigeneration* plants [BoK01][EDU01][HeB02][HeS03] [RDC03][LiF06][WuW06][CTwww], also identified with the terms Combined Heat Cooling and Power (CHCP) or Combined Cooling Heat and Power (CCHP). Trigeneration represents an extension of cogeneration, and consists of the production *in situ* of the threefold energy vector requested by the user. In many cases, energy is produced from a unique source of fuel (*e.g.*, natural gas). From this outlook, one of the most practical incentives to extend cogeneration to multi-generation applications has come from the evolution of various technological solutions for *cooling power* production around the world, whose performance is fast improving. The development of these technologies also responds to the growth of comfort needs, recently felt as relevant and urgent by a significant part of consumers in industrialised countries. Indeed, in the last few years the uprising demand of seasonal air conditioning has drastically boosted the cooling equipment market [CTwww]. In warm climates and seasons the need for cooling is higher during the hours of electrical peak load. This may cause electrical network congestions and may lead to cascaded failure events in the power systems, as it occurred in specific situations worldwide [USC03]. Furthermore, peak load hours typically correspond to high electricity rates for tariff-based consumers, and in these hours high electricity prices are very likely to occur also in the electricity market. The adoption of multi-generation (and in particular trigeneration) solutions, in the cases in which it is possible to identify simultaneous usage of different energy vectors, can improve the energy efficiency of the overall system and reduce the technical and economic burden (also from the network side) required to satisfy the needs of more and more demanding users. In addition, availability of technological solutions supplied by different types of energy or fuels can limit the exposure of the multi-generation system to fuel price volatility for specific types of fuel and to possible inefficiencies of operation of some energy markets.

Simultaneity of production is one of the key aspects to be considered when planning and analysing a multi-generation system. There may be situations in which the demand of the various energy vectors is not simultaneous. Hence, the discriminating factors in multi-generation system assessment are the relative duration of time in which there is actual simultaneity of different types of demand, referred to a specified time horizon, and the corresponding amounts of such demand. In this respect, in classical trigeneration plants a CHP prime mover is coupled to an absorption chiller fired by cogenerated heat in order to produce cooling power [MaP00][MaT02][MiL03]. This makes it possible to produce thermal power also in the summertime to obtain from the absorption chiller the amount of cooling needed to satisfy the cooling load. This solution is particularly helpful to circumvent the lack of adequate thermal request throughout the whole year, which represents one of the biggest shortcomings that often make cogeneration unprofitable. In fact, by *transforming* the cooling demand into thermal demand, the prime mover can run for longer time at averagely higher load. Regarding the energy amount, adequate cooling demand in the summertime and thermal demand in the wintertime (*seasonal* trigeneration) are needed to make trigeneration solutions economically feasible.

In a more general situation, cooling generation is needed not only for air conditioning, or in any case not only for the summertime span, but there is a simultaneous energy demand for electricity, heat and cooling throughout the year. In this respect, potential trigeneration applications on a small-scale basis include hospitals, hotels, food industry, schools, department stores, commercial buildings, offices, residential districts, and so forth [MaT02][BaK02][WuW06][ZiP06][LiW07][CTwww]. Other applications at larger scale, such as airports [CaP06b][CaS06], have been reported in the recent literature. In such cases, the analysis of different trigeneration alternatives becomes even more complex, considering the variety of possible components and schemes. On the other hand, availability of various alternatives, in which different technologies may be fed by different fuels to produce different energy vectors, can be an effective option to exploit at both the planning and the operational stages. In particular, the further degrees of freedom could be challenging to address, for instance in terms of optimising the system design and the control strategies [ChM09b], but could lead to several advantages concerning better energy and environmental performance, eventually leading to shorter investment pay-back times [ChM06]. In general, specific objective functions can refer to energy saving, emission reduction, profit maximization, or multi-objectives. The different objectives may often be in conflict with each other. However, the higher flexibility of operation allowed by availability of multiple trigeneration equipment may lead to synthesize plant design and operation strategies that converge towards pseudo-optimal solutions or compromise solutions of different objective functions [CaP06].

As far as available technologies are concerned, heat and/or cooling generation equipment can typically be heat-fired or electricity-fired. Various schemes are then possible to satisfy combined needs of electricity, heat and cooling. The most widespread solution is the coupling of a cogeneration technology to a *thermally-fed* system for cooling production, such as absorption/adsorption chillers, which indeed represents the *classical* trigeneration scheme discussed above [WuW06]. In addition, a number of alternatives, among which high-efficiency electric heat pumps (EHPs) [Dan06], are now emerging to fit similar needs and enhance the overall energy system performance, while the potential economic benefits have to be evaluated more thoroughly.

A further aspect to take into account is that in several cases the cooling generation equipment is *reversible* and can actually operate under either cooling mode or heating mode (and sometimes simultaneously under both modes). In this case, the plant schemes could be simplified by avoiding the purchase of some pieces of equipment, and plant profitability could be enhanced by better exploiting the CHP production [ChM06][CaP06][ArF07]. In addition, the cooling solutions may have a *heat recovery* option, so leading to potentially set up a cooling-and-heat cogeneration system to be coupled to a "classical" electricity-and-heat cogeneration system.

The complexity of the possible multi-generation options and schemes has led the authors to identify a suitable framework and generalized models for trigeneration planning [Man06][ChM06][ChM06b][ChM07a], which can be set up by looking at the plant as a *black box*, with an array of inputs and a manifold output. The trigeneration planning problem can thus be extended to encompass the analysis of different equipment for electrical, thermal and cooling generation, besides the classical schemes with absorption chillers fed by cogenerated heat [RDC03][BrV05][LaN06b][WuW06]. Each component of the trigeneration system can be characterized by means of its black-box characteristics [Man06], that is, by its energy input-output interactions with the other plant components, typically described through conventional energy efficiencies [Hor97][Dan06], and without going into further details with the equipment internal description. In turn, combination of various pieces of equipment into aggregated block-components can be described by composite black boxes, up to model the whole multi-generation plant itself as a black box, with the aim of limiting the number of variables into play [ChM09b]. In this way, given the required energy outputs, it is possible to identify the solutions with limited energy inputs and thus more efficient use of energy [Sch07]. The models can be further improved in order to take into account the interaction with and the impact on external energy networks [ChM08c][ChM09b].

Following the line of cogeneration and trigeneration applications, a number of combined generation systems can be envisioned and applied to cope with *diversified* energy needs. For instance, considering the combination of CHP systems with different technologies for thermal/cooling generation, the enthalpy levels at which thermal/cooling power can be produced can be manifold. More specifically, it could be possible to use heat generation equipment such as electric heat pumps to produce hot air for space heating, while hot water for domestic use and steam for feeding an absorption chiller could be produced through a cogenerator, *e.g.*, an ICE. Likewise, one of the potentially most useful applications in the future refers to cogeneration of electricity and hydrogen, in the light of using hydrogen as a storage energy vector. This possibility is of specific interest to better exploit uncertain electricity production from renewable sources such as wind or sun.

More extensively, the number of simultaneously needed products can be higher, including for instance electricity, heat at different enthalpy (temperature/pressure) levels, cooling, hydrogen, dehumidification, desalination, or other chemical products used in specific processes [GaJ04][LiG04][UcS04][HeZ07][WaL07][GaW08]. Hence, from the point of view of considering the physical carriers that are generated to carry different forms of energies, it is possible to identify cases of tri-generation, quad-generation, and so forth, depending on the number of outputs. In these cases, the conceptual line followed leads to introduce the rationale of *multi-generation* (or *poly-generation*) [LiN06][WeM06][ChM08b][ChM09].

This approach calls for formally devising planning and assessment models able to properly account for the manifold energy vectors that can be potentially generated within a combined framework. Indeed, extending the above concepts relevant to trigeneration, the energy system designer and planner can benefit from the variety of technologies available for multi-generation in order to endeavour to develop optimal solutions for the specific cases under study. Exploration of various alternatives can provide enhanced outcomes referred to different objectives such as energy efficiency, emission reduction, economic profitability, and reliability and quality of the service provided to the user [BuT03][CaP06][LiN06] [ArF07][ChM07a]. Suitable planning methodologies and evaluation metrics dealing with specific objectives are addressed in Chapters 4÷8.

1.5. The Distributed Multi-Generation (DMG) Paradigm

The concepts discussed in the previous sections allow the introduction of a specific categorization of the applications developed in the decentralized energy production area. In particular, the main aspects include the number of fuels used to supply the thermal side, the number of energy vectors required at the system output, and the number of sites considered, where a *site* corresponds to a unique interface with the energy networks, and may contain various generation units and loads.

Taking into account the structure and the operational aspects of the energy systems discussed, such a categorization can be expressed in increasing order of complexity as [ChM09]:

a) *classical cogeneration*, in which the energy production refers to a single site, and the generating unit(s) is supplied by a single input fuel, and provides two types of energy at the output;
b) *multi-generation* (MG), again referred to a single site, with generating unit(s) supplied with single/multiple input fuels to provide manifold energy vectors at the output;
c) *distributed multi-* (DMG), referred to multiple sites, each of which generally contains an MG system, linked together from the system analysis viewpoint through technical, environmental or economic relations.

In particular, in single sites the local plants perform the conversion from various types of energy with the objective of covering the local demand and, if allowed, to exchange energy with the external energy networks at the interconnection point(s). However, all aspects of the interaction with the external energy networks can be studied by using quantities known or assessed at a local level. In multiple sites, the location of the individual sites connected through various energy infrastructures [ShF05][SöP06], as well as the characteristics of the energy infrastructures themselves, become relevant information to be used for carrying out specific analyses. In today's systems, the main infrastructures are the Electricity Distribution System (EDS), the Gas Distribution System (GDS), District Heating networks and District Cooling Networks (DCN). On the fuel side, besides classical solutions, thermal equipment can be supplied by using alternative fuels such as bio-masses [DeD07][FiC07] or bio-fuels [KoK07]. More in general, hybrid solutions may be set up by coupling the local generation equipment to solar or wind energy systems [KhH05][CrT06][BaJ07]. Further studies aim at assessing the possibility and effectiveness to adopt a Hydrogen Distribution System (HDS) as an energy carrier, to be used either at the input side to supply the plant, or at the output side to deliver locally generated hydrogen to external or storage systems [Pen06][McE06].

Part of the bulk of research activities in progress worldwide are framed into specific projects [ChM09], such as the *Integrated Energy Systems* [ZaP06] started in 2001 [DOEww], focused on the integration of distributed generation equipment with thermally-activated technologies [DOE04], with particular reference to laboratory applications and to set up test protocols and perspective standards for the development of integrated energy solutions. Another relevant project is the long-term based "*Vision of Future Energy Networks*", which has led to develop the *Energy Hub* model [GeK07][GeA07] and aims at studying the evolution and perspectives of the energy system structures in scenarios extended over 30 years in time. More extensively, these evolutions encompass the development of *Multi-Source Multi-Product* energy systems [HeZ07].

The interactions among the local units due to the presence of distributed generation scattered in the distribution system are a traditional issue of electrical systems. Indeed, the crucial aspect is the *link* or *coupling* among the local generations connected to different sites in the energy networks. In practice, in many cases, even in the presence of scattered MG plants connected through energy networks, there is no true coupling variable to be used for the assessment of DMG system characteristics and performance, and the calculations on the energy system can be carried out by taking into account the plants at each single site independently of the characteristics of the other sites, resulting in single-point models of the multi-generation systems.

In relatively few cases the presence of coupling information among the sites makes it impossible to carry out a separate analysis of the various sites. Some examples are:

- The need for taking into account the network-side effects of the local inputs to (or outputs from) the various sites, such as network losses or costs associated to energy wheeling rates [MoM07].
- Composite environmental constraints defined on a global basis, for instance referred to the implementation of emission trading schemes.
- Global limits on resources that could be dedicated to the individual plants connected to the energy networks in a mutually exclusive way, such as budget limits, limits to the overall incentives, or limits in the provision of fuel in congested or critical situations.
- Coordinated control: in the present status of the energy production, many local resources are scattered within the system, but there is little or no coordinated control of these resources. In perspective, DMG systems could be subject to coordinated control and possible combined optimisation of the local plant operation according to specified objective functions, for instance formulated in economic terms and including network- and environment-based constraints. This aspect is for instance addressed in the studies aimed at establishing the potential of forming and managing micro-grids [LaA02][HeG05][AbA06][KaI06][HaA07] with interconnected DER units, adding to them the multi-generation dimension.
- The management of market aspects related to energy production from a consortium-based perspective, in which the constraint can be the overall economic transaction between the consortium and the supplier. The degrees of freedom for diversification of the energy production are provided by the local loads and available network interfaces, also taking into account possible coordinated control options.

The combination of DER *and* MG opens new perspectives to decentralized electricity generation, depending on availability of local thermal power plants allowing multi-generation applications. In particular, considering the multi-generation dimension can lead to substantial changes in the global vision of the energy sector, calling for formulating a comprehensive framework of analysis. Thus, the DMG approach qualifies itself as the most innovative view in the comprehensive "*Triple E*" scenario encompassing *Energy, Environment and Economics*. In this scenario, the flexibility of design and applications leaves considerable room for several effective developments, also including uptaking technologies [KoW04].

The strength of the new developments is becoming clearer to all the subjects and entities operating in the energy sector. On the scientific side, topics such as energy resources, power systems, environmental impact, economic and financial analysis are becoming more and more interconnected, calling for an interdisciplinary view for bridging the gaps among the various disciplines. The relevant publications produced since the beginning of the new millennium are accordingly moving towards comprehensive energy modelling [JeI06], introduction of large-scale approaches and methods to decentralized energy planning [HiS07], deployment of various technologies for combined cooling, heat and power production [WuW06], development of structured frameworks for addressing the DER diffusion within electrical systems [CaF06][AlP07][MaF07][PeH07], and so forth. All these aspects find a global synthesis within the DMG approach illustrated in this book.

The relevant research results are expected to enlighten the way to understand the complex relations among the various fields involved in the present development of effective solutions of the energy sector, and to provide useful hints for the application of these solutions. In terms of technologies, the implementation of the various solutions depends on a host of subjects. The overall contours have to be fixed by appropriate legislation promulgated by the governments and rules set up by regulatory bodies, Authorities and local communities. Manufacturers are expected to propose advanced technological solutions with more and more efficient and economic equipment. Energy distributors and suppliers are assigned the challenging task of analysing and solving specific problems arising in the new structure of the energy delivery system. Overall, consumers' choices act to a large extent as key drivers for the development of small-scale applications. The possible success of actual developments in the DMG direction will highly depend on the effectiveness of creating and exploiting synergic actions among the various entities participating in the energy system evolution.

Chapter 2

DISTRIBUTED MULTI-GENERATION SYSTEMS: STRUCTURES AND SCHEMES

This chapter contains an overall illustration of the structure of multi-generation plants interconnected through various energy networks according to the distributed multi-generation framework. The plants are classified by using a synthetic functional block structure, with specific distinction between two main blocks, namely, a more classical combined heat and power block, and an additional generation plant block with different types of connections with the combined heat and power block, the local user, and the external energy networks. Within the overall scheme, the presence of various types of plant components and the definition of specific control strategies may determine significant differences in the energy flows of the manifold energy vectors involved in the plant operation. In addition, some openings to the possible exploitation of distributed energy storage, supply from renewable energy sources, and hybrid systems are briefly summarized.

2.1. MULTI-GENERATION PLANT STRUCTURE

2.1.1. Overall block structure

According to the definitions introduced in Chapter 1, an MG plant encompasses a number of components installed in a single site connected to the external networks. The variety of possible components calls for a conceptual interpretation of their role within the plant and with external interconnections. In this respect, an MG plant can be considered as composed of a number of local subsystems producing the various energy vectors such as electricity, heat, cooling, and so forth. The basic layout of a multi-generation plant is shown in Figure 2.1 (modified from [ChM09]). The various energy flows indicated in the figure depend on the plant components included in the block structures and are not necessarily all simultaneously present. Furthermore, other possible products or effects not explicitly indicated in Figure 2.1 among the MG outputs include chemical compounds, as well as desiccant or dehumidification effects.

The general MG plant structure is characterized by two main physical blocks, namely, the CHP block and the Additional Generation Plant (AGP) block.

The *CHP block* contains all the plant components forming the cogeneration side. The basic component is a cogeneration prime mover. Furthermore, thermal back-up and/or peak shaving are ensured by a Combustion Heat Generator (CHG) (or Auxiliary Boiler, AB). The CHP block mainly produces electricity W and heat Q to various possible final uses, among which there are the user's loads and energy supply to the AGP, as well as possible outputs towards the EDS, the DH network, the HDS, the hydrogen user, and so on. Besides the fuel supply, additional inputs to the CHP block may come from external connections to Renewable Energy Sources (RES), EDS, DH network, HDS, and from specific outputs of the AGP block. Possible direct links from RES to the local load are managed as pass-through connections within the CHP block.

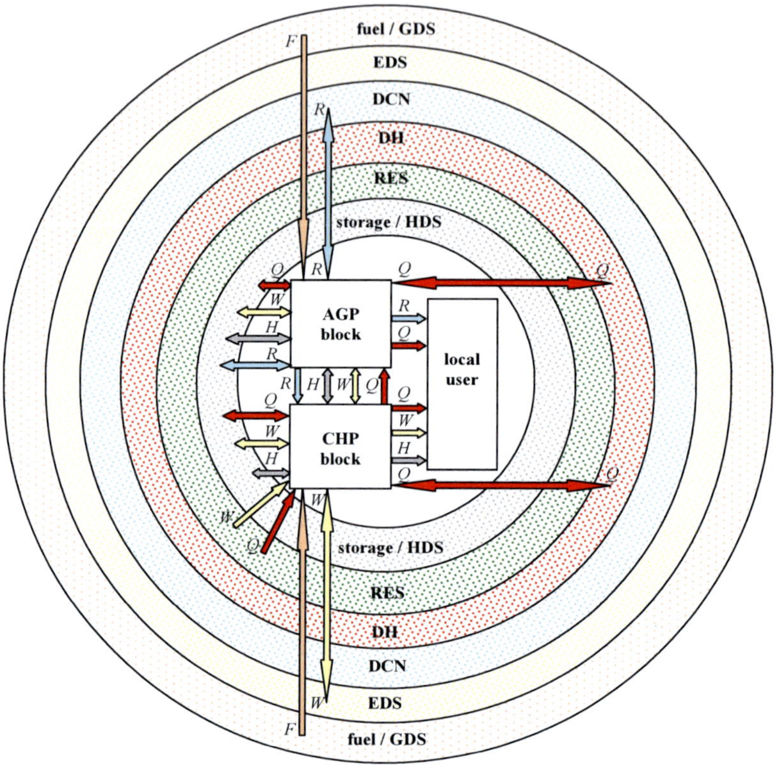

Figure 2.1. Layout of a multi-generation plant with distinction between the CHP and AGP blocks and interconnections to external systems.

The *AGP block* can contain different types of equipment for possible production of cooling, heat, hydrogen, and other products. From a general viewpoint, and focusing on CCHP systems, it is possible to identify two different *linking modes* with respect to the CHP side [Man06][ChM07a][ChM09]:

- *separate* (or *parallel*) linking mode (Figure 2.2): the AGP is only supplied by energy vectors not produced by the CHP plant; these energy vectors come from the fuel supply (natural gas is a typical energy input) and from other external networks or storage systems; as a matter of fact, the AGP is totally "decoupled" from the cogeneration side;
- *bottoming* (or *series*) linking mode (Figure 2.3): the AGP is totally or partially supplied by the CHP side outputs; in specific cases, it is possible to distinguish between an "*electrical bottoming*" case, in which the AGP is fed by electricity coming from the CHP side, and a "*thermal bottoming*" case, in which the AGP is fed by heat coming from the CHP side.

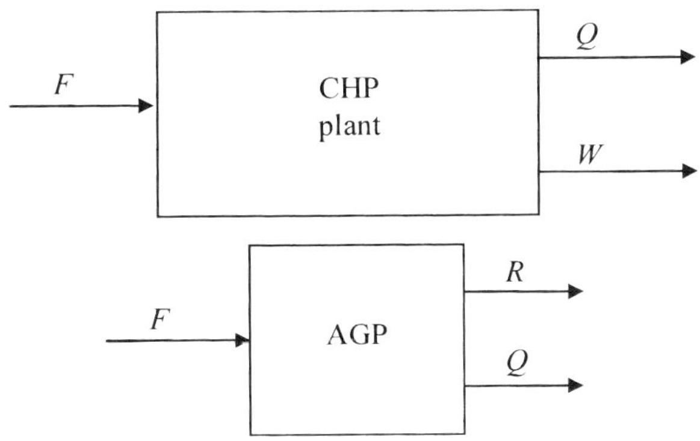

Figure 2.2. Multi-generation configuration with separate AGP.

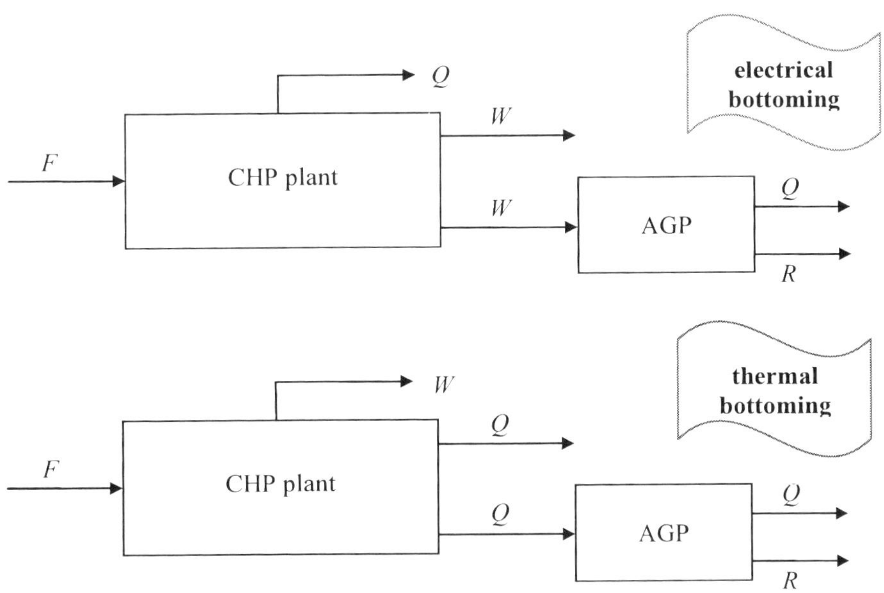

Figure 2.3. Multi-generation configurations with bottoming AGP.

Some equipment in the AGP block can be reversible, that is, able to operate under *cooling mode* or *heating mode*, typically producing heat for air conditioning at enthalpy levels different from the one produced in the CHP side [Afo06]. As a matter of nomenclature, within the specific trigeneration context, the AGP containing cooling production equipment can also be referred in the sequel as Cooling Generation Plant (CGP).

2.1.2. Energy vector description

With reference to Figure 2.1, the energy vectors are represented by the entries *F*, *W*, *Q*, *R* and *H*. More specifically:

- The thermal energy contained in the fuel *F* is typically based on the fuel Lower Heating Value *LHV* [EDU01][EPAww]. The type of fuel to be used depends on the fuel available on-site, or on the fuel distribution network options available at the DMG plant inlet. Concerning energy networks, the GDS is widely spread in urban areas, while other options such as the HDS are under analysis in different countries [ArS00] and could be envisioned for future applications [AkY06][McE06][Pen06]. For small-scale cogeneration applications based on MTs or ICEs, the CHP block can be directly fuelled by using natural gas, diesel, dual-fuel [BoK01][EDU01], and bio-masses [ChM05] in case co-fired with natural gas [DeD07][FiC07]. Natural gas fuel can also be used for further hydrogen production in order to increase the hydrogen supply of a fuel cell [McE06][Pen06][HeZ07]. In the separate linking mode, the AGP could be fed with different fuels. The most typical one is again natural gas. Generally, it is possible to use different types of fuel to supply different equipment (*e.g.*, fuel oil for the heat generators, natural gas for separate cooling generation in the AGP, and hydrogen for FCs). In case of multiple fuels, the different types of input thermal energy should be assessed on the basis of the *LHV* of the corresponding fuel.

- Electrical energy *W* can be purchased from the EDS, produced by the cogeneration plant, or generated by RES available on-site, such as small hydro plants, PV systems, or wind systems. Electricity can be used to supply the local user, as well as to feed the AGP in the *electrical bottoming* case, with the aim of producing thermal/cooling energy, or hydrogen from electrolysis. Possible surplus of electricity production can be sold out to the EDS. Profitability of this operation generally depends on the prices for energy consumption and local generation at different hours of the day.
- Thermal energy *Q* can be produced at different enthalpy levels, typically in the form of hot water and/or steam. Usually thermal energy is produced within the CHP block to supply the local user. Thermal energy is used to feed the AGP in the *thermal bottoming* case, with the aim of producing additional heat for the user at different enthalpy levels (*e.g.*, hot air by means of heat pumps), or to generate cooling power through absorption chillers. Thermal power could also be exchanged with DH networks, or might be recovered from a chiller condenser to supply part of the thermal load. An additional amount of heat can be produced from RES, in particular from solar thermal systems.
- Cooling energy *R* is produced within the AGP to supply the local user. The typical production is in form of chilled water. Part or all of the cooling power can be fed back into the CHP plant, for instance to pre-cool the intake air of a turbine or MT in order to improve the electricity generation capacity and efficiency of the overall system [HeS03][AmH04][Hwa04]. Furthermore, cooling energy exchanges with a DCN are possible.
- Hydrogen energy *H* could be produced within the CHP block from methane steam reforming or from electrolysis typically starting from renewable sources [MdE06][Pen06][HeZ07]. Furthermore, it could be stored within hydrogen storage units to be used at a later time [KoH06][GrK07], or exploited by the local user, for instance, for transportation or industrial applications [Dun02][Win05][Win05b]. The amount of hydrogen energy generated can be used to supply a FC in the CHP plant. In the presence of an HDS, hydrogen energy can be exchanged with the HDS. In particular, excess of local production not used in the local hydrogen storage system could be injected into the HDS for further distribution (for instance, for future automotive applications [GoL08][CaM09]).

2.2. THE CHP BLOCK

2.2.1. Equipment and characteristics

The basic component of the CHP block is the cogeneration prime mover, whose capacity is in case dwindled over more units. The prime mover is sided by one or more auxiliary boilers, for alternative or backup supply. Electricity can be usually exchanged through the connection to the EDS, with the exception of cases in which the MG plant is used in stand-alone operation [ZoG06]. The presence of the EDS guarantees higher reliability and flexibility to the plant owing to the possibility of selling to the EDS the electricity produced in excess of the local demand (for instance, because adopting a heat-led strategy, or as a result of analyses showing the convenience of increasing the production to score economic profits). The option of exploiting a DH network as a source to cover lacking thermal production or as a sink to inject exceeding production opens further operational options for profitably managing the plant.

Figure 2.4 presents a detailed view of the CHP block components. The energy vectors are indicated in a dedicated way, by adopting the subscript y to denote cogeneration. The prime mover is the core of the plant. Its only input is the cogeneration fuel thermal power F_y, which is transformed into cogenerated electricity W_y and cogenerated heat Q_y. In turn, the cogenerated heat Q_y can be supplied at different thermal levels, depending on the prime mover and on the recovery heat generation equipment used.

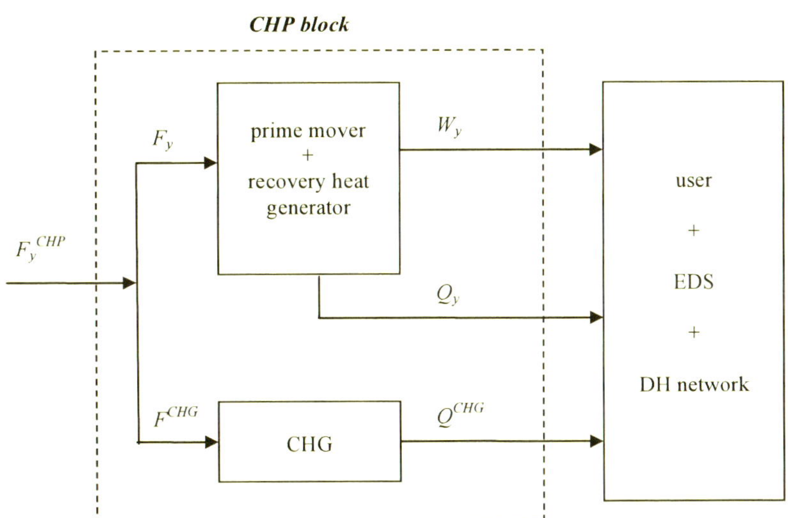

Figure 2.4. CHP plant scheme.

Some indications on possible technological solutions are provided below. Further details on the specific technologies are illustrated in Chapter 3.

On a small-scale basis, the cogeneration technology nowadays mostly adopted is the ICE [WiS00][BoK01][EDU01], that can be considered as a consolidated benchmark. However, the MT, the breakthrough technology of the last decade, is rapidly increasing its market shares [BoK01][EDU01]. More specifically, the growing success of MTs is due to their excellent modularity and flexibility characteristics (above all if used in clusters [DaG99]), that make them suitable for a wide range of applications and loads, as well as fuels [BoK01].

In addition, in terms of NO_x and CO pollutant emissions, the full-load performance of natural gas-fuelled MTs is in general very good with respect to natural gas-fuelled ICEs [EDU01][EPAww], so that MTs can be more and more encountered where there are specific binding environmental constraints (for instance, urban areas [AkO05][AkO06]). However, several studies show that the off-design emission characteristics of a number of MTs could significantly worsen, so that the local environmental pressure could be higher than the one estimated from a full-load analysis [MaC09].

Referring to small-scale DG, if the prime mover is an MT [WiS00][EDU01][EPAww], usually the heat is recovered in the form of either hot water or steam. The typical fuel adopted is natural gas, while bio-masses or a mix of bio-mass and natural gas are spreading as valuable alternatives [ShF05][AsM07], above all in the perspective of CO_2 emission reduction [ClI04][AsM07]. If the prime mover is an ICE [WiS00][EDU01][EPAww], hot or superheated water can be produced, as well as steam. Typical fuels adopted are natural gas, diesel, dual fuel, and bio-masses; however, more and more stringent emission constraints might limit the use of diesel-fuelled units to back-up applications, although with a large potential of being turned into dual-fuelled systems [WiS00]. Future development might include co-utilization of natural gas and hydrogen with higher performance and lower emissions [Bys07].

On a micro-scale level (below 10 kW_e), typical household-targeted products based on Stirling are gaining interest on the market, particularly for their smooth and silent operation, and their relatively high heat-to-electricity ratio suitable for several residential applications [OnU06][KuK08][Tho08]. Stirling engines are external combustion engines, and as such can in theory burn a number of different fuels [OnU06]. However, in practice natural gas is typically used, while solid fuels would further reduce the electrical efficiency performance [KuK08]. Heat occurs in the form of hot water, with temperatures up to about 85°C. However, lower supply temperatures may as well be sought in order to avoid possible electrical efficiency penalties due to thermodynamic reasons [Tho08].

Cogenerative FCs are still too expensive for a broad market exploitation, but a number of examples available in the literature hint that they might play an important role in the next future as a high-efficiency low-emission local energy generation source for various capacities and applications [SiL01][DoW05][KhH05][AkY06][Ob06][WeM06]. In particular, the possibility of adopting FCs in combination with MTs or gas turbines could lead to promising perspectives for setting up CHP hybrid schemes with excellent energy and environmental performance [MaC01][BuT03][GrM03][BiM04][SeP06]. Hydrogen is typically locally generated from natural gas through steam methane reforming [Dun02][McE06], with potential for co-production of electricity, heat and useful chemicals such as nitrogen by means of an internal reforming fuel [HeZ07]. In case, the steam needed for the reforming process can be supplied by the cogenerator itself (the FC in case coupled to a turbine in a hybrid cycle), so leading to a high-efficiency multi-generation scheme [LuU01][MaC01] [GrM03][LuB03].

The auxiliary heat generation is usually provided by a CHG group [EDU01][OnU06], very often gas-fired, or alternatively oil-fired or diesel-fired. It is in practice a group of that can produce hot water or steam depending on the user's applications. The thermal capacity is often dwindled over more units, for higher overall partial-load efficiency, reliability and flexibility.

As discussed in detail in Chapter 4, normally the user's needs do not match the production, due to the plant sizing or to specific operating conditions (for instance, the CHP group is switched off when it is not profitable to produce electricity). In this way, ABs are needed for thermal back-up, besides possible peak-shaving applications. Indeed, even if the plant is in theory sized to satisfy the electrical and/or thermal peaks, electrical and thermal production in the prime mover are correlated, whereas thermal and electrical load are relatively more independent of each other. Because of this, positive and negative thermal and electrical unbalances between production and loads are commonplace in cogeneration plants. As far as electricity is concerned, when the prime mover production fails the load or exceeds the load, it is the EDS (under the hypothesis of grid-connected plant) to make up for the unbalanced load, by allowing injection/drawing of energy to/from the plant. When the thermal production exceeds the load, the heat surplus is wasted to the ambient (solution to avoid, when possible, due to loss of environmental benefits from cogeneration). When the heat production fails the load, instead, the unbalanced share is produced on site by means of the ABs. Of course, the possibility of being connected to a DH network may change the operational options of the MG plant.

2.2.2. Prime mover control strategies

While planning the set-up of an energy system, taking into proper account the possible control strategies for all the plant equipment can play a fundamental role. Indeed, control strategies are directly related also to the sizing itself of the various components that can be differently interconnected and operated.

Among the possible control strategies for the CHP prime mover, it is possible to mention:

- *Electrical/thermal base load*: the prime mover is designed to cover only the constant part of the electricity/heating load; the machine is thus operated at a fairly constant load, which optimises the production efficiency.
- *Electrical/thermal load-following*: the cogenerator is operated to follow the evolution of the electricity (heating) load patterns. This kind of approach requires good dynamic performance, which is usually not a stringent constraint for ICEs and MTs [ZhT02][ElS07][UzO07]. For the sake of terminology, load-following is typically used by the operators in the power systems and markets; the same concept is often denoted as *load*, or as the likes of *electricity/heat-led strategy*, especially in documents prepared by thermal system operators.
- *Peak shaving*: the system is operated not continuously, but only to cover part of the load during peak conditions (typically electrical load, whilst thermal peaks are instead covered by boilers).
- *Off/on operation*: the prime mover is forcibly kept off or on, in the latter case at the rated capacity (*always-on* control strategy) or according to one of the above controls, usually on the basis of economic considerations (typically, the machine is switched off in the periods in which the electricity costs are lower).
- *"Spark-spread" control*: this kind of approach is a generalization of the previous one; in this case, if the plant operates in a competitive market environment, the price of electricity (variable according to a certain time step, for instance hourly) is compared to the production cost of electricity from the plant, and then the unit is run (at a certain load level) or not according to economic considerations.

Considering small-scale CHP units, the performance of ICEs and MTs often becomes poor and more uncertain below 50% of the rated electrical load, so as to make it necessary (and in many cases advised by the constructors themselves) to turn the group off.

2.3. THE AGP BLOCK

The AGP can be fed by different energy vectors and can thus be characterized by different components, also according to the *separate* or *bottoming* linking modes outlined above. In this respect, besides different plant layouts, adopting different cooling (and/or heating) machines could lead to significantly different interactions among the energy flows. Normally, the refrigerators are run under a cooling load-following control strategy, although the operation of some equipment such as the heat pump can also be mutually interconnected with the CHP one. Modelling of these interactions, with special reference to the interface between CHP side and AGP, is one of the key drivers to efficient planning of the whole MG system.

A first classification of the solutions and of the corresponding acronyms used in this book is indicated below. Specific technological details are provided in Section 3.3.

2.3.1. AGP equipment in separate linking mode

If the AGP is *separate* from the cogeneration side, typically the following technologies of chillers/heaters directly fed by fuel can be adopted:

- GARG (Gas-fired Absorption Refrigerator Group) [ToJ98][FoD00][ASH00][Kre01][DOE04][Man06][WuW06]: this type of equipment, fed by gas, belongs to the category of *direct-fired* chillers (DFCs). The gas thermal content is transformed into cooling effect directly in the machine by exploiting the absorption process. The chiller is most of times double-effect, while the higher-performance triple-effect technology is being lately introduced in the market [ToJ98][Dan06][WuW06]. The heat, usually discarded by means of a cooling tower, could be in case recovered (for instance, by means of an electric heat pump to increase the heat temperature to levels suitable for user's applications [Hav99][MaC01]). Often, the machine is reversible and could be used as a heat pump, in case avoiding purchase of additional boilers.
- GAHP (Gas-fired Absorption Heat Pump) [FoD00][DOE04][Dan06][CoP07]: these machines are capable to draw thermal power from a "free" heat source (such as air, water, ground water, or the ground itself) and to transfer it to a hotter ambient after increasing its temperature. GAHPs are also directly-fed by fuel, and from the outside they can be basically seen as boilers, but with efficiency possibly higher than unity. However, the output heat temperature is often limited to less than 70÷80 °C.
- EDC (Engine-Driven Chiller) [ASH00][RDC03][LaN06b][Dan06]: in this case, a conventional vapour-compression chiller, instead of being driven by an electrical compressor, is driven by a mechanical compressor, whose shaft is directly connected to a conventional internal combustion engine. Seen from the outside as a black box, the system, also directly fed by fuel (and as such belonging to the DFC category), is completely equivalent to a gas-fired absorption chiller, although from the energy point of view an engine-driven chiller has the advantage that part of the fuel input can be more easily recovered, as in normal cogeneration ICEs [ASH00][Dan06][WuW06]. Thus, the machine can provide at the same time heat and cooling power.
- EDHP (Engine-Driven Heat Pump): the engine-driven chiller is often a reversible machine able to work as a heat pump; in this case, the possibility of recovering thermal power also from the driving ICE allows enhanced performance as a total energy thermal generator [LiZ05][Dan06][LaN06][LaN06b].

2.3.2. AGP equipment in bottoming linking mode

Considering a *bottoming* AGP, the equipment mostly used is:

- WARG (Water Absorption Refrigerator Group) [ToJ98][ASH00][FoD00][Kre01] [RDC03][DOE04][Dan06][Man06]: in the classical trigeneration case, differently from a direct-fired chiller, the absorption machine is indirect-fired, supplied by heat produced in cogeneration (*thermal bottoming*). For this reason, this machine type is also referred to as Indirect-Fired Absorption Chiller (IFAC). The chiller can typically be single-effect (usually fired by hot water) or double-effect (usually fired by super-heated water or steam) [ASH00][Kre01][Dan06]. Triple-effect chillers now appearing on the market typically require higher pressures and temperatures than double-effect ones, which could limit their applications in bottoming schemes and make them more suitable to be direct-fired [WuW06]. However, promising solutions could be represented by exhaust gas direct-firing from an MT or, for larger applications, from a Gas Turbine (GT) [MaC01][KoW07]. Again, the thermal power discharged to a cooling tower could be in theory recovered [Hav99][MaC01]. As for a gas-fired absorption chiller and in analogy with a gas-fired heat pump, also in this case the machine is often reversible and could be used for heating purposes as a Water Absorption Heat Pump (WAHP), in case co-fired by natural gas [FoD00][BaK02][BrV05].
- CERG (Compression Electric Refrigerator Group) [ASH00][VoD03][Dan06] and EHP (Electrical Heat Pump) [Hor97][ASH00][VoD03][MiM04][PoM04][Dan06]: the electric chiller is the classical solution to produce cooling power, and in a multi-generation system the feeding electricity could be produced in the CHP unit (*electrical bottoming*). In most cases, however, an electrical heat pump is adopted, since it is a reversible machine capable to work in both heating mode and cooling mode (as a chiller, then). In this way, the CHP installed electrical capacity and the working hours of the plant throughout the year are optimised, exactly as for the thermal power with an absorption chiller fed by cogenerated heat. When adopting an electrical chiller, the discharged thermal power could be sometimes recovered by means of a heat recovery condenser [Wul99].

2.3.3. Other equipment

Further technologies, typically heat-fed and thus of possible interest for setting up thermal-bottoming schemes cascaded to CHP systems, can be encountered in the AGP, above all for building applications. In particular, *adsorption* [Zie02][Dan06][WaO06][HuW07] [Wan08][WaO08] and solid or liquid *desiccant* [Dan06][DaW06][LaC07] systems, as well as *ejector cycles* [InI05][GoB07], are more and more often adopted for MG of electricity, heating, cooling, and dehumidification [LiG04][ZaP06][HeP07]. Often, combination with low-temperature thermal sources such as solar power results attractive [FIT02][MaB02] [GoG07][Hen07], including emerging technologies with enhanced performance such as the ones based on the so-called Generator Absorber heat eXchange (GAX) absorption cycles [VeB02][Zie02][GoV08][PaK08] for both cooling and heating generation.

2.4. Interactions with External Systems

2.4.1. External networks

The possibility of interacting with different energy networks is a peculiar feature of a DMG system. This aspect gives the energy system high potential in terms of energy and economic performance, and at the same time calls for adequate modelling and planning tools, as discussed in the next sections.

In a DMG system, several local multi-generation units like the one in Figure 2.1 can be scattered over the territory and can interact with each other by means of the available energy infrastructures. A conceptual sketch is represented in Figure 2.5, where the arrows exemplify some possible links. It is important to notice that the absence of connection to a specific external network (*e.g.*, DCN) does not limit the possibility of serving a local demand of the corresponding energy vector (*e.g.*, cooling load), since this energy vector could be produced on-site by transforming the input from another energy vector/network.

In the presence of different characteristics of the local sites, and in terms of plant components and demand requirements for the different energy vectors, it could be possible to envision some kind of supervised control of the single sites, leading to coordinated management of the production of the various energy vectors and of their energy exchange with the external networks. This possibility follows the rationale of creating a suitable portfolio of energy customers to be managed in a consortium-based fashion by a single provider or energy service company, (ESCO) optimising the plant operation and economics according to specified objective functions [AkO05]. Determining the characteristics of such a portfolio of customers is one of the present research challenges. Some activities in this direction can be assisted by means of classification or clustering techniques able to group together customers exhibiting similar characteristics in their manifold energy use [ChN03][ChN05][EnA07].

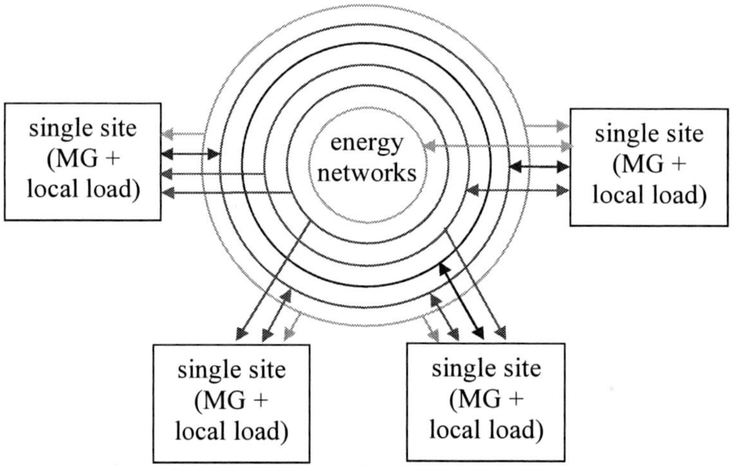

Figure 2.5. Conceptual scheme of multi-generation system interactions with the external energy networks represented as concentric circles. For instance, from the inner to the outer circle: storage/HDS, RES, DH, DCN, EDS, and fuel/GDS.

A group of MG sites to be managed in a coordinated way could be connected through networks with different owners (or public networks) that supply all or part of the energy vectors. In this case, the entity managing the MG sites has to interact with the various owners, each of which has its specific interests from both the technical and the economic points of view. On the operational side, the DMG system could be subject to coordinated control, thus enabling combined optimisation of the local plant operation. Defining the objective functions of such an optimisation may depend on the purpose of the study and in particular on the goals of the entity addressing the problem. These goals differ significantly if considering the network operator (aimed for instance at reducing the network losses), a possible ESCO representing a portfolio of clients, or the profit-based owner of a local MG system. Common objectives can be searched for by formulating multi-objective problems with agreed objective functions. In alternative, the specificity of the operators can be taken into account by applying distributed computing techniques, for instance agent-based applications. In this case, each operator is modeled as an individual agent with decision-making capability [KuK03][MaV04][DiH05][BaE07], and the interactions among the agents are represented by the linking structure describing the various interfaces of technical and economic type, as well as the ICT providing the communication infrastructure among the multi-site energy systems.

2.4.2. Distributed storage

The various MG schemes can be completed by various types of *storage systems*. Some of these systems act directly on the *electrical*, *thermal*, or *cooling* side, storing energy to be used at a later time. In other cases, the energy stored could be of different type with respect to the one to be used, thus calling for specific conversions (*e.g.*, from electricity to hydrogen and *vice versa*). The transformations adopted in the process of storing and re-using energy necessarily produce technical losses, thus making the whole process generally not convenient in energy terms. However, availability of stored energy can become profitable in other terms. For instance, thermal/cooling storage applications [Has98][Has98b][Wul99][Dan06][ZhG07] are based on the creation of a corresponding energy buffer, to be used for thermal load shifting and control purposes [AsB03][KhR04]. These applications prove to be effective to improve (or even change dramatically [ToS01]) the plant design, management, economics, and environmental performance [LiC01].

On the electrical side, there is a variety of *distributed electrical storage* devices [RiJ01][CII04][IsK05][BoB07]. However, the use of such devices is mainly limited by their relatively low capacity. Hence, electric storage systems are typically used to enhance the continuity of supply of particularly sensitive users (for instance hospitals, computer centres, critical industrial applications in food industry or paper mills, and so on). The capacity of these systems is exploited in transient conditions and for short time durations (*e.g.*, from a few seconds up to a few tens of minutes), while backup generation is needed to ensure the system supply for longer outages. In addition to these uses, electrical storage devices can play an important role in enhancing the convenience of producing electricity through RES, especially in the cases of highly uncertain availability of the primary source (e.g., sun or wind). The effectiveness of applying DS devices to deal with highly variable and unpredictable renewable-based MG can be improved by storing *hydrogen* produced by electrolysis as an equivalent electricity DS means [KoH06][WiH06]. Suitable control strategies can be developed to manage the generation-load balance, with the possibility of improving the plant flexibility and economics [VaP05][WaN07]. Potential benefits of the integration of small-scale wind or PV systems with hydrogen have already emerged for stand-alone applications [ZoG06]. Further benefits are expected from the inclusion of DS solutions within DMG applications. These benefits could refer to the possible interaction with electricity markets [KoH06][GrK07], as well as to environmental aspects [GrD07].

2.4.3. Renewable energy sources and hybrid systems

Classical CHP applications are based on production of electricity and heat starting from gas of fossil fuels. However, various other sources can be used for cogeneration purposes, among which RES play a relevant role due to their environmental benefits [ElC03]. More specifically, various solutions adopting solar technologies [Jäg07], such as Photovoltaic (PV) modules, thermal collectors, and hybrid Photovoltaic/Thermal (PV/T) systems, are emerging to provide effective integration with cooling/heating equipment in MG plants.

In principle, electricity can be produced in a PV system and then used to supply an electric chiller. However, from the energy and economic viewpoints the adoption of heat-fired cooling technologies (*i.e.*, , absorption or desiccant systems [FlT02][Hen07]), to be fed by cogenerated heat in a PV/T solar system (solar trigeneration) or by heat produced in a solar collector, generally proves to be even more effective. In fact, in PV/T solar units the heat recovery system brings additional energy benefit by decreasing the module temperature [TrN02][VoC06] and thus increasing the PV electrical generation efficiency. In this way, the almost inverse correlation among temperature and solar irradiance of the PV modules is beneficially exploited. In particular cases, such as MG with concentrating PV/T systems, the thermal power produced at temperatures higher than 100 °C can be effectively deployed in applications such as steam generation or supply of double-effect absorption chillers [MiK07].

The production of solar thermal power for trigeneration applications has the advantage of being effective throughout the whole year [Hen07][Sar07]. In fact, in the wintertime, the produced heat can be used in addition to the heat cogenerated through classical solutions. In the summertime, the thermal power can be used for cooling generation purposes. In this sense, the advantages are the same as those relevant to CHP plants in CCHP applications.

A further promising field of application refers to *hybrid* DMG systems, in which controllable CHP plants of various sizes are coupled to renewable sources. Notwithstanding the fluctuating power supply provided by uncertain resources such as wind or sun, it is possible to obtain enhanced performance regarding the overall energy efficiency and especially the environmental impact of the integrated energy system [ElH02][Lun05][BeD06] [LuM06][AnL07][MeC08]. An example of application to a CCHP system with PV is provided in Section 8.1. Furthermore, the plant economic profitability could be positively improved from selling electricity to the grid and benefiting from feed-in tariff structures [MuO07] established in compliance with specific environmental regulations [SwM06]. Further benefits could be brought by adopting reversible EHPs to exploit electrical output from RES for heat or cooling generation within the composite DMG system [Kil99][BaL01][LuM06].

The key aspects concerning the use of RES refer to energy efficiency and environmental impact [Din99][VoB00][Hon05][RiD06]. In this respect, the environmental aspects should be addressed in a comprehensive way through Life Cycle (LCA) techniques [KhH05][FtA06][Peh06][Wei07]. The same occurs for biomass-supplied energy systems [ChM05][FiC07]. For comprehensive analyses, above all in comparison with traditional fossil thermal solutions, the LCA-based approaches should take into account the *external costs* associated to the entire life cycle [KaG03][FtA06][Gul06][ExE07].

Considering DMG solutions with RES (mainly solar systems) applied to buildings, the energy used for system operation (heating, cooling, ventilation, electrical uses, etc.) is the largely prevailing part of the energy consumed during the whole life cycle [SaH07]. Hence, such DMG applications are efficient even in an LCA context. Potential life cycle energy savings can also be obtained by combining PV and heat recovery for MG [CrT06].

Chapter 3

MULTI-GENERATION COMPONENTS: CHARACTERISTICS AND MODELS

Multi-generation can be effectively implemented by adopting a number of different technologies. A summary of the characteristics of various plant components, with special focus on cogeneration and trigeneration systems, is provided in this chapter. The illustration starts from the case of cogeneration and proceeds with further components for cooling and heat production, which can be used according to the combined production principles. The description is orientated to provide the basic information that can be used to develop suitable models of these components to be applied within an energy system-based approach. The plant components considered refer to available solutions that have already reached the commercial stage. In particular, for cogeneration applications the main components addressed are internal combustion engines and microturbines, backed by combustion heat generators. For cooling plants, the focus is set on various types of chillers, namely, compression chillers, absorption chillers, adsorption chillers, engine-driven chillers, and heat pumps. For the components analysed, the main aspects highlighted refer to thermodynamic characteristics, performance indicators to be adopted within a black-box modelling approach, and relevant off-design models aimed at representing the component operation under generic conditions. Further aspects such as heat recovery, operational constraints, and additional means to enhance the plant component performance are described in specific cases.

3.1. COGENERATION PRIME MOVERS

3.1.1. General aspects

The possible combinations of CHP technologies and schemes are numerous and various. For instance, classical applications for steam turbines refer to topping cycles (with downstream steam available to a thermal user) or to bottoming cycles (in which the thermal input is recovered from a high temperature topping cycle gas turbine [Hor07]). Similarly, an MT can be applied as cogeneration heat source for a thermal user, but also used as a thermal sink in combination with an plant in a hybrid model [BoK01][ZhT01]. In the sequel, after a general comparative overview of widespread DG thermal technologies, further details are given on the available technologies most suitable for the small-scale applications considered in this work, that is, ICEs and MTs. A synthetic view on Stirling is also included, since the market for SEs is increasing for domestic cogeneration applications. Further potential applications could refer to trigeneration cases [KoW04]. Other technologies such as FCs, not yet mature at the commercial application stage, but of possible interest for the future, are not addressed in this book. However, the general models developed in this book can be readily applied to these technologies as well.

Addressing GTs of large size and steam turbines is outside the scope of this book. The interested reader is referred to the technical literature, for instance [Hor97][Loz00][Hor07].

An example of general data for the most used distributed CHP technologies, also including gas turbines as an upper-size reference (considering 30 MW_e as the upper bound for DG), is reported in Table 3.1. These data are representative of average commercial figures, and are meant to be useful for a first comparative look. For many fields of Table 3.1, the entries are presented with large spread due to manifold variables, such as equipment size, manufacturer, application, and finally availability of meaningful information. In particular, some MT data (such as the useful life) are reasonable extrapolations from little available data, due to lack of sufficient statistical information for this relatively new technology. The ranges of figures shown are consistent with those indicated in various references, among which [BoK01] and [PiW05]. However, the numerical values of the data indicated may present significant differences and even divergences according to different sources and market researches. Indeed, the fast development of some CHP technologies in recent years is making any survey on market availability very likely to be soon outdated.

Table 3.1. Characteristic data for some cogeneration technologies

	Diesel ICE	*gas ICE*	*gas turbine*	*microturbine*
size [kW_e]	10÷20,000	10÷20,000	500÷30,000	25÷300
electrical efficiency (at *LHV*) [%]	30÷45	25÷45	25÷40	20÷35
total efficiency [%]	85÷88	85÷88	80÷90	75÷85
partial-load performance	best	good	poor	good
start-up time [s]	10	10	600 ÷3600	60÷600
fuels	diesel, residual oil	natural gas, biogas, propane	natural gas, biogas, propane, distillate oil	natural gas, biogas, propane, distillate oil
uses for heat recovery	hot water, low pressure steam, district heating	hot water, low pressure steam, district heating	hot water, low/high pressure steam, district heating	hot water, low pressure steam
CHP output [Btu/kWh]	3500	1000÷5000	3500÷12,000	4000÷15,000
usable temperature for CHP [°C]	70÷450	130÷250	250÷600	190÷330
availability [%]	90÷94	95÷97	92÷96	92÷97
NO_x emissions [ppm]	400	30÷95	10÷25	10÷45
noise [dB at 1m]	70÷20	70÷120	75÷90	70÷80
useful life [h*1000]	20÷60	50÷60	90÷150	60÷80
cost [€/kW]	500÷1000	600÷1200	500÷700	700÷1300

The prime movers used for cogeneration applications have different potential for thermal energy recovery, as a consequence of different electrical and thermal efficiencies. In order to describe at the same time the relative output of heat and electricity, let us consider the electrical power output W, the thermal power output Q and fuel thermal power input F.

The prime movers are usually characterized by [Hor97]:

- the *cogeneration ratio* λ, or Heat-to-Power Ratio (*HPR*), ratio of thermal to electrical power generated:

$$\lambda = \frac{Q}{W} \tag{3.1}$$

- the electrical *efficiency*, ratio of the electrical power output to the fuel thermal power input:

$$\eta_W = \frac{W}{F} \tag{3.2}$$

- the *thermal efficiency*, ratio of the thermal power output to the fuel thermal power input:

$$\eta_Q = \frac{Q}{F} \tag{3.3}$$

- the *overall efficiency* or Energy Utilisation Factor (*EUF*), conventionally calculated as the sum of the electrical and thermal efficiencies:

$$EUF = \frac{W+Q}{F} = \eta_W + \eta_Q \tag{3.4}$$

A conceptual framework directly related to the cogeneration ratio is formulated in Section 4.2. A systematic definition of the above efficiency indicators is also provided in Section 4.4.2 in a multi-generation modelling context.

Figure 3.1 shows the commercially available electrical sizes for the most used CHP technologies, along with their typical ranges of rated cogeneration ratio. Given the prime mover, the rated λ can change depending on the heat recovery system used. This is the reason why there can be quite wide ranges for the cogeneration ratio. Indeed, different technologies have different systems to recover the waste heat, which can be more or less complicated and prompt a variable thermal output. For instance, the waste heat from an ICE can be recovered from only the cooling water and lubricating oil, or also from the exhaust gases (as detailed below), leading to the possibility to get different cogeneration ratios from the same prime mover. However, it has to be pointed out that the cogeneration ratio is usually defined at rated conditions, but it changes at partial load and generally in off-design conditions, similarly to what happens to the other parameters of the prime movers, as shown in the sequel.

The technologies reported in Table 3.1 include well-consolidated CHP solutions such as reciprocating engines and gas turbines, as well as relatively innovative microturbines.

Reciprocating engines can generally be diesel-fuelled or gas-fuelled (with the latter typically following an Otto cycle [BoK01]), and can be found in sizes up to few tens of MW_e. The available smaller sizes and larger sizes overlap respectively the microturbine and gas turbine application fields. Reciprocating engines exhibit very good electrical efficiencies (more than 40% for diesel engines, almost as much for Otto engines, owing to recent impressive technology improvements) and excellent performance at partial loads (down to

50% of nominal load, diesel engines boast almost constant electrical efficiency, whereas the efficiency drop for gas engines is to only 85÷90% of the nominal one). This level of performance makes reciprocating engines suitable for load-following control strategies and applications with even highly variable loads. Diesel engines are relatively cheap, but their use becomes critical because of the high emission level; exhaust treatment requires expensive equipment and the results are not always satisfactory, so as to often make them a less viable choice with respect to the less polluting gas engines.

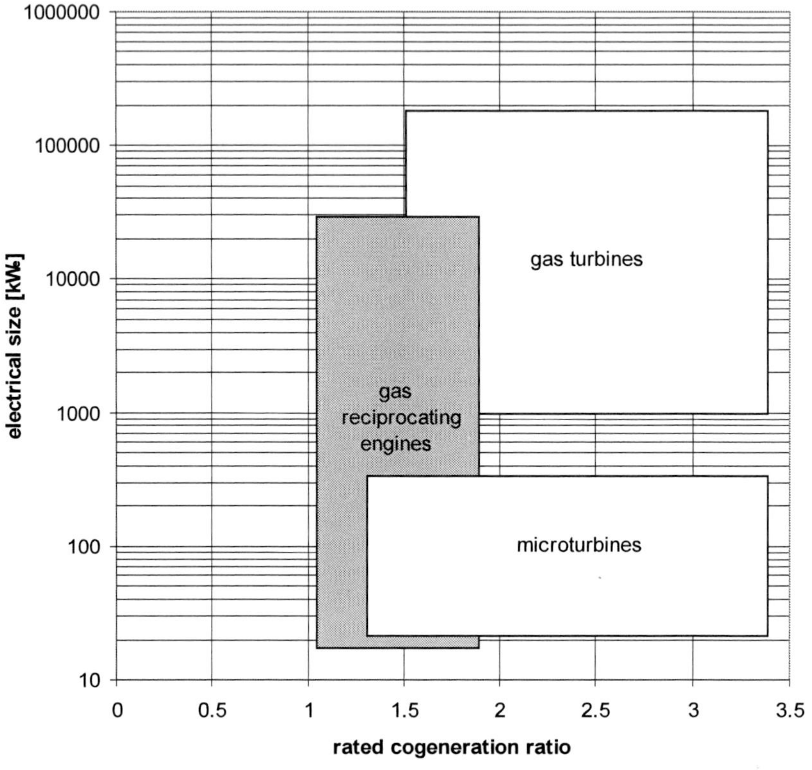

Figure 3.1. Cogeneration ratio for different technologies and sizes.

Gas turbines are highly reliable, have low emission levels, and can easily produce high pressure steam, besides boasting low maintenance costs and high power/size ratios. Their use for DG CHP applications starts from about 5 MW_e; lower sizes are used only for reserve applications where the NO_x emission constraints are very severe, while for continuous generation diesel engines are preferred. Below 2 MW_e, however, at present the capital costs make this technology not competitive in the market. Compared with reciprocating engines in the overlapping sizes, penalising factors for GTs are the partial-load efficiency drop (under the constant rotation speed constraint) and the need for high-pressure fuel compressors.

Microturbines [Pil02] represent an evolution of GTs thought for small-scale applications. Commercially available sizes range from few tens of kW_e to 300÷400 kW_e. Some manufacturer is thinking of rising the size up to about 1 MW_e, in order to compete with the leading ICEs. MTs use a single-stage centrifugal compressor and usually a recuperator to improve electrical efficiency. Their polluting emissions are lower than for ICEs, alongside GTs, and the major difference with respect to the performance of their "larger relatives" is the partial-load characteristics: microturbines are not bound to the constant speed constraint, so that the efficiency drop at lower loads is about 10÷15% of the rated value. On the other hand, the variable rotation speed requires electrical power conditioning equipment for grid connection. Costs are higher than for ICEs (especially due to the electrical converters and the recuperator), but increasing sale volumes are likely to make their cost drop soon.

Figure 3.2 shows an example of comparison among the partial-load characteristics for the different technologies presented. The chart stops at 50% of the electrical rated load, because below this point the efficiency for some CHP technologies may drop significantly so that several machines (in particular, MTs and gas ICEs) are in general advised not to be operated.

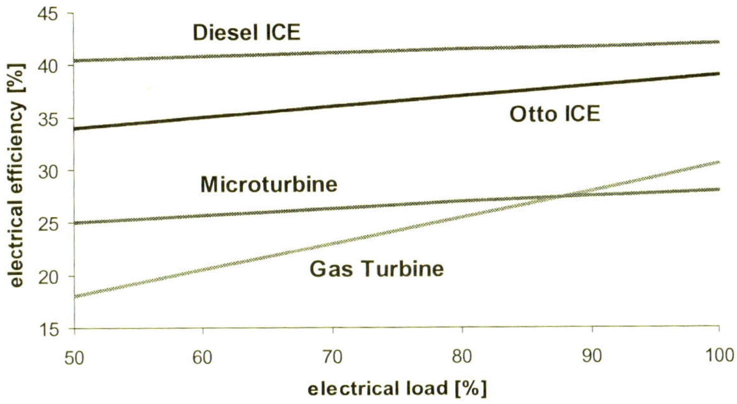

Figure 3.2. Partial-load performance comparison for some CHP technologies.

3.1.2. Internal Combustion Engines

The *Internal Combustion Engine* (*ICE*) technology is the most mature and well-proven one among CHP prime mover options. Upsides for ICEs include low investment cost, very good efficiency (also at partial load), suitability for on-off cycling operation, and high temperature exhaust flow rate for CHP applications [BoK01][Kre01]. Moreover, they are a well-known technology, and their maintenance procedures are well-defined. Hence, carrying out the scheduled maintenance on a regular basis ensures good availability and reliability. ICEs represent the overwhelming technology in the size range below 1 MW_e and are adopted in an array of applications ranging from rural electric cooperatives to educational facilities. However, ICEs represent a viable option also in the range 1÷5 MW_e (for instance, for medium-scale DH applications), while for larger capacities GTs units become the leading technology. For small-scale capacities, the ICE market is now facing the competition of newer technologies such as MTs and, for the lowest capacity range, of SEs. The main characteristics of ICEs are presented in the sequel, with special focus on their CHP applications and on gas-fed units.

3.1.2.1. Cooling and heat recovery systems

ICEs can be air-cooled or liquid-cooled. Air is used mostly for very small and mobile units, with fins being adopted to increase the radiation and convection area and with blowing fans to improve the heat exchange. Liquid-cooled engines are usually characterized by internal conducts through the shell and are more silent owing to the absence of fans. They are widespread also owing to the possilbity of an easy heat recovery, as discussed below. In addition, it is common to have a lubricant cooling circuit, whose aim is to keep the lubricating fluid (normally oil) at about 10 °C above the coolant temperature, but however below 125 °C [BoK01], in order to avoid *cracking* phenomena.

For cogeneration applications, heat is normally recovered from the oil cooler (small amount) and the engine coolant, as well as from the exhaust gases. Small amounts of heat can also be available from the turbocharger intercooler and/or aftercooler. All these recovery systems practically allow heat recovery without significant modification of the engine operation (thermodynamic cycle) as an electricity generator [Ang98], which represents a major upside.

Table 3.2 shows a typical breakdown of the recoverable heat (in percentage of the fuel thermal input) into the single contributions.

Table 3.2. Typical breakdown of the single contributions to heat recovery in a gas ICE

source	*recoverable heat* [%]	*temperature* [°C]
exhaust gas	10÷30	350÷600
cooling water	20÷30	90÷125
lubricating oil	1÷4	80÷100
exhaust gas condensation latent heat	5÷10	55÷75
turbocharger air	0÷5	50÷150

Approximately 50% to 70% of the fuel thermal input can be recovered in thermal CHP applications, depending on the engine thermal efficiency and on the heat exchangers used. The upper bound of energy conversion today for ICEs is about 90% of the fuel input. Indeed, regardless of the quality of the recovery system, small thermal losses will always be present, mainly due to radiation incomplete combustion and pumping [Gan96]. Moreover, a part of the enthalpy contained in the exhaust gases is not always recovered (gases kept at a temperature above the condensation threshold [Ang98]).

Figure 3.3 shows a *closed-loop heat recovery system* with recovery from jacket coolant, oil circuit and exhaust gases, sometimes dubbed as *total energy system*. In this case, the cogeneration system is sold in a "package", the heat exchangers are contained inside the CHP shell, and the installation requires only the main hydraulic and electrical joints. This is the typical case of a thermal user with a unique enthalpy level requirement, consistent with the total recovery. The different heat exchangers are set in series giving priority to the lower temperature ones, so as to optimise heat recovery. Such one-level recovery systems usually supply hot water at 85÷90 °C, and the water returning from the cooling circuit must often be kept below 70 °C in order to avoid dangerous heating of pistons, cylinder head and liners. For this purpose, a radiator (or a cooling tower for the bigger plants) is needed in all the cogeneration installations to ensure dissipation of the heat exceeding the thermal use, sized on the worst-case basis to dispose of all the heat in case of (even temporary) lack of thermal request.

Ebullient cooling systems may also be found in some CHP applications. In these systems, a boiling coolant (usually water) is made circulate through the engine. In particular, the water enters as a pressurized liquid at its boiling point and starts evaporating when receiving heat from the engine. Not all the liquid boils, though, so that the double-phase mixture remains at uniform temperature along the circuit, hence improving heat transfer and combustion efficiency, as well as lengthening the engine life [EDU01]. In alternative, forced-circulation systems may operate above the usual pressures, in a temperature range of 120÷130 °C, and can also produce low temperature steam [EDU01].

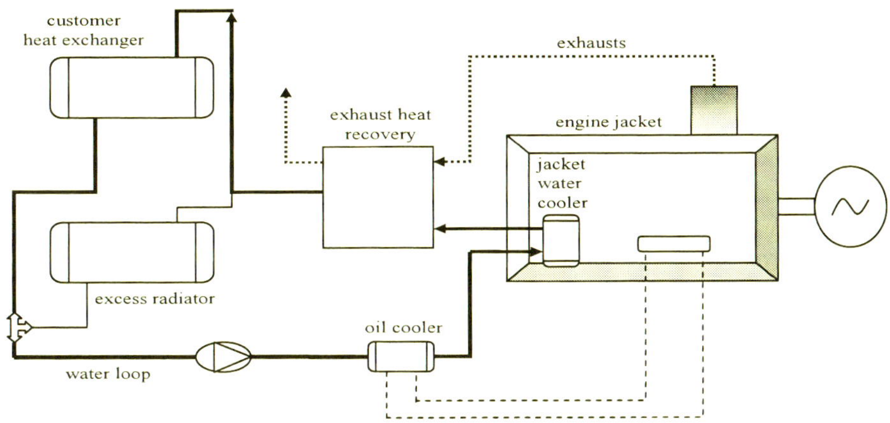

Figure 3.3. ICE closed-loop hot-water heat recovery system.

One of the main upsides of ICEs is that they can supply heat at different temperature levels (Table 3.2) for different uses, which often determine the recovery system configuration (the heat exchangers at the different temperature levels are placed in cascade in order to maximise the thermal recovery, as mentioned above) and the temperature parameters of the heat exchangers.

A particular role is played by the heat recovered from the exhausts, which may account for 10% to 30% of the thermal input. Usually, air-to-water heat exchangers are used. Exhaust gases can be cooled from 400÷500 °C at the engine outlet to about 130÷150 °C, and can produce, for instance, steam at 10÷20 bar (180÷210 °C), pressurized hot water, or hot water at a temperature as low as 95 °C. However, in the smaller engines, which also exhibit specific needs for adequate jacket cooling, the heat contained in the exhaust gases may be not sufficient to raise steam [EDU01]. In general, the higher is the temperature of the media used for the thermal production, the higher must be the minimum exhaust temperature, and thus the lower is the amount of thermal energy that can be recovered.

Not all the heat available from the exhausts is usually recovered, since the temperature of the flue gases is often kept above the condensation threshold (to avoid potential arising of corrosive acids in the exhaust piping and stack). The minimum temperature at which exhaust gases can be cooled is about 100 °C for natural gas engines (natural gas is normally sulphur-free, but however the exhaust gases are not cooled further in order to allow easy discharge and dispersion into the atmosphere), and about 140 °C for diesel fuels. In general, for gas ICEs the minimum exhaust outlet temperature is usually kept above 120 °C, which allows heat recovery of up to 70÷80% of the enthalpy content.

When lean-burn engines [BoK01][Gan96] are used, the content of oxygen in the exhaust gases can be up to about 10÷15%: this may allow supplementary firing in a Heat Recovery Steam Generator (HRSG), in case with additional fresh air, as in turbine post-combustors [Loz00].

Recovered heat can be used for a number of applications, such as space heating, reheating, domestic hot water, absorption cooling, and so on. More specifically, it is possible to identify the following major schemes:

- separate production of steam with pressure up to 20 bar using the heat contained in the exhaust gases, and hot water at 85÷90 °C from the engine cooling systems;

- production of hot water or superheated water by means of cascaded heat exchangers that make up a "total energy" system;
- direct recuperation of the gases, that can be used in certain processes such as drying or to fire absorption chillers;
- generation of hot air through air-to-air heat exchangers, for instance for heating purposes; in this case, special care should be paid to the maintenance of the heat exchanger at the exhaust side.

3.1.2.2. Efficiency and off-design performance of ICEs

Today's ICE electrical efficiencies may reach up to about 45% at rated conditions (the best values available today, apart from combined cycle groups), and drop slightly at off-design (*i.e.*, lower) loads. The partial-load performance of ICEs, especially diesel units (efficiency in practice constant until half of the rated load), are the best of all the technologies today available for electrical generation. Also the partial-load electrical performance for gas engines is quite good, especially if a "quality" control is carried out by valve control of the doses of the inlet air and fuel, as opposed to a throttling valve control with poorer performance [Gan96]. At 50% load, the electrical efficiency of a gas engine is about 85÷90% of the rated one.

Along with electrical efficiency, the ICE thermal efficiency may change significantly at partial load, as all the conditions for which combustion and heat transfer have been designed change. Moreover, also the heat transfer in the heat exchangers used to recover the thermal power changes, in such a way often not easy to work out [BeK03].

Manufacturers can give numerical information or charts to describe the partial-load characteristics of the machine (basically, electrical and thermal efficiency, or similar information from which these efficiencies can be calculated). The technical literature on ICEs (for instance, [Gan96][EDU01][OnU06]) may also provide plots to describe the *heat balance* at partial load, *i.e.*, electrical output and thermal energies available at the different circuits, representing the potentially recoverable heat (apart from the efficiency of the heat exchangers, usually in a range 85÷95%).

It is not infrequent that only few performance points are given from the manufacturer (for instance, efficiencies at 50% and 75%); in this case, usually linear interpolation or second-order model interpolation is carried out on the given data, also on the basis of analogies with other engines or literature models. If sufficient information is known about the mechanical characteristics, it is sometimes possible to approximately estimate partial-load behaviour of the engine through *analytical models* [Gan96].

Typically, the engine characteristics below 50% of the rated load are known with higher uncertainty; moreover, the steep drop of electrical efficiency below this level makes it practical to switch off the unit; for lower loads, it is commonplace to resort to grid supply or to dwindle of the installed capacity over parallel smaller units.

Figure 3.4 shows the partial-load characteristics for a gas-fed 836-kW$_e$ Otto ICE, drawn after processing data available in [BuC03]. The thermal power is produced in two different ways and thermal levels: 1) 70÷85 °C hot water in a circuit that recovers the heat from jacket coolant, lube oil coolant and intercooler; 2) 10-bar steam produced in an unfired HRSG by from exhaust gases (inlet temperature at the recovery boiler of 500 °C, outlet at 220 °C). At 50% partial load the electrical efficiency drops less than 10% with respect to the rated value. The heat recovered through the hot water circuit increases slightly, due to the larger amount of heat rendered available by the lower electrical efficiency. The heat generated through the HRSG follows the profile of the electrical production: the exhaust temperature at the engine outlet increases slightly at partial load, which balances out the heat transfer decrease within the recovery boiler due to decreased mass flow rate and thus exhaust speed. As an overall effect, the exhaust temperature at the stack keeps fairly constant, so that the steam produced decreases proportionally with the mass flow rate. Similar values for a gas ICE of the same size are reported in [EDU01].

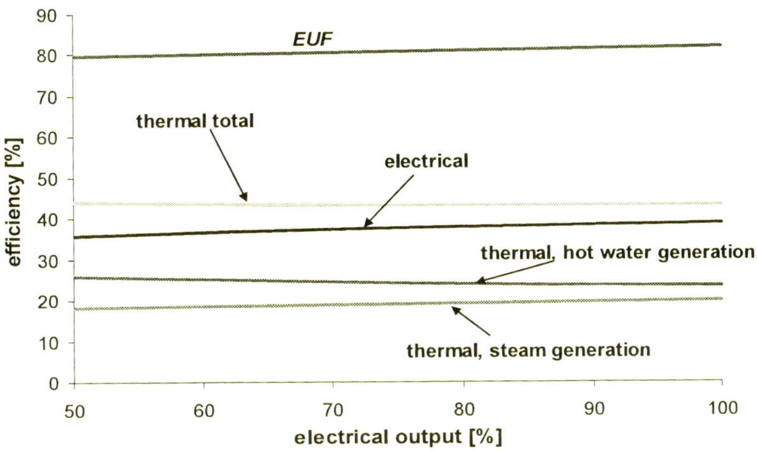

Figure 3.4. Partial-load characteristics for a 836-kWe Otto ICE.

In general, analogous considerations can be carried out in similar plant configurations, although actual figures depend on the specific case, given the wide range of possible structures and operating conditions.

Concerning the engine dynamics, diesel engines are slightly better than Otto engines; however, both technologies perform very well for load-following applications (as an indicator, the start-up time is about 10 s, Table 3.1) and even for on-off cycling [BoK01], with no significant dynamic constraints (such as ramp rates, on/off duration limits, maximum number of cycles in a given time period, and so forth) to limit the range of operations the engine can perform.

A last consideration refers to the possible de-rating of ICEs due to atmospheric conditions. As a general reference, according to [EDU01], internal combustion engines may have a decrease of electrical output of 2 to 3% for each 300 m of altitude, also depending whether turbo-charged or not, while the power output may decrease by about 1% per 5.5 °C increase in the ambient temperature with respect to ISO condition (15 °C). However, detailed data on every specific unit should be provided by the manufacturers.

3.1.3. Microturbines

3.1.3.1. Generalities on microturbines

Microturbines (MTs) [Pil02] are today's breakthrough technology for small-scale DG applications, with units available on the market from 25÷30 to 300÷400 kW_e and intention to extend the application range up .to 1 MW_e. MTs are conceptually very simple, based on the open Joule-Brayton cycle [CaG96][Hor97][Bej97][Loz00] as the bigger gas turbines, although with better performance at partial load, due to the possibility of running at non-synchronous speed [Cam00]. With respect to ICEs, MTs exhibit lower emissions at full load and, from a cogenerative point of view, higher cogeneration ratio (at the cost of lower electrical efficiency), which might make them more suitable for specific applications.

In this section, the major aspects of MTs for small-scale cogeneration applications are presented, focusing on those issues concerning their characteristics and performance as a part of an MG energy system.

Whereas GTs with sizes of few MW_e are just a rescaling of the bigger ones that use axial flow turbo-groups, the MT architecture has been completely rethought. Indeed, MTs use radial flow compressors and expanders or turbines and rotate at high speed, of the order of 100,000 rpm. Constructive and economic reasons make it suitable to have only one stage of compression, so that the typical compression ratio is 3:1 to 5:1, very high for a single-stage compressor but low in absolute (compared with 12:1 to 15:1 for GTs). The lower compression ratio leads to higher exhaust temperature and lower efficiency (less than 20%), so that it is common to recover part of the exhaust thermal energy to pre-heat the turbine inlet air before it enters the combustor by means of an air-to-air heat exchanger [BeK03] called *recuperator* or *regenerator* [Loz00][McD00]. A *recuperated cycle* allows consistent fuel saving and efficiency increase compared to a simple cycle. Hence, most MTs available commercially have a recuperator and boast electrical efficiencies of about 25÷30% [BoK01][EDU01]. In the sequel, special focus is set on recuperated single-shaft MTs, which represent the most widespread application.

3.1.3.2. Off-design characteristics

For most applications it is commonplace that a DG CHP prime mover is required to supply partial loads for a relatively long portion of its operation time. An MT is not bound to running at fixed speed as most GTs: this is in fact one of the key factors to justify its wide use for load-following or stand-alone applications, as its dynamics and performance at partial load are very good [Cam00].

For a conventional open cycle gas turbine, the *heat rate* (defined as the inverse of the electrical efficiency, see Section 5.2.2) may increase considerably with decreasing load. Instead, recuperated low pressure ratio cycles with the ability of maintaining a high *Turbine Inlet Temperature* (*TIT*) also at partial load exhibit far better performance [BoK01]: an MT is indeed designed with these characteristics. In particular, when the load decreases the shaft speed and the fuel supply are adjusted so that the machine slows down and decreases the air mass flow operated. At the same time, the control system changes the fuel flow so that with a significant speed (and therefore mass flow) reduction the *TIT* can be kept fairly constant over a large range of partialisation. Meanwhile, thanks to the non-synchronous coupling to the load/grid, a Power Conditioning Unit (PCU) makes up for the speed (and therefore frequency) change [BoK01][Hwa04]. With the *TIT* kept high enough, it is possible to get to as low as 50% of the rated load with electrical efficiency drop of only few percentage points (not more than 10÷15% drop at half load), unlike GTs bound to constant speed constraint [Cam00]. Below the 50% load threshold, the efficiency degradation may become significant [Hwa04]. From a practical standpoint, it is advisable to switch the machine off at loading levels smaller than 50% of the rated load (as for ICEs). Sometimes the switch-off is done automatically by the MT itself [PeL04].

For cogeneration applications, it is also important to evaluate the partial-load characteristics of thermal efficiency. Indeed, not much information is available on this subject, and the relevant performance may actually strongly depend on the specific machine characteristics and heat recovery system.

For instance, an illustrative experiment-based model of thermal efficiency at partial load was presented in [Per05]: there, the characterization of a 105-kW_e MT (100 kW_e, net of fuel compressor parasitic power) is carried out by setting the electrical load (in a range 50÷110% of full load) and measuring both thermal and electrical efficiencies. The results are given for different temperatures of the cogenerated hot water, in a range between 60 and 80 °C.

However, the MT exhausts might in general be able to produce hot water at a temperature as high as 130 °C. Whereas the electrical efficiency is rather independent on the water outlet temperature and recalls the characteristics drawn in Figure 3.2, the thermal efficiency tends to be lower for higher outlet temperatures, due to higher exhaust temperature and therefore higher thermal waste towards the environment. Figure 3.5 is aimed at providing a simplified idea of the complexity of the experimental results found.

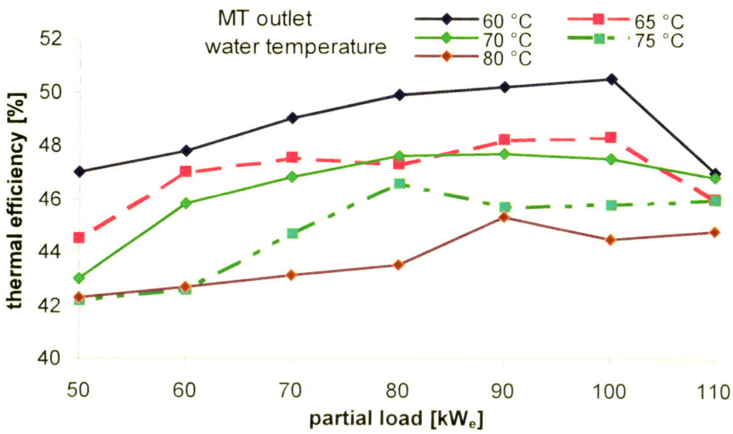

Figure 3.5. Partial-load thermal efficiency for a 105-kW$_e$ MT.

Another model for the off-design characterization of an MT is reported in [BaC99]. In that work, the 45-kW$_e$ MT analysed is equipped with a *multi-step add-on recuperator*. At rated electrical load, the possibility of modulating the flow levels in the recuperator and the cogenerative heat exchanger through a by-pass valve enables to obtain different cogeneration ratios and overall efficiencies at the cost of electrical efficiency drop. Table 3.3 summarizes the main characteristics of the MT described in [BaC99] (by-pass equal to 0 means that all the exhaust gases circulate through the recuperator).

Off-design characteristics like the one shown should be taken into account in all real situations when evaluating the energy system performance and the coupling system between energy source and user (which could in case be represented by a bottoming device such as an absorption chiller, in CCHP applications).

Actual efficiency figures and partial-load profiles depend on the specific type of MT, its size, and its design characteristics (for the recuperator, in particular). However, for modelling purposes the electrical efficiency of an MT at partial load could be typically characterized through a quadratic curve, with the maximum generally at about full load, and the efficiency practically constant down to about 90% of rated load. The thermal efficiency is more complicated to model, being dependent on the design of the specific machine even more than the electrical efficiency. However, once known the characteristics of the exhaust gases at partial loads from data sheets (if available), it is possible to derive partial-load efficiency models considering the minimum temperature to which the exhausts can be cooled down (depending on the applications, and however typically not below 110÷120 °C) and the heat exchanger efficiency (usually higher than 90% [EPAww]).

Table 3.3. Characterization of an MT equipment with a multi-step add-on recuperator

by-pass flow level	[%]	0	20	40	60
electrical power	[kW_e]	45	45	45	45
water flow	[kg/s]	1.24	1.65	2.06	2.49
water-.in temperature	[°C]	70	70	70	70
water-.out temperature	[°C]	90	90	90	90
fuel consumption	[m^3/h]	22.12	24.44	27.37	30.96
overall efficiency	[%]	69	77	82	85
electrical efficiency	[%]	21	19	17	15

Other major concerns to account for in an MT-based cogeneration system are the efficiency and the power output variation with the ambient conditions, such as ambient temperature and altitude [Loz00]. The rated figures that can be found in data sheets usually refer to ISO conditions (15 °C, 1.0123 bar, 60% humidity [Loz00][Kre01]).

Typical temperature-variable characteristics of power output and heat rate (Section 5.2.2) of an MT [Loz00][Hwa04] are shaped as in Figure 3.6. As the ambient temperature increases, both power output and efficiency decrease (that is, the heat rate increases). Indeed, fixed the rotational speed and therefore the volumetric flow rate aspired by the compressor, the intake air mass flow rate goes down since the air density decreases with increasing temperature. At the same time, modification of the thermodynamic cycle occurs, which brings about consequent efficiency worsening. The same holds true for the fuel flows to the combustor [Loz00][Kre01]. Similar output capacity drops occur due to altitude increase and consequent ambient pressure decrease [Hwa04], whereas the relevant impact on electrical efficiency is negligible [Loz00]. However, this matter definitely depends on the design approach and control system implemented on the specific machine. For instance, temperature controls can be set to take into account the ambient temperature. Adjustments on the turbine speed and fuel gas compressor speed could be arranged somehow so as to keep flat capacity for temperatures higher than about 25 °C [Hwa04].

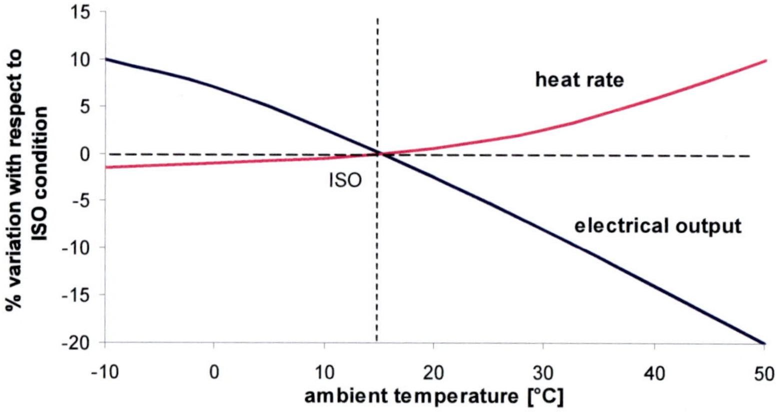

Figure 3.6. Electricity output and heat rate characteristics at different ambient temperatures for a typical MT.

Figure 3.7 shows the electrical and power output and electric efficiency curves in function of the outdoor temperature for a 60 kW_e MT [MTJ06]. Also the thermal power undergoes variations with respect to ISO conditions. Indeed, both exhaust temperature and exhaust mass flows are modified by changes in outdoor characteristics, so that also the energy content of the exhaust gases changes. Similar plots are for instance reported in [PeL04].

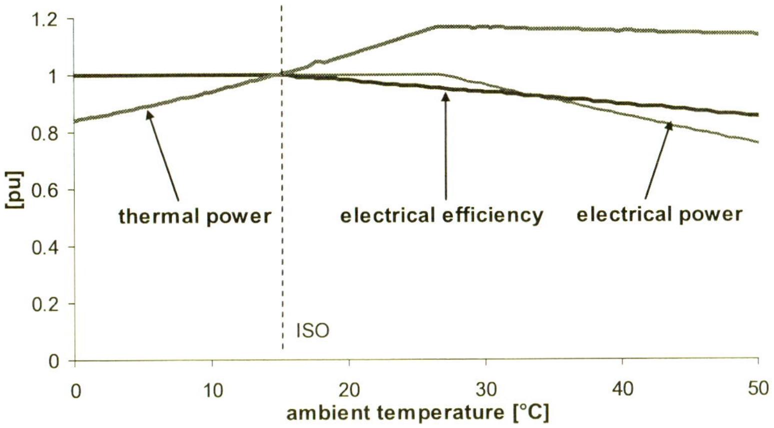

Figure 3.7. Electricity and heat output and electrical efficiency characteristics at different ambient temperatures for a 60-kW$_e$ MT.

Sometimes, the efficiency and power output drops for relatively high inlet air temperatures can be significant. In this respect, a number of techniques might be explored to pre-cool the intake air, which often proves to bring consistent efficiency and economic benefits [GaR04]:

- *Evaporative humidification* [Loz00]: by spraying water over the inlet air, the temperature drops to the wet-bulb one [Ang98][Kre01]; this solution is economic but often inefficient (in humid climates, for instance, where the temperature is already close to the wet-bulb temperature) and however requires consumption and availability of water.
- *Absorption chiller* [MoA98][FoD00][Loz00][Hwa04][AmH04]: in this case, an absorption chiller, often fed by cogenerated heat, is used to generate cooling power to cool the turbine inlet air. In practice, a particular case of trigeneration is set up [HeS03]. Although potentially very effective, this solution can be unattractive in terms of plant costs, so that adequate economic assessment must be run [GaR04]. In addition, an aspect not to be neglected is the counter-pressure due to the absorption chiller and cooling-air heat exchanger, which might balance off the power improvements gained with pre-cooling the inlet air. Also in this case, humidity plays an important role [KiR00][HeS03].
- *Electric chiller* [Loz00]: this solution is in general not very effective, because of the efficiency and power output drop due to the additional electricity consumption.

For an off-design model of an MT equipped with an add-on recuperator, all the issues considered in this section should be taken into account. It is apparent that some approximation in a model for code implementation may be someway needed. However, simulations can provide better results if more experimental data on the specific machine are available.

3.1.3.3. Cogeneration applications

The typical components and configuration of a single-shaft MT with add-on recuperator for cogeneration applications is shown in Figure 3.8 (elaborated from [BaC99]). Quite often the prime mover, the recuperator, the cogeneration heat exchanger, and the PCU are all contained in the MT package. The footprint is thus relatively limited (also due to the actual machine size, much smaller than for ICEs, for instance) and the machine can be sited with minor efforts. As an illustrative example, a 100-kW$_e$ MT package has size of about 3x2x1 m. In this case, usually just pipeline terminals to the fuelling system and to the user's hot water distribution system need to be arranged, besides electrical connections.

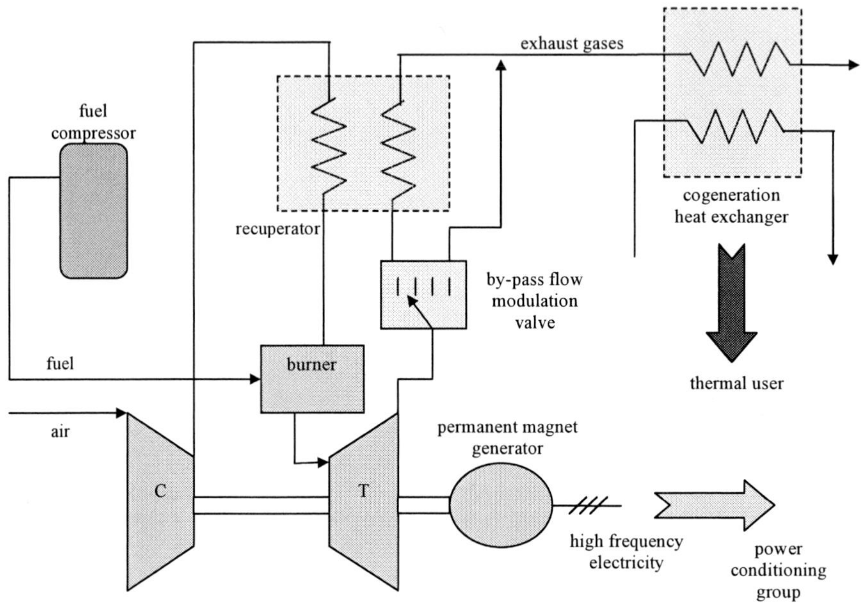

Figure 3.8. Typical configuration and components for a CHP recuperated-cycle MT.

The MT lends itself naturally to operate in cogeneration, since it is technically quite easy to recover thermal energy from the exhausts, without altering the electrical performance of the machine (apart from the small counter-pressure at the turbine outlet [Loz00]). The exhaust gases, after leaving the microturbine, flow through the recuperator (if any) and the cogeneration heat exchanger. The exhaust temperature at the turbine outlet can be from 500 to 700 °C, while at the recuperator outlet can be of about 300÷350 °C, and is usually cooled down to as low as 100 °C in the cogeneration heat exchanger.

The exhausts, after pre-heating the compressor outlet air, can be utilized for several cogeneration applications, similar to the ones possible for the heat recovered in a HRSG from ICE exhausts:

- They can flow into an air-to-water recovery boiler, to produce hot water or low-pressure steam. The hot water supplied to the thermal user usually ranges between 60 °C and 150 °C [BoK01][Per05]. However, normal figures for the in-out temperature at the heat exchanger (user side) are 70÷90 °C.
- They can flow into an air-to-air heat exchanger to provide space heating or air drying.

- They can be directly utilized in ovens or similar industrial and manufacture applications.
- They can drive absorption chillers (directly-fed by the exhaust gases, or fed by hot-water or low-pressure steam in turn generated in a HRSG).

The overall MT efficiency can typically reach up to 80÷85% [BaC99][Per05], also depending on the temperature at which the hot water is supplied (see also Figure 3.5) and the exhaust gases are cooled down to. The recovery system can be controlled in such a way as to accommodate the user's needs over the time. This can be done by means of fan radiators to dissipate the excess heat [BoK01][Per05] and a step recuperator (in case) [BaC99]; of course, this kind of modulation is made at the cost of increasing the overall energy losses.

Rated values of the cogeneration ratio range from about 1.5 (recuperated cycles) to as high as 5 (non-recuperated cycles). Moreover, given a unit equipped with an add-on step recuperator, by modifying the position of the flow modulation valve it is possible to obtain a cogeneration ratio spanning from 0 (all the thermal power is wasted off by means of the radiators) to 6 or even higher. In this way it is possible to endeavour to *simultaneously* follow the electrical *and* thermal load, thus improving the energy system management.

3.1.3.4. Considerations on single-shaft MTs and comparison with ICE technologies

Diesel and Otto ICEs are the kings in the market in which MTs have entered recently. ICEs are well known and reliable technologies, with higher electrical efficiency and lower investment cost than MTs, therefore with essentially lower investment costs and fuel costs for electricity production. However, there are several advantages from MT systems with respect to natural gas and diesel ICEs, especially when considering cogeneration applications. In fact:

- microturbine NO_x emissions at rated conditions may typically be 2÷5 times less than for gas ICEs and 10 times less than for diesel ones (however, for the sake of completeness it must be noticed that off-design emissions might increase consistently [MaC09]);
- the outlook for future electrical efficiency improvement is much higher for MTs than for ICEs;
- the partial-load electrical efficiency drop is slightly better than for gas ICEs;
- weights, size, and footprints are smaller, with subsequently higher installation ease and possibility to find adequate plant room;
- noise and vibrations are much less, and the produced noise is at high-frequency, so that is relatively easy to control [Kre01]; this makes MTs very suitable for installations also in residential districts and commercial buildings, with no need of expensive abatement systems;
- the possibility of adopting several fuels with minor adjustments is wider;
- maintenance tends to be cheaper [PiW05]; in particular, diesel groups require more frequent and onerous maintenance (working hours to maintenance hours ratio of about 20, against 500÷1000 for MTs [BaC99]);
- the cogeneration ratio is higher and, in case, adaptable, which can make MTs fit to some specific CHP applications better than ICEs;
- the MT dynamic response for load-following applications and islanded operation is generally better: ICE synchronous generators have high internal reactance, along with large time constants of the excitation system; MTs, instead, thanks to the electrical load/grid coupling through DC bus and controllable inverters, boast more effective, more stable and faster voltage and frequency control [BaC99];

- controllability and operation flexibility are much better, thanks to the hi-tech PCU;
- cluster-based operation and control of several units is far easier, thanks to the DC bus and the PCU, which allows better dispatching and deployment of the single units, which leads to higher efficiency when the overall load is partialised.

3.1.4. Stirling engines

A Stirling engine [BoK01][Kre01][OnU06][Peh06b][KuK08] is an *external* combustion heat engine, operating on a closed Stirling thermodynamic cycle. This is a major difference with respect to "classical" *internal* combustion. External heat is supplied at a high temperature to the engine heater head by a continuous combustion burner, and waste heat is rejected at ambient temperature. However, SEs are also based on a reciprocating motion principle (with an intermediate medium typically being hydrogen or helium) in all similar to the one produced in an ICE, and as such can be coupled to an alternator to produce electricity. Although conceived more than one century ago, in the past reliability problems had limited its use mainly to hobbyists, while today the overall technology improvement trend that has brought *in auge* DG has allowed development of units with commercial potential.

With adequate recovery systems, SEs can in theory reach up to 95% of overall efficiency, since almost all the heat energy can be recovered. In practice, values of overall efficiency close to about 90% are feasible, provided that hot water generation is within certain ranges [Tho08]. Electrical efficiency may also be relatively high even for micro-units (below 1 kW_e), in principle as high as 30% to 40%. In practice, smaller units reach electrical efficiencies of the order of 10÷15% also due to efficiency-of-scale reasons, while larger units (but anyway below 50 kW_e) exhibit values close to 25%.

As the combustion is external, it does not require a specific fuel. Indeed, SEs may be operated using propane natural gas, gasoline, diesel, wood, biomasses, and even solar energy. Nevertheless, most of the units that are being commercialized are fed on natural gas, whereas efficiency penalty might arise in the presence of solid fuels [KuK08]. Typically, owing to the high amount of heat generated, Stirling engine application is mostly associated with household products that substitute conventional gas boilers and produce *also* electricity. Hence, although also ICEs may be used for such applications, the so-called *micro-CHP* systems [Peh06b] are often associated to SEs with electrical capacity within the range 1÷10 kW_e and thermal one in the range 7÷50 kW_t. Indeed, the external combustion allows a relatively better control of the combustion process, and the fewer moving components with respect to ICEs reduce vibrations and wear. Thus, SEs may be preferred to classical ICEs owing to negligible operation noise level and small pollutant emissions, as well as potentially excellent reliability, with little maintenance required relative to ICEs and even heat-only boilers.

Today, there are several residential micro-CHP applications run mostly by utilities and manufacturers, in both pilot plants and actual household installations. However, although commercial diffusion is starting in these years, their uptake is still limited with respect to the initial forecast. Indeed, due to their high cogeneration ratio (typically between 5 and 8), such systems are suitable for buildings with very high thermal requirements with respect to the electrical ones, so that the engine does not work at partial loads or on/off cycle too often. Indeed, discontinuous and partialised operation can worsen the engine performance (as well as reliability and lifetime) consistently, thus decreasing the estimated energy and environmental benefits [CaT07]. In addition, in current regulatory frameworks incentives to micro-CHP units (relatively more expensive than ICEs or MTs) may be not sufficient to allow widespread adoption.

Nevertheless, SEs represent a promising distributed technology for DMG applications in the next years, also owing to enabling technologies such as thermal storage systems [HaL07][HoA08] or to potential coupling to low-temperature absorption/adsorption systems for CCHP applications [HuW07][HuW07b][DeW08]. In fact, both options could flatten out the engine operation and thus improve the energy and economic performance across the year.

3.2. Combustion Heat Generators

The core of a cogeneration plant is the cogenerative prime mover that produces electricity to deliver to the electrical user/electrical grid and heat to deliver to the thermal user/district heating network. However, in nearly every cogeneration plant also auxiliary heat generators for thermal peak-shaving or back-up supply are usually needed and thus present. In general, the heat generator is combustion-based, as it allows higher reliability and versatility, especially if it can be fed by a wide range of fossil fuels (natural gas, fuel oil, wood, coal, and so forth) [Ang98][Kre01]. This section is dedicated to illustrate the main aspects concerning combustion heat generator design and applications to cogeneration plants. Many general considerations on combustion and burner characteristics can be also applied to ICEs and MTs. Other types of heat generators such as heat pumps, which are typically electricity-fed and are often used also in a reversible way as chillers, are addressed in Section 3.3.6.

3.2.1. General aspects of heat generation groups

Combustion heat generators can be ideally and conceptually seen as a two-side system at two temperature levels (Figure 3.9):

- the *hot side*, composed of a burner, a combustion chamber and the hot-side of a heat exchanger, in which the thermal energy is generated and transferred;
- the *cold side*, which basically is the cold side of a heat exchanger that receives from the generator hot side the thermal energy to be delivered to the user.

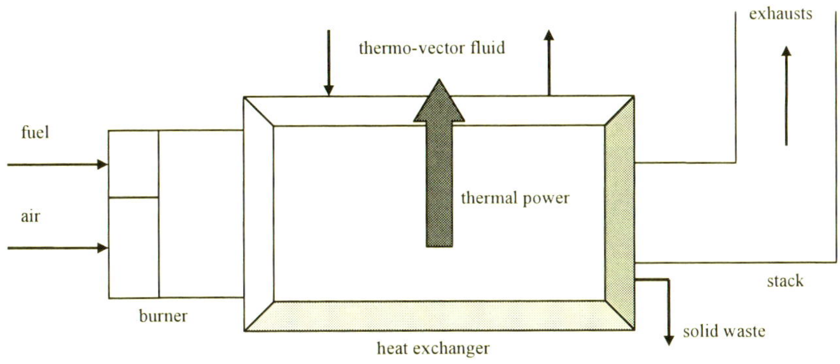

Figure 3.9. Combustion heat generator schematic model.

In the *hot side*, the borders of the "intermediate" combustion chamber are often not defined, and actually overlap the two subsystems. The thermodynamic model refers to an air-fuel-exhausts system operating under steady-state conditions and at ambient pressure. Following the air-fuel-exhausts path, the following stages can be pointed out [Ang98]:

- the circuit begins in the atmosphere, where the combustion air is drawn from;
- in the *burner*, the combustion between the two reactants, *i.e.*, air and fuel, gets started and a steady flame is generated, which develops in the *combustion chamber*;
- the combustion reaction continues in the hot side of the *heat exchanger*, where the heat generation properly occurs; the flame from the combustion chamber and the combustion products transfer heat to the cold side of the heat exchanger, through which the thermo-vector fluid delivers the thermal energy to the user flows;
- the exhaust duct (*stack*) leads the flue gases outside into the atmosphere, releasing them at a non-dangerous height (so that the polluting concentration at the ground level is low enough); the exhaust circuit can be pressurized, under-pressurized, or both, depending on the fan system used to overcome the resistance to the mass motion in the burner, heat exchanger and stack;
- when coal or fuel oil are adopted, a part of the combustion products is also dumped out of the generator shell as solid waste.

3.2.2. Boiler efficiency and losses

Boiler performance is reported in manufacturer's data sheet and is fundamental for device selection. However, often there is the need, the willingness or the requirement to assess "on the field" the actual efficiency. Therefore, a first way to simply evaluate it from field measurements (*direct method* [Ang98]) is by means of the expression

$$\eta^{\mathrm{CHG}} = \frac{Q}{F} \tag{3.5}$$

where η^{CHG} is the efficiency of the combustion heat generator; Q is the thermal power rendered available from the combustion to the heat distribution circuit and that in general may be not straightforward to measure; F is the fuel thermal power delivered from the chemical energy contained in the fuel mass flow $\dot{m}_F$ through the combustion process:

$$F = \dot{m}_F \cdot LHV \tag{3.6}$$

The *LHV* can be calculated once known the fuel composition [Loz00][Kre01].

It is worth mentioning that *condensation boilers* allow the recovery of also the condensation latent heat from the water contained in the exhausts. In this case, the *HHV* (Higher Heating Value) should be used when evaluating the thermal energy available from the combustion reaction [Loz00][Kre01]. Therefore, when using the *LHV* for evaluation of the performance of a gas-fired condensation boiler, efficiencies higher than 100% could occur, which is the pride of several manufacturers. Typical *LHV*-based efficiencies may thus be in the practical range of 0.95÷1. Most condensation boilers are gas fired, since exhausts from oil and coal must be kept at a temperature high enough (150 °C [AlF97]) to avoid condensation. In the liquid, in fact, there might be sulphur oxides generated during the combustion, which are corrosive acids when dissolved in liquid water and would require costly stainless steel stacks [Kre01]. However, sulphur oxides also contribute to acid rains, so that their emissions are subject to stringent controls.

The same considerations apply to the stack discharging the flue gases from MTs or ICEs.

A more theoretical way to define the boiler efficiency is based on subtracting from the fuel thermal energy F the overall Q_L occurring in the boiler (*indirect method* [Ang98]):

$$\eta^{\mathrm{CHG}} = \frac{F - Q_L}{F} \tag{3.7}$$

This approach to boiler efficiency definition has the upside of making it clear that losses need to be reduced in order to improve the performance. Despite its appearance, the term Q_L is often easier to measure and/or evaluate than Q in (3.5). A full understanding of the single terms that origin Q_L is also important for the understanding of boiler partial-load characteristics. The term Q_L can be therefore broken down into the sum of the following components [Ang98][Kre01]:

- *Dispersion heat losses* Q_{L_δ}, due to the unwanted heat exchange between the boiler shell and the environment, at first approximation (considering convection-only heat transfer, neglecting a small amount due to radiation) proportional to the difference between the thermo-vector fluid temperature (for instance, T_o at the heat exchanger outlet) and the ambient temperature T_A. This term is usually small, thanks to the good thermal insulation of modern boilers. Usual ranges are between 0.1% and 5% of the rated output, with lower figures for bigger devices, for which the dispersing surface to heat production volume ratio is smaller.
- *Sensible heat losses in flue gases* Q_{L_σ}, due to the fact that the exhausts are released at a relatively high temperature ("*stack*" *losses*).
- *Latent heat losses in flue gases*, due to combustion of hydrogen into water (when it is not recovered in condensation boilers).
- *Heat losses due to incomplete combustion of carbon*, to avoid because they imply the presence of toxic carbon monoxide in the flue gas, besides energy waste; current emission constraints in most countries make this loss negligible, with volume concentration of CO usually kept below 0.1%.
- *Heat losses from hot solid waste*, due to unburned carbon in ash from coal and oil.

The most significant losses are associated to flue gases and depend on their stack temperature. In traditional boilers, exhaust temperature at the stack outlet is 150 to 180 °C, with a lower limit to avoid problems with acid condensing of about 120 °C. Typical figures are reported in Table 3.4. As mentioned above, when the reference for fuel energy calculation is the *LHV*, for condensation boilers the rated efficiency may become even higher than unity.

Table 3.4. Typical boiler efficiencies (*LHV* reference)

boiler	*exhaust temperature at the stack outlet* [°C]	*rated efficiency*
conventional	150	0.8÷0.85
conventional	120	0.9÷0.95
condensation	40	1.05

3.2.3. Partial-load characteristics

Combustion heat generator operation varies over time in order to follow the user's needs and depends on the plant control strategy. Because of the wide variations that the thermal power usually experiences (for instance, due to seasonal and night/day cycling), it is indeed common for heat generators to operate at partial loads. As a consequence, neglecting partial-load behaviour might bring to biased efficiency and economic evaluation of the CHG plant [Kre01].

Information on partial-load characteristics and efficiency from manufacturers' data is in general not readily available. Furthermore, examples found in the literature do sometimes report contrasting information on the partial-load behaviour (for instance, with efficiency dropping rather than increasing at partial loads). The approach followed here is taken from [Ang98]. This approach is sound from a theoretical standpoint and well matches experimental characteristics.

The bigger contributions to efficiency drop in a CHG are due to dispersion losses and stack losses, so that its partial-load behaviour can be assessed with good approximation by means of the simplified formula

$$\eta^{\mathrm{CHG}} = 1 - L_{\delta} - L_{\sigma} \tag{3.8}$$

where L_{δ} and L_{σ} are, respectively, the dispersion losses and stack losses expressed in *per unit* (abbreviated as pu) of the fuel thermal power *F*.

The amount L_{δ} rises when the thermal load decreases: indeed, the associated thermal losses $Q_{L_{\delta}}$, that are a function of the operating temperatures, remain roughly constant since the thermo-vector fluid temperature often does not depend on the load. On the other hand, the power *F* supplied through the fuel mass flow rate is directly related to the thermal power to deliver, regardless of the control system adopted.

The amount L_{σ} depends on the *control system* implemented on the heat generator [Ang98], usually a thermostatic control at the burner. In general, the feed-back variable is the temperature of the thermo-vector fluid at the outlet of the heat generator; the controlled variable is the fuel thermal power *F*:

- adopting a *modulating burner*, the fuel mass flow is controlled according to (3.6) following the required thermal load; with this type of control, usually if the fuel mass flow decreases the exhaust temperature also decreases, and so does L_{σ};
- adopting an *on-off burner*, the CHG operates only at rated $\dot{m}_F$, and the control is carried out by on-off switch cycling for adequate time intervals controlled by the thermostat settings. This control is rougher than the modulating one, and forces the heat generator and the distribution system to fatigue thermal cycles; however, it is also cheaper and therefore is adopted in smaller and/or older devices. With this control, stack losses are roughly constant: in fact, when the burner is on, the exhaust temperature and composition does not change significantly from rated values (apart from negligible inertial phenomena), and therefore nor does L_{σ}.

The above considerations are summarized in Figure 3.10 (elaborated from [Ang98]), which shows typical partial-load characteristics (stack losses, dispersion losses and corresponding efficiencies) for boilers with on-off burner control and modulating burner control. The modulating field is supposed to be in the range 0.3÷1 (usual values): below this interval, on-off control is performed. Below 20÷30% of the rated load, however, actual partial-load characteristics may drift from theoretical ones and efficiency results very poor, so that usually the boiler is not operated with these load levels when possible. Commonly, boilers can also be operated in over-load by typically 10% [Kre01] with efficiency very close to the rated one.

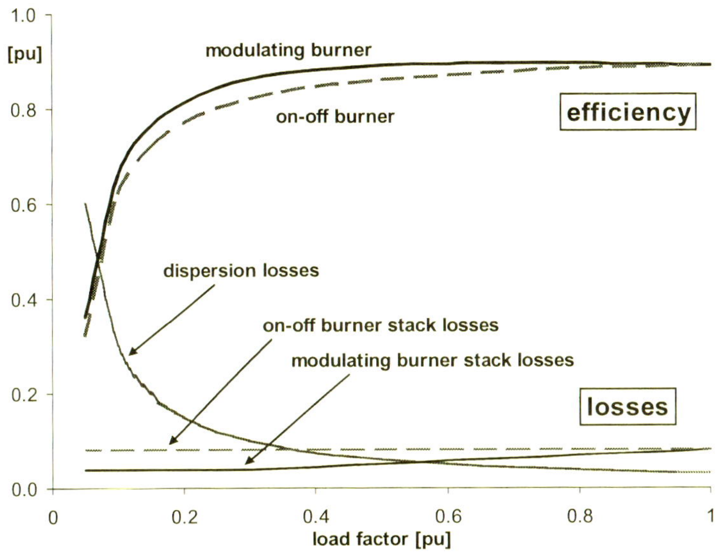

Figure 3.10. Typical partial-load characteristics for boilers with on-off and modulating burners (pu values are referred to rated base values).

A physical-mathematical model, suitable for numerical implementation, can be formulated by expressing the boiler efficiency as

$$\eta^{CHG}(Q) = \frac{Q}{Q + Q_L(Q)} \tag{3.9}$$

which is just another way to express (3.8) pointing out the thermal losses Q_L in function of the thermal output Q. In general, thermal losses in a thermal device, regardless of the control technique, depend on models related to the heat transfer (radiation, convection and conduction), thermodynamics and fluid-dynamics laws [BeT96][Bej97][Kre01][BeK03].

According to the considerations carried out above and to practical numerical aspects of commercial boilers, the thermal losses can be modelled fairly well as a polynomial function (typically of the second order, or even of higher order) of the output power:

$$Q_L = \nu_2 \cdot Q^2 + \nu_1 \cdot Q + \nu_0 \tag{3.10}$$

where ν_2, ν_1, and ν_0 are the model parameters. The relation (3.10), applied in (3.9), yields an efficiency model as a function of the load Q that often fits very well data available from commercial catalogues or found in field tests, as well as characteristics like the ones plotted in Figure 3.10.

A "blinder" model can be worked out just considering that, in practice, boiler efficiency at partial loads is fairly constant within the control range. Fit-to-data models of order as high as the second or the third, therefore, can well link the output variables (thermal load, *Q*) to the input variables (fuel thermal power, *F*) at partial loads, such as, for instance,

$$F = \nu_2' \cdot Q^2 + \nu_1' \cdot Q + \nu_0' \qquad (3.11)$$

where *F* and *Q* are expressed in per unit of the rated values [Kre01] and again ν_2', ν_1' and ν_0' are fit-to-model parameters.

For CHGs, it is also possible to define an indicator similar to the Integral Part Load Value (*IPLV*) used to characterize overall energy performance of a chiller over a certain period [Pal04] (see also Section 3.3.3.5). However, as the efficiency for a boiler is fairly constant in a wide range, in several European countries directives on the minimum efficiency values (for energy saving purposes) are often given only at the boundaries of the modulation field, *i.e.*, typically 100% and 30% of full load (see [Ang98] for Italian values).

3.3. Cooling Generation Plant Equipment

3.3.1. Generalities on cooling plants

The first and main purpose of a CGP is to "*produce cooling*" (or *cold*), that is, in other words, to subtract heat from a system (in a thermodynamic sense) whose temperature is lower than the one of the environment where this heat is then discharged.

In order to produce cooling, an *operating fluid* goes along a thermodynamic inverse cyclic transformation in specific machines, called *chillers* (usually when the desired cooling effect is produced by means of *water* at a temperature above 0 °C, typical of air conditioning applications) or *refrigerators* (usually when the desired cold effect is produced at temperatures below 0 °C, typically for industrial processes). In this work, the two terms are considered as synonyms, and the general considerations that are carried out apply to any kind of cold temperature levels. However, specific examples mostly refer to air conditioning applications [Afo06].

There are several applications for which it is desirable to produce cooling. First of all, air conditioning has rapidly grown in recent years, passing from being a niche luxury only few could afford to a standard included in most of residential, tertiary and commercial applications, such as hospitals, school buildings, offices, and so on. Moreover, a wealth of industrial applications (food processing, cold storage, manufacturer plants, just to mention some) require the production of thermal energy in the form of cooling at various temperature levels, even up to -60 °C.

In the present work, the interest is mostly on the interaction of the cooling plant, seen as a *generator* of thermal energy in the form of cooling, with other energy sources which can be cascaded with it, at a topping, bottoming or also parallel level, and which all together make up a *trigeneration* or a *multi-generation system*. Within the general DMG framework, cooling plants suitable for small-scale applications (below 1 MW_c) are addressed in detail, although all the models and the methodologies developed can be extended to larger components outright. In the range of capacitities considered, the refrigeration or air conditioning needs can be satisfied by cooling production from a large variety of electric (or, more generally, mechanical) or heat-activated equipment, as well as by a combination of them. In particular, electric vapour compression chillers, direct-fired and indirect-fired absorption chillers, and engine-driven chillers are analyzed. Hints are also given on adsorption chillers. All these chiller types can often operate in reversible mode as heat pumps. In this respect, special care is dedicated to electric heat pumps (that can in turn operate also in cooling mode). In fact, EHPs represent a powerful tool for enhancing the DMG plant overall performance, thanks to their relatively high efficiency and to the possibility of improving the heat recovering rate.

The focus is set on those aspects connected with small-scale generation of cooling power and heat recovery within MG energy systems, especially at the interface with other energy sources, giving only few hints on those aspects more tightly related to the user's side.

3.3.2. Cooling plants characteristics

A CGP is aimed at keeping an ambient at a given temperature T_2, lower than a reference (typically the outdoor) temperature T_1, by dragging form the ambient a thermal power R_2 (cooling power), while going along an *inverse thermodynamic cycle* [CaG96][Bej97][AlF97] [Ang98][Kre01]. In order to do so, a certain amount of energy is needed to lift the energy level of an intermediate fluid from the "cold" state (at a temperature lower than T_2) to the "hot" state (at a temperature higher than T_1). The energy input can be in various forms, giving birth to different categories of chillers. The main energy inputs can be thermal energy (absorption/adsorption chillers, for both IFAC and DFC categories, according to the definitions in Section 2.3) or mechanical/electrical energy (vapour compression refrigerators, which can be framed in the EDC and CERG categories, Section 2.3).

Some data referring to *typical* figures for vapour compression chillers and absorption chillers [AlF97][Ang98][Kre01] are shown in Table 3.5.

Table 3.5. Characteristics of cooling equipment used for refrigeration and air conditioning

vapour compression chillers			
lowest temperature achievable [°C]	pressure [bar]	cooling power [kW_c]	compressor
-25	> 1	0.1÷30	reciprocating hermetic
(basic cycle)		30÷500	reciprocating semi-hermetic
		250÷1500	reciprocating open
-60		150÷500	screw semi-hermetic
(double		500÷3000	screw open
compression cycle)		3÷350	scroll
-150		300÷6000	centrifugal semi-hermetic
(cascade cycles)		300÷30000	centrifugal open
absorption chillers			
lowest temperature achievable [°C]	pressure [bar]	cooling power [kW_c]	refrigerant/absorbent
0	0.01	50÷5000	$H_2O/LiBr$
-150	0.2	5000÷10000	NH3/H2O

For vapour compression chillers, the most important element that distinguishes a type of equipment (and therefore its performance) from another is the compressor used to lift the energy level in the cycle (as well as the type of cycle used to perform this energy lifting). For absorption machines, instead, the main distinction is based on the fluids used in the cycle. However, also for the vapour compression machines the refrigerant used in the thermodynamic cycle play a fundamental role, especially after the latest regulatory requirements in matter of environment, as it influences both the compressor and the cycle.

Besides the technologies analysed in this work, others are available on the market. However, these technologies are limited to special or niche applications, such as water vapour compression plants for water refrigeration and baby food industry, gas compression plants for aeronautic applications, thermoelectric devices (without refrigerant fluid) for electronic and aerospace applications, and so forth [Ang98].

Special mention deserves a particular category of cooling equipment whose cycle is based on vapour compression, but whose mechanical compressor (in general screw or reciprocating) is "thermally-driven" by a combustion engine. Hence, in terms of energy balances the primary source can be considered to be thermal, as for absorption chillers, so that they also belong to the DFC family (Section 2.3.1). This aspect will be relevant for formulating suitable energy and environmental models, as it will be clearer in the sequel. This equipment type, called *engine-driven chiller*, is particularly interesting because of potential recovery of the rejected heat (as in the case of cogeneration through ICEs).

3.3.3. Vapour compression chillers

Most of the systems used for Heating Ventilation and Air Conditioning (HVAC) as well as industrial applications such as food process (refrigerating and freezing) use vapour compression cycle chillers to generate the desired cooling effect. The first patent on a mechanically driven refrigerator dates back to as far as 1834, while the first commercialized equipment was produced in 1857 [Kre01]. In spite of that, today vapour compression chillers are undergoing large modifications, re-designs and re-thoughts, especially because of the banning of most refrigerant fluids used so far due to environmental issues [MoR74][SaS92][Kre01]. This section goes through the main characteristics of the vapour compression refrigerators, with special regard to the equipment suited for small-scale applications and electrically-fed (CERG).

3.3.3.1. Thermodynamic aspects and components

Vapour compression chillers are typically electricity-fed or, more generally, mechanically-driven. They are based on a *bi-thermal cycle*, with two energy interfaces ("reservoirs", "sinks", "sources") at the temperatures T_1 and T_2, as shown in Figure 3.11. The subscripts i, 1 and 2, already used in the previous paragraph, refer to typical notations used in thermodynamics for, respectively, the input, the higher and the lower temperature, as well as all the energy flows exchanged with the thermal sinks/sources at those temperatures.

In terms of notation, R_2 is the desired cooling effect (in the picture, the R_2 flow direction points out heat *drawn* from the coldest sink), W_i is the electrical (or mechanical) energy input, Q_1 is the energy discarded. Indeed, the "hot" system receives the *heat rejected* from the cycle, sum of the heat drawn from the source to cool down, plus the energy input needed to perform the compression, so that the first principle of thermodynamics applied to the system in Figure 3.11 yields:

$$Q_1 = R_2 + W_i \tag{3.12}$$

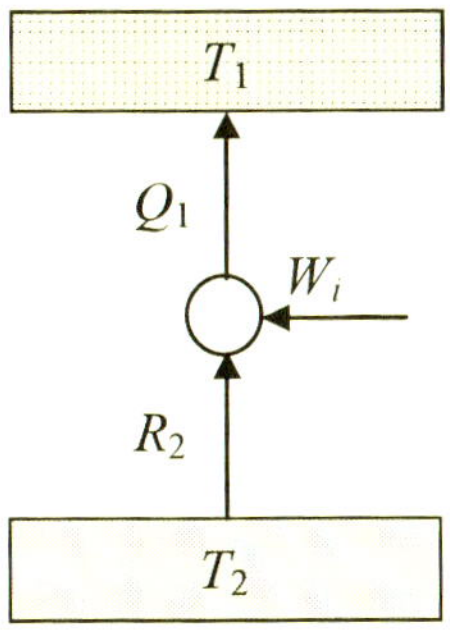

Figure 3.11. Vapour compression chiller general scheme (T_1>T_2).

A conventional efficiency can be defined as the ratio of the desired output to the input cost necessary to obtain it (see also Section 5.2.1). For a vapour compression cycle, the *cooling effectiveness* (or *efficiency)* ε, or *Coefficient Of Performance* (*COP*), is defined on the basis of the obtained *useful effect* as ratio of the cooling energy (desired output) to the energy cost (input):

$$COP = \frac{R_2}{W_i} \tag{3.13}$$

The upper performance reachable in an inverse cycle is given by the Carnot efficiency ε_C from an ideal inverse Carnot cycle [CaG96][Bej97][Kre01][Dan06]:

$$\varepsilon_C = \frac{T_2}{T_1 - T_2} \tag{3.14}$$

Of course, in real equipment the actual efficiency is far lower than the theoretical one due to energy losses and the technical impossibility of operating along a Carnot cycle.

In a vapour compression cycle [AlF97][Ang98][Kre01], a working fluid (the *refrigerant*) evaporates and condenses at temperatures and pressures suitable for cost-effective equipment design given the cooling needs. The four main components in this kind of cycle are compressor, condenser, expansion device, and evaporator. More specifically:

- The *compressor* raises the pressure of the refrigerant to such a level that the corresponding saturation temperature is slightly above the temperature of the cooling medium used in the condenser. The type of compressor depends on the application (cooling power and minimum temperature at which this cooling power is needed).
- The *condenser* is a heat exchanger used to reject heat from the refrigerant to a cooling medium, typically air (especially residential-sized applications) or water (larger chillers).
- The *expansion device* (a throttle valve) makes the refrigerant expand (after leaving the condenser) down to the evaporator pressure.
- The *evaporator* is a heat exchanger circulated by the refrigerant (after the expansion) and by the medium to cool (air or water); the refrigerant is now at a temperature lower than the temperature of the medium to cool and therefore absorbs energy from this latter one, so closing the thermodynamic cycle.

A simplified diagram of a vapour compression cycle is shown in Figure 3.12, where also the (desired) cooling effect and the disposed heat, as well as the input energy, are indicated. The input energy is supposed to be electricity, as for a CERG, but more in general it could be mechanical energy, for instance from the mechanical shaft of an EDC or a turbine. The direction used for R_2 in this case points out that R_2 is a cooling flow, with a useful effect, in kW_c, delivered *to* the user (in terms of heat flow, heat would be ideally drawn *from* the user).

The heat Q_1 must be rejected through the condenser to the environment, for instance directly to air-cooling the condenser, or indirectly to water and then to a cooling tower (for larger powers and most of absorption equipment applications). Sometimes, depending on the temperature the heat is rejected at, it is possible to re-use the otherwise wasted heat to provide simultaneous heating to the buildings or for sanitary applications (chillers with heat recovery condenser, as detailed in Section 3.4).

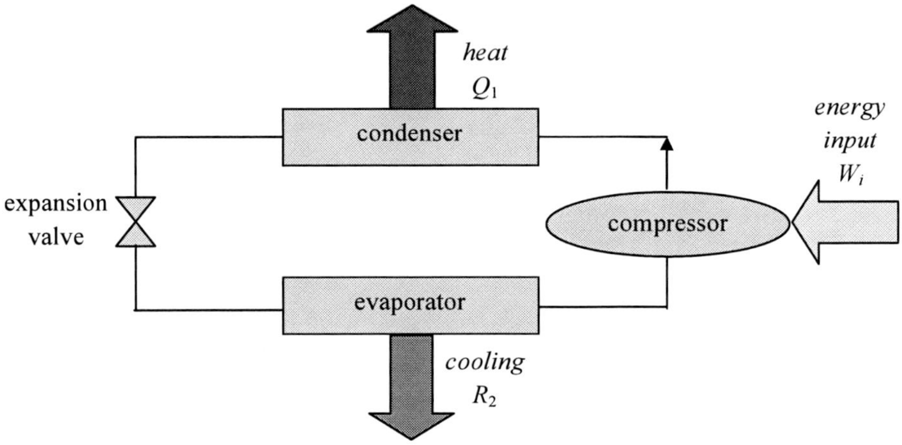

Figure 3.12. Simplified schematic diagram of a vapour compression cycle.

3.3.3.2. Refrigerants

The refrigerants used in cooling plants are indicated by symbols composed of the letter R ("refrigerant") followed by a series of numbers and, in case, letters, related to their chemical structure [CaG96][AlF97][Ang98][Kre01][Dan06].

The first "classic" refrigerants used in cooling plants since last century's beginning were natural, among which water (H_2O, *i.e.*, R-718), ammonia (NH_3, *i.e.*, R-717), CO_2 (which requires high condensation pressures, about 150 bar), and Hydro-Carbons (HC) (although flammable and explosive), such as propane (R-290), propylene (R-1270), and isobutane (R-600a). Ammonia has been for years the king of refrigerants, with optimum thermodynamic and thermo-physics qualities [AlF97][Kre01]. However, since it is moderately flammable and toxic, today its application is mostly limited to industrial refrigeration.

From the 30's till the 80's *Chlorofluorocarbons* (CFC) (and as a sub-family also *Hydro-Chlorofluorocarbons* - HCFC) were largely used in HVAC plants (R-11, R-12, R-13, and above all R-22). Although very stable, non-toxic, non-flammable, and odourless, today CFC production is banned, according to several international agreements [Kre01], and new refrigerants are substituting the traditional ones. Indeed, the presence of chlorine makes CFCs capable to combine with molecules of O_3 and convert it into O_2, therefore contributing to the ozone layer depletion and greenhouse effect [MoR74][SaS92][ExE05].

Refrigerants are classified according to their capability of attacking the ozone layer in the stratosphere through an index called *ODP* (Ozone Depletion Potential), and to their contribution to the greenhouse effect through an index called *GWP* (Global Warming Potential) [AlF97][Ang98][Kre01]. Both ratings are normalized to the value of R-11. Also toxicity and flammability represent a feature according to which refrigerants can be rated, being divided into groups from A1 (non-flammable and least toxic) to B3 (flammable and most toxic) [Kre01].

Refrigerants widely used in the past were the R-11 ($ODP = 1$), R-12 ($ODP \approx 1$) and R-22 ($ODP \approx 0.05$). Today, they are being substituted, in the newer machines, by R-123 ($ODP \approx 0.02$) and R-134a ($ODP \approx 0.25$) [Ang98][Kre01]. R-123 provides efficiency advantages over the R-134a, but it is slightly more toxic so that special guidelines have to be followed in its applications.

An important issue raised in the recent years is to define new indicators that could allow more exhaustive evaluation of the global warming impact due to the use of refrigerants, also including the contribution of system efficiency besides the release of the refrigerant charge itself [Kre01]. For trigeneration systems, this issue is addressed in [ChM08].

3.3.3.3. Compressors

The compressor is the most important component in a refrigerator, so that it often gives the name to the whole machine. Vapour compression chillers normally utilize electrically-fed compressors (some can be also engine-driven), and usually are divided into groups according to their *functioning principle*, namely:

- reciprocating,
- screw,
- scroll,
- turbo-centrifugal.

Other categories into which compressors can be grouped are the type of *construction* (hermetic, semi-hermetic, open) and the *size*. Often these conventional categories are used at the same time to specify a given compressor/chiller.

Selection of the best type of chiller/compressor depends on various factors, namely:

- *Application*: it determines the size of the plant. Moreover, the quality of the service (for instance, tertiary HVAC or industrial one), as well as the influence the refrigerant to adopt (due to safety reasons, for instance), can play an important role.
- *Rated cooling power*: in general, reciprocating chillers, as well as screw chillers and scroll chillers, are more suitable for smaller flow rates and higher compression ratio, compared with centrifugal turbo-compressors.
- *Chilled water temperature* to be reached: turbo-compressors, for instance, are mostly used in large plants where the medium to keep cooled is at relatively high temperature (large air conditioning applications, for instance, where the chilled water is produced at about 7 °C), so that their compression ratio is relatively low.
- *Refrigerant* used: some refrigerants, due to safety reasons, cannot be used or are preferred not to be used in specific applications (for instance ammonia, which is slightly toxic and inflammable, is typically not used for residential air conditioning applications).
- *Compression ratio* (as a consequence of the previous variables): for instance, screw compressors are widespread in industrial applications. In fact, the compression can

often be inter-cooled, since the lubricating oil is able to draw heat from the refrigerant. In this way higher compression ratios (and so lower temperatures) can be reached.

- Expected operation at *partial loads*: as a cooling plant is usually rated on the base of the summer peak, the chiller is bound to work for most of times at partial loads. Hence, specific care is needed when selecting the chiller, also taking into account this aspect and the control strategies it can be subject to.

For the power range of interest in this work, reciprocating chillers and screw chillers are the most suitable equipment. These machines are further investigated in more details. Other general characteristics and information about chiller compressors can be found in [AlF97][Ang98][Kre01][Dan06].

3.3.3.4. Considerations on reciprocating and screw compressors for cooling plants

This section takes a closer look at the characteristics and performance of reciprocating and screw compressors, the leading categories for small-scale electric chillers. In general, there is no clear preference for one type rather than the other, and the selection often depends on the specific case.

Both screw and reciprocating compressors are *positive displacement* compressors [Kre01], in which the pressure increase is caused by a volume reduction. The mechanism for volume reduction basically differentiates the two categories.

The reciprocating compressor is a well-known technology, with a fairly constant efficiency also with different compression ratios (relevant to different condenser and evaporator temperatures) and different cooling outputs; both these aspects are relevant to partial-load operation, which is indeed very good. In addition, reciprocating compressors require relatively simple maintenance. Proper design and siting help avoid annoyances such as liquid blows, vibration and noise. It is a technology in its prime, seeming to have reached the mature stage.

Screw compressors are more in evolution. They perform better for higher capacities (open and semi-hermetic types, above 500 kW_c), with rated efficiency comparable with centrifugal chillers, whereas smaller groups (semi-hermetic, from 100 to 500 kW_c) exhibit efficiency lower than reciprocating compressors. For all capacities, the working process (slide valve and fixed compression ratio) is such that off-design operation is penalised with respect to reciprocating compressors. However, larger exploitation of inverter drives in the future could be able to make up for this shortcoming. Installation of screw compressors is advised where the compression ratio varies within a narrow range over the whole working season (water-cooled groups are then preferred). If the load were to drop under 75% quite often (for instance, for highly variable HVAC applications), dwindling of the capacity into smaller units should be carried out. If noise constraints are a primary issue, then screw compressors may be more indicated. Other upsides are the absence of vibrations, the high reliability, and the possibility of working with variable rotation speed (flexibility). Margins for technological improvement of screw compressors in the upcoming years are large (especially if driven by inverters). However, reciprocating compressors remain the leading technology for small-scale and middle-scale applications.

3.3.3.5. Off-design models

The performance of a chiller may change consistently in off-design operation, due for instance to variations of the outdoor conditions (related to the condenser temperature), or indoor conditions (related to evaporator temperature and cooling load). The *COP* of an electric/mechanical chiller, in fact, depends on various factors, *in primis* the efficiency of the compressor. Considering a simplified thermodynamic model (with no sub-cooling, no super-heating, and no pressure losses [CaG96][Ang98]) with condenser temperature T_1 and evaporator temperature T_2, the *COP* can be formally expressed as

$$COP = \varepsilon_C \cdot \varepsilon_{II} \cdot \eta \tag{3.15}$$

where

ε_C = efficiency of the inverse Carnot cycle operating between the same temperatures;
ε_{II} = exergy efficiency (thermodynamic 2nd law efficiency [DiR07]) of the cycle: in a simplified model, thermodynamic irreversibilities are mainly due to the throttling process in the expansion device, and depend only on the refrigerant used, condensation temperature and evaporation temperature;
η = compressor efficiency (including the electrical drive).

All the parameters in (3.15) may change in off-design conditions, so that the full-load *COP* is often an insufficient indicator even to compare different chillers. In this respect, various indices have been proposed in order to encompass more general situations. The most adopted of such indices is the *IPLV* [Kre01][Dan06], which takes into account the performance at different operational points (100%, 75%, 50%, 25% of the rated load) by weighting the relevant *COPs* with different weights. However, sampled points and weights might change depending on the specific regulation. Hence, apart from the rated *COP*, the *IPLV* or similar indicators allow fairer evaluation of the likely performance of different chiller types when put to operate in real conditions.

From the general expression (3.15), there is a tight interaction among different parameters, namely, the temperatures of the ambient to keep cooled and of the hot sink (impacting on the evaporator and condenser temperatures, and therefore on the cycle performance), as well as the loading level (particularly impacting on the compressor performance, also depending on the regulation system adopted [Man06][Dan06]). In addition, loading level and condenser and evaporator temperatures are correlated [Dan06]. For instance, in general once fixed the required indoor temperature, if the outdoor temperature decreases also the cooling loading level decreases (less cooling power is required to keep the desired comfort level). Although depending on the specific compressor [Kre01][AsB03], as also discussed above, partial-load operation may often mean worse chiller performance; on the other hand, lower outdoor temperature favours the condenser operation, increasing the chiller performance [Dan06]. At first approximation, these two effects may counterbalance each other, so that in most cases the *COP* keeps roughly constant while the load changes [YuC05]. Hence, at first approximation it is possible to formulate off-design models that are only a function of the outdoor temperatures (fixed the internal one), with no partial-load efficiency penalty. This is also consistent with the data usually provided by manufacturers (in table form as a function of chilled water and cooling water/air temperatures). In addition, for relatively bigger groups often the capacity is dwindled over several units in order to better track the load and guarantee higher reliability. In this case, the overall *COP* can be considered constant at partial load and varying only with outdoor conditions with even better approximation. The same also applies at first approximation in the presence of variable speed drive, as in most of new relatively large applications [Dan06]. However, if the manufacturer provides detailed off-design information, of course more precise models can be derived case by case for the specific chiller.

3.3.4. Absorption chillers

3.3.4.1. General characteristics

In order to generate cooling power, *absorption chillers* [Kre01][Dan06] use a heat source [ToJ98], instead of electricity (or more in general mechanical energy) as for traditional chillers. Hence, the major incentive to using an absorption chiller has always come from the possibility of exploiting thermal waste available on-site. A variety of applications can be proposed for absorption machines, although the main market in most countries is relevant to production of chilled water for HVAC in residential, tertiary and commercial buildings [MeB08], as well as the production of low-temperature cold (below 0 °C) for industrial applications [CoG03]. As economic conditions of energy supply (namely, gas and electricity, above all) as well as climatic characteristics vary from country to country, absorption systems may represent, at the same time, a niche technology in one country and a very profitable one in another. On these premises, this section is dedicated to the major aspects concerning performance and operation of absorption chillers, with special focus on their features for DMG applications.

The first absorption machine was patented in 1859 and used an ammonia-water solution. Lithium bromide and water were not used until 1940. After the 60's, when this equipment well competed with electric chillers for large building air conditioning, the rising of prices of oil and gas led absorption machine operations to be more expensive than electric ones, until the 80's, with both decrease of gas and oil prices and the introduction of the more efficient two-stage technology [Kre01]. Today absorption machines are competitive again with electric machines for a large range of applications. Indeed, they offer some advantages over conventional electric chillers, among which:

1. they do not use refrigerants causing concerns about ozone depletion or global warming (namely, CFC or HCFC refrigerants);
2. they can be fed by a variety of sources, such as natural gas, steam, solar-heated water, waste heat (typically cogenerated), which can be positively exploited to reduce the electrical demand when electric peak rates are high;
3. they are far less noisy, due to the absence of rotating parts.

In order to perform the absorption and "thermo-compression" operations that characterize absorption machines, two fluids are needed, namely the *refrigerant* (or *absorbate*) and the *absorbent*; refrigerant and absorbent together constitute what is called a *working pair*.

The first absorption machine classification can be made according to the working pair employed. Many pairs have been proposed through the years but only two of them have been widely used:

1. *H_2O-LiBr systems*: water as the refrigerant and a solution of lithium bromide in water as the absorbent;
2. *NH_3-H_2O systems*: ammonia as the refrigerant and water as the absorbent.

The ammonia-water pair is mostly found in refrigeration applications, with evaporation temperatures lower than 0 °C; pressure levels are usually above atmospheric pressure. The water-lithium bromide pair, instead, is widely used for air conditioning applications, where it is not necessary to reach temperatures below 0 °C; of course, in order to produce chilled water the evaporation temperature must be low enough, so that the refrigerant (water) is brought to operate in partial vacuum (less than 1 kPa at the evaporator, 5 to 7 kPa in the condenser [Kre01]).

The second important classification for absorption chillers refers to the method used to supply the thermal input [Kre01]:

1. *Indirect-fired* absorption chillers (IFAC, Section 2.3) use steam or hot liquids as the energy source, such as hot water or steam from a boiler (at least in principle), GT, MT, ICE, and so on. The most natural application of these chillers is thus to recover heat cogenerated in CHP systems.
2. *Direct-fired* absorption chillers (belonging to the general category of DFC, Section 2.3) use the thermal energy produced by the direct combustion of fossil fuels: the combustion section is basically as for a boiler, and the fuel used is usually natural gas; to some extent, this equipment could be considered as a boiler that produces cold instead of heat. Reversible versions of absorption chillers can indeed be switched from cooling to heating mode so as to obtain an *Absorption Heat Pump, AHP* [AlF97][KeP08].
3. *Heat-recovery* or *exhaust-recovery* absorption chillers use waste gases, typically exhaust gases from CHP systems, as the heat source [MoC01][ViB07]. Often, this kind of systems are also classified as "direct-fired", with the exhausts circulating in the "hot side" of an internal heat exchanger and providing (through the "cold side") the thermal energy needed in input by the chiller.

Moreover, absorption chillers can be classified according to whether the cycle is single-effect or multi-effect [ToJ98][Kre01][Dan06][WuW06]. In particular:

1. In a *single-effect* absorption machine [KiI08], all the condensing heat cools down in the condenser and is rejected to the atmosphere usually by means of a cooling tower.
2. In a *double-effect* absorption machine, the generator (see below) is split into a high-temperature generator, fed by external heat, and a low-temperature generator, fed by internally recovered heat, so improving the overall cycle efficiency (exactly in analogy to turbine regeneration or power combined cycles). By cascading this process, triple-effect chillers, quadruple-effect chillers, and so on, can be obtained [Dan06], with two main schemes for the internal heat exchanges, namely, a series-flow cycle and a parallel-flow cycle [KiZ02].
3. *Triple-effect* machines [DeM90][MoC01][KiZ02] are under commercialization and represent a promising technology for the next future, with energy performance potentially better than for double-effect chillers by about 30%, although maintenance problems related to corrosion in H_2O/LiBr chillers operating with very high temperatures might arise [KiZ02][Dan06][WuW06].

Since the fluids most commonly used as the working pair in the commercial, tertiary and residential sector (suitable for small-scale applications) are by far water and lithium bromide, in the sequel these systems are analysed in further details. However, apart from the numerical values and corresponding specific operating conditions, all the general considerations relevant to DMG applications can be extended to all the other possible working pairs, in particular NH_3-H_2O.

Some data available from manufacturers are summarized in Table 3.6. For triple-effect chillers, only few data are provided, based on likely estimates. For double-effect chillers, the bulk of them are steam-fired or direct-fired (from exhaust gases, in case), although also some hot water-fired units are available, with superheated water temperature in the range of 155 to 205 °C, and efficiency in the same range of steam-fired machines. With the increasing interest towards heat recovery and energy efficiency in the last years, manufacturers are more and more offering single-effect "low-temperature" hot water equipment, where hot water can be supplied to the absorption machines at about 90÷95 °C [ECD01][Per05] or even less [HeB02][HoC04][Dan06]. These units are designed for a variety of applications, among which supply from solar energy and trigeneration with direct connection to the jacket cooling circuit of an ICE (at temperatures of around 80 °C [LoG05]) or to the water circuit at the outlet of an MT package, and are becoming a market standard.

Table 3.6. Typical characteristics of absorption chillers available on the market

single-effect			
firing source	capacity range [kW_c]	temperature [°C]	*COP*
steam	300÷5,000	110÷130	0.6÷0.8
hot water	300÷5,000	115÷150	0.6÷0.8
"low-temperature" hot water	35÷1,800	80÷90	0.6÷0.8
double-effect			
firing source	capacity range [kW_c]	temperature [°C]	*COP*
steam	100÷5,000	120÷185	0.9÷1.3
direct-fired	50÷5,000	---	0.9÷1.3
triple-effect			
firing source	capacity range [kW_c]	temperature [°C]	*COP*
steam	---	> 200	> 1.5
direct-fired	---	---	> 1.5

3.3.4.2. Thermodynamic aspects

Absorption chillers exploit, instead of a mechanical energy input, mainly thermal energy at relatively low temperature. Also mechanical energy is required to drive internal pumps and auxiliaries, but the consumption of such devices is negligible in system analyses. This is the reason why these systems are told to be based on a *tri-thermal* thermodynamic cycle (as seen from the outside), with two energy interfaces at the temperatures T_1 and T_2, and a third input thermal energy source Q_i at the temperature T_i (Figure 3.13).

The *heat rejected* from the cycle in this case is:

$$Q_1 = R_2 + Q_i \tag{3.16}$$

and the *cooling efficiency* (or *COP*) for an absorption unit is defined as

$$\varepsilon = \frac{R_2}{Q_i} \tag{3.17}$$

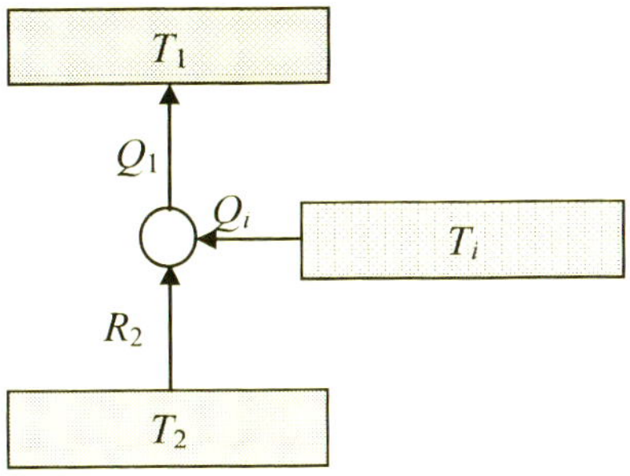

Figure 3.13. Absorption chiller general scheme (T_1>T_2).

The Carnot efficiency for a totally reversible tri-thermal machine is given by [AlF97]

$$\varepsilon_C = \frac{\frac{1}{T_1} - \frac{1}{T_i}}{\frac{1}{T_2} - \frac{1}{T_1}} \tag{3.18}$$

and depends only upon the temperatures of the sources used in the cycle.

The relation (3.18) indicates that if the temperature of the "firing" source T_i rises to infinite, the formula tends to be the same as (3.14). This means that, given the temperatures at the operating conditions (condenser and evaporator), theoretically the efficiency of an absorption machine is always lower than the one of a vapour-compression machine (operating along a bi-thermal cycle). Also in practice, the *COP* of an absorption machine is far below the one boasted by electricity-fed equipment, ranging from about 0.6 for single-effect units to about 1.2 for double-effect ones. However, comparison between the efficiencies (3.13) and (3.17) (defined conventionally as ratio of the desired output to the necessary input) does not provide adequate indications. In fact, absorption chillers may be powered, for instance, by low-grade waste heat, whereas vapour compression chillers must be driven by high-quality energy such as electricity or shaft work, so that the performance evaluation needs to be framed within an overall energy system analysis (see in particular Chapter 5).

Interesting theoretical and engineering details on heat powered refrigeration cycles are provided in [ToJ98] and reported in [Dan06], also for absorption chillers coupled to CHP systems.

The schematic structure of an absorption chiller is similar to the one of a classical electric chiller, apart from the compressor, substituted by a "thermal compressor" with an inner thermo-chemical subcyle [Dan06], the core of the absorption cycle. The basic cooling absorption cycle can thus be synthesized as follows [CaG96][AlF97][ToJ98][Kre01][Dan06]:

- the low-temperature *liquid* refrigerant absorbs heat from the water to be cooled and is converted into a vapour phase (*evaporator* section);
- after mixing with the absorbent in the *absorber* and being pumped into the *generator* by a *solution pump*, the refrigerant *vapour* is driven off to a higher pressure by exploiting heat supplied to the generator (*thermal compressor* section);
- the refrigerant is converted back into a *liquid* by rejecting heat to the outside (*condenser* section);

- the refrigerant expands to a low-pressure *mixture of liquid and vapour* (*expander* section) that goes back to the evaporator section, and the cycle repeats over and over again.

Seen from the outside and for system energy analysis purposes, the main difference between absorption and electric machines is in the compressor, as mentioned earlier. An electric chiller uses an electric-driven motor to drive the compressor that lifts the energy level from a low temperature to a higher temperature. Instead, in an absorption cycle chiller, compression of the refrigerant vapour is carried out by a set of elements (the *absorber*, the *solution pump* and the *generator*) in combination, fed by heat and a negligible amount of electricity, which all together make up the thermal compressor, also called "termo-chemical compressor". Thermodynamic and practical details on absorption cycles (for both cooling and heating applications) can be found in [Kre01][Dan06].

From (3.16), the heat to dispose is the sum of the heat supplied and the cooling effect produced. From the relatively low figures given for absorption *COPs*, it is apparent that absorption chillers need to get rid of large amounts of heat, which is commonly obtained by means of cooling towers or evaporative cooling condensers [AlF97][Ang98][Kre01].

3.3.4.3. Absorption chiller off-design characteristics

If cooling power is needed for air conditioning and no thermal storage system is implemented, it is highly likely that cooling loads undergo large seasonal and daily variations. In these conditions, off-design efficiency of cooling production assumes a key importance.

In general, according to several models presented in the literature (see for instance [AlF97][ECD01][Soz01][BaK02][LoG05][KiK07][GoH08]) and from elaboration of data available from manufacturers, H_2O-LiBr absorption chillers show good features when operating at partialised capacity. In a simplified way, it is possible to say that the efficiency keeps quite constant, or even improves slightly, when the load decreases up to about 30÷40% of the rated capacity; below this load level, the *COP* falls more or less in a linear fashion.

As a theoretical justification of the fact that the *COP* of an absorption chiller has a maximum at intermediate load, it can be noticed that at lower cooling loads there are *internal* due to dissipation from fluid losses and mechanical losses, while at higher loads *external* losses play a prevalent role [HoC04]. However, in a wide range the *COP* variation is of few percentage points.

A detailed map of the dependence of *COP* and capacity would involve many variables and diagrams, depending on the specific type of chiller and manufacturer, to whose catalogues the reader is referred if willing to figure out further details. In addition, also condenser and evaporator temperatures must be considered while addressing the partial-load model issue, as all the absorption cycle characteristics (and in particular temperatures) are closely intertwined, as further discussed below.

As an example, from some data from manufacturers and thermodynamic considerations and assumptions, it is possible to plot *COP* curves against the partial cooling load (the most useful variable for energy system simulation purposes) with condensation or evaporation temperature as a parameter. In general, the performance improves if the cooling water temperature at the condenser decreases, or the temperature of the chilled water (cooling effect) increases [Kre01], as intuitively inferable by the theoretical model (3.18). Detailed explanations of *COP* models in dependence of the cycle temperatures can be found for instance in [Soz01][LoG05][KiK07][GoH08]. Figure 3.14 shows a general possible model of the partial-load *COP* for a double-effect absorption chiller (for instance, a DFC) and a hot-water single-effect absorption chiller (a WARG), both adopting H_2O-LiBr. It is interesting to notice how the plots recall the partial-load efficiency characteristics of a boiler. Figure 3.15

shows the dependence of the partial-load *COP* of a single-effect H_2O-LiBr absorption chiller on the condenser cooling water (usually requiring a cooling tower). Usual temperature values for the cooling water in summertime are about 30 °C or even more, with typical drop of 5 °C in the cooling tower [AlF97][Kre01].

Curves as in Figure 3.14 and Figure 3.15 can be obtained by fitting real data by means of polynomial models [LiN06]. If a *black-box* fit-to-data model is implemented to track manufacturers' data, good fitting is given by a second-order curve, such as

$$R = \nu_2 \cdot Q_t^2 + \nu_1 \cdot Q_t + \nu_0 \tag{3.19}$$

where R is the cooling power, Q_t is the thermal input, while ν_2, ν_1 and ν_0 are model parameters. Similar curves hold in general true also for reversible utilization of the absorption machine as an AHP [LiN06].

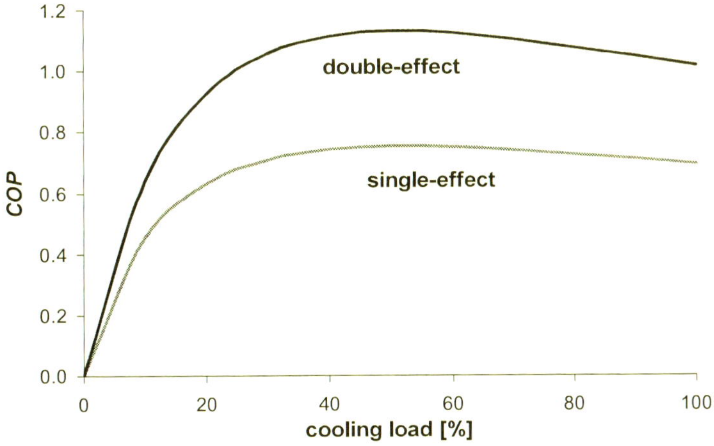

Figure 3.14. Comparison of partial-load characteristics for double-effect direct-fired and single-effect hot water absorption chillers.

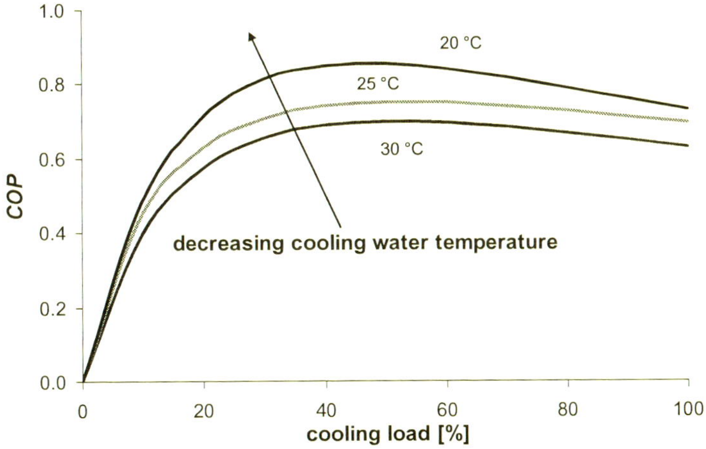

Figure 3.15. Dependence of partial-load characteristics for a single-effect absorption chiller on the condenser cooling water.

However, it is up to the designer or the analyst to use the models they believe to be the most adequate to estimate off-design characteristics, according to the data rendered available by manufacturers' and often extrapolating "simple" analytical models (linear or polynomial) from tables and graphics, in order to implement them in simulation programs.

3.3.4.4. Temperature constraints for heat sources

Once the evaporator temperature and condenser temperature are chosen, the lowest temperature at which the thermal input can be supplied to the generator is also determined, setting a system constraint on the heat sources that can be adopted [AlF97][CoG03]. For instance, for a single-effect H_2O-LiBr chiller with evaporation temperature equal to 2 °C and condensation temperature of 37 °C, the energy supplied to the generator must be at the lowest about 90 °C [ECD01]. Practical and economical design considerations on heat transfer rates and heat exchanger areas, though, may increase this lowest level to more than 100 °C (typically about 110 °C for H_2O-LiBr [AlF97] and 120÷130 °C for NH_3-H_20 [CoG03]), so requiring, in case, superheated water or steam.

In the last technological developments, compact single-effect absorption chillers can be fired by temperatures in the range 80÷90 °C. Temperatures required by double-effect chillers would be much higher, in the range 170÷180 °C, although actual figures in the most recent units could be below 150 °C. Triple-effect chillers would typically require temperatures above 180 °C.

Heat sources at these temperatures can be readily available for direct-fired equipment. Of course, in this case the quality of the heat (its high temperature and exergy content [DiR07]) may somehow be seen as "wasted", but it is paid back with a more compact and efficient design of equipment. However, since the great upside of absorption chillers over electric chillers lies in the possibility of exploiting low-temperature waste heat, problems might arise if the heat source were available at a temperature lower than the rated one needed at the absorption generator. Hence, it is important to characterize the dependence of the machine performance on the temperature at which heat is supplied to the generator. In this respect, Figure 3.16 shows typical characteristics on this dependence for a single-stage ammonia refrigeration plant [CoG03]; however, except for the numerical values (firing temperatures are lower for H_20-LiBr chillers), the same general profiles apply also for water-lithium bromide units [Dan06][KiI08]. The *COP* profile is highly nonlinear in dependence of the heat source temperature, and might experience a dramatic fall below a certain generator temperature, while no significant improvement can be achieved beyond a certain temperature threshold. In addition, such profiles change with changes in the condenser and evaporator temperatures, so that in each operating condition a typical generator temperature with maximum *COP* can be found (see [LoO05] for a detailed explanation on this issue). Indeed, the absorption cycle temperatures are closely intertwined with each other, and the required generator firing temperatures for proper chiller operation increase with the number of stages [Dan06]. As a consequence, the coupling of the energy source to the absorption machine has to be thoroughly studied. In particular, in every operating point (including off-design conditions) a CHP system must be able to supply heat to the chiller at an adequate temperature in order to avoid getting cooling efficiency (as well as capacity [KiI08]) below the stated one and uneconomical energy system operation. Furthermore, also crystallization problems in the absorption solution might arise if the generator/evaporator/condenser temperatures are not opportunely controlled [Dan06], above all for air-cooled H_2O-LiBr systems [LiR07], which might lead to system breakdown. On the other hand, too high firing temperatures would bring no benefit to the chiller performance, whereas they would decrease the CHP thermal performance (as for instance discussed in Section 3.1.3 for MTs), thus worsening the overall performance of the CCHP system [CoG03].

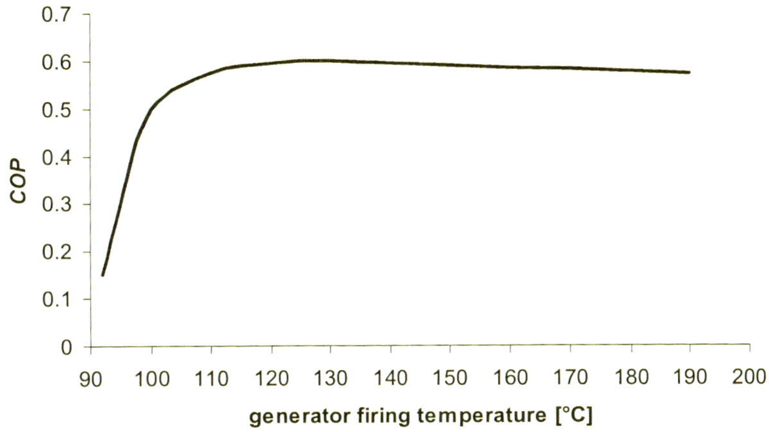

Figure 3.16. Absorption chiller *COP* typical dependence on generator firing temperature.

On the above premises, it is possible to summarize the characteristics of suitable coupling between absorption chillers and small-scale CHP systems [WuW06] as follows:

- *Triple-effect*: the chiller could be directly fired by fuel (natural gas), with internal combustion to provide energy to the generator, or by exhausts from GTs, MTs, and in perspective FCs, through an internal heat exchanger [MaC01][MoC01][ViB07]. Firing temperatures are typically above 200 °C.
- *Double-effect*: the chiller could be fired through steam, usually at a temperature between 120 and 185 °C. The steam may be produced in a HRSG fed by an ICE, an MT, or a GT. In case, also (pressurized) hot water might be used. Alternatively, the chiller could be fired directly by exhausts from a GT or an MT [MoA98]. As reported in [HoC04], a double-effect chiller could also operate with a combination of an MT (or ICE) exhaust heat and direct gas firing (the generator works in this case as a fired HRSG), or could be set downstream the post-combustor of an MT [BrV05].
- *Single-effect*: this machine could be fed by steam in a range about 110÷120 °C, or by superheated water at 115÷150 °C. More and more units are available on the market with generator firing temperature much lower, and which might actually be fired by hot water at 80÷90 °C, although with higher specific costs. Hence, small-scale MTs and ICEs, as well as micro-scale SEs, could generate hot water suitable for these low-temperature applications. For practical examples, the 105-kW_c absorption chiller presented in [Per05] is fed by hot water at 95 °C produced by an MT, and operates with a rated efficiency of about 0.7, while the 35-kW_c chiller analysed in [HoC04] can reach a *COP* of about 0.6 with firing hot water at about 80 °C.

The issues related to CHP-absorption chiller coupling are further pointed out throughout the numerical assessment examples provided in Chapter 7.

3.3.4.5. Comparison between absorption chillers and vapour compression electric chillers

There are several upsides and downsides for both absorption and electric chillers. For a synthetic overview comparison, the main aspects to take into account may be:

- *Driving energy and operating costs*: absorption chillers are driven by heat, which can be at very low cost if cascading as residual from topping cycle. On the contrary, electricity often implies high operating costs.
- *Initial costs*: electrical units cost yet quite less than absorption ones.
- *Performance*: *COP* for electrical chillers is far higher, even though the comparison is not "fair", given the different quality of the energy input. Partial-load performance is generally good for both types, namely, quite steady for electrical units, and with significant reduction only at low loads for absorption chillers (whereas performance typically increases at intermediate loads).
- *Maintenance costs and reliability*: absorption chillers have relatively few parts in movement, so that the corresponding maintenance costs are lower and reliability is higher.
- *Availability*: as a consequence of more frequent maintenance stops, redundancy may be needed for electrical units in order to obtain the same availability as for absorption chillers.

However, a comprehensive assessment cannot in general be carried out only on the basis of a direct comparison, but the two different types of chillers need to be analysed within an energy system perspective (see for instance [MaC08b]). Suitable models and analyses of the impact of different cooling alternatives (including electric and absorption chillers) on the overall plant performance within a DMG framework are discussed in the next chapters.

3.3.5. Adsorption chillers

As for absorption systems, the interest towards adsorption systems is driven by the possibility of recovering waste heat (even at lower temperature, and thus of lower thermodynamic values, than in absorption machines), thus improving the overall generation efficiency, as well as by the absence of ozone-depleting refrigerants [WaO08]. Hence, research about adsorption chillers has witnessed consistent increase in the last years, although commercial units are still quite more expensive than absorption and mechanical units [Dan06].

Both absorption and adsorption cycles are grouped within the generic term of "sorption" cycles [Dan06]. In synthesis, *absorption* corresponds to absorbing and releasing a substance with a chemical change, whereas in an *adsorption* cycle no chemical change occurs. In both cases, thermal power is used to separate the refrigerant, and this operation represents the major difference with respect to mechanical chillers, as already discussed. Hence, also for adsorption chillers the same thermodynamic assessment (based on heat-fired cycles) as for absorption ones holds true [Hen07].

A basic adsorption chiller is composed of a single bed of solid material (*adsorbent*) that cyclically absorbs and releases a certain substance (*adsorbate*). The most typical adsorbent-adsorbate working pair is silica gel-water. The silica gel desiccant absorbs water, but, differently from the absorption cycle, it does not undergo chemical changes, apart from gaining weight due to water addition. However, the two main adsorption cycle stages closely recall the characteristics of absorption cycles. In fact, at first the adsorbent is cooled by means of the *adsorption* process, bringing about refrigerant evaporation in the evaporator. This brings the useful cooling effect. Then, the adsorbent is heated up by means of the *desorption* process, during which the refrigerant condensates in the condenser and heat is released outside [WaO08]. However, with a single adsorbent bed, cooling effect might be provided only in part of the cycle. More continuous refrigeration operation can be obtained by setting up more beds, for which suitable design trade-off between cost and performance/capacity increase must be sought [Dan06].

The desorption process occurs thanks to heat externally supplied, so that it is also called *generation* process, in analogy to absorption chillers. This heat can be supplied by a low-grade source, at a temperature typically lower than for absorption units (as low as about 50 °C), so that a higher level of heat recovery may be carried out in the presence of low thermodynamic quality heat. In addition, low-temperature hot water produced from solar energy can be effectively utilized for cooling applications, with potentially high energy and economic benefits due to the high correlation between solar radiation and cooling needs [Hen07]. However, the main drawback of adsorption chillers is represented by the relatively low *COP* (defined as for absorption chillers) due the cycle thermodynamic characteristics [Dan06], with typical values in the range 0.3÷0.5. In this respect, a number of researches are taking place worldwide, with the aim of getting further knowledge of and thus improving adsorption chiller performance by enhancing the physical-chemical characteristics of the cycles for given working pairs, as well as by looking for new working pairs [WaO08]. On the other hand, it has to be pointed out that adsorption chillers generally perform better than absorption ones for temperatures below 80 °C. At this temperature level, single-effect absorption chillers may experience dramatic performance fall (and even crystallization), as aforementioned, whereas for decreasing firing temperatures the performance curve in adsorption units slopes down more gently [Dan06][Hen07].

3.3.6. Heat pumps

The term "heat pump" derives from an analogy to hydraulic pumps, as this device performs the operation of "pumping" heat into an ambient at higher temperature from a source at lower temperature. Typically, a heat pump extracts heat from an environmental source (ground, groundwater, water from lakes or rivers, air, and so on) and raises its temperature to a level high enough to satisfy the user's needs (space heating, hot water service, and so on). Energy consumption is needed to *lift* the energy level from lower to upper temperature, so that the model is completely analogous to that of a cooling generator. Indeed, the thermodynamic cycle is the same, and most of modern devices can operate reversibly to provide also cooling power. This is why, although heat pumps are someway more similar to heat generators according to the heating production purpose, they are dealt with in this section. Traditional applications refer to air conditioning and heating, by producing hot water (also for building services) or hot air.

In this section, the main characteristics of electrical heat pumps are presented, focusing on those aspects of major interests for applications within a multi-generation energy system. System models and applications within DMG architectures are widely illustrated in the next chapters.

3.3.6.1. Classification of heat pumps

Today, the bulk of heat pumps are based on a vapour compression cycle and are electricity-fed (*Electric Heat Pumps – EHPs*). The interest towards heat pumps, especially electric ones, from an energy standpoint is very high, as they can supply to the hot ambient an amount of energy far higher than the one needed to operate (although of different thermodynamic value). Typical efficiencies range from 2 to 5, with practical maximum of about 7, when they are used for heating purposes from cold sources at 5÷10 °C. This high energy effectiveness is someway due to the fact that the energy from the cold source, typically available from the ambient, is available "for free". Furthermore, scenarios envisaging the utilization of EHPs are promisingly arising in the perspective of new combined cycle technologies that might reach electrical efficiencies up to 70% (for instance, high-temperature FCs combined with bottoming combined cycles [Nor03]).

A number of absorption chillers can operate reversibly as absorption heat pumps [LiN06][LaN06][KeP08], although their use is yet limited compared with the overwhelming EHPs. They are basically absorption chillers that use available firing thermal input (hot water or steam) for heating purposes, with efficiency values up to about 2 [LaN06][KeP08]. Reversible absorption heat pumps that can operate in both chilling and heating mode coupled with a firing boiler or direct-fired are also available [AlF97]. Finally, also (reversible) *adsorption* heat pumps [KeP08] are gaining market shares.

Also the interest towards Gas *Engine-Driven Heat Pumps* (*EDHPs*) is lately increasing [ZhL05][LiZ05][LaN06][LaN06b][HeE09]. Their use could be profitable in particular applications, especially thanks to the possibility of exploiting heat recovered from the exhausts. However, weight and footprint are often a major concern, as well as noise level often unacceptable in building applications [AlF97].

Specific technology issues related to AHPs and EDHPs can be essentially inferred from the ones discussed in Section 3.3.4 for absorption chillers and in Section 3.3.7 for engine-driven chillers. Hence, without losing generality, in the sequel the heat pump concepts are discussed with specific focus on EHPs owing to their widespread application.

3.3.6.2. Thermodynamic aspects of EHPs

EHPs can be addressed in analogy to electricity-fed vapour compression chillers. In fact, if operated reversibly, EHPs actually *are*, to every extent, CERGs. As such, they are based on a *bi-thermal inverse cycle*, with two energy interfaces ("reservoirs", "sinks", "sources") at the temperatures T_1 and T_2 (Figure 3.17).

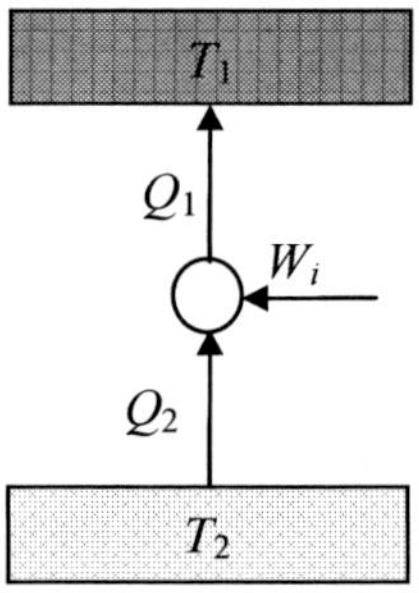

Figure 3.17. Heat pump general scheme (T_1>T_2).

The subscripts i, 1 and 2 usually refer to the input electricity, and the higher and lower temperature, respectively. Furthermore, the subscripts 1 and 2 are used to represent the energy flows exchanged with the reservoirs at these temperatures.

The same concepts as for cooling cycles essentially apply. In fact, the main difference lies in the *purpose* of the inverse cycle: whereas in a cooling plant the purpose is to draw energy from the lower temperature reservoir, in a heat pump the purpose is to inject power into the higher temperature reservoir. Therefore, Q_1 is the desired heating effect, W_i the electrical energy input, and Q_2 is the (free) energy source.

The first principle of thermodynamics applied to the system in Figure 3.17 yields

$$Q_1 = Q_2 + W_i \tag{3.20}$$

i.e., the "hot" system receives a heating effect sum of the free heat drawn form the lower temperature system and the electrical input.

As a consequence of the change of purpose, also the definition of efficiency changes with respect to chillers. Given that a general efficiency is defined as the ratio of the desired output to the input cost necessary to obtain it (see also Section 5.2.1), for an EHP it is possible to define the *heating effectiveness* or *efficiency* (ε'), or *COP'*, as the ratio of the heating power (desired output or *useful effect*) to the *energy cost* ("needed" electrical input):

$$COP' = \frac{Q_1}{W_i} \tag{3.21}$$

The *prime* indication on *COP'* in (3.21) points out that the efficiency refers to the one of a heat pump, although calculated in an inverse cycle as for chillers. Similarly, when referred to the *COP* of a reversible EHP, the subscript *c* (cooling) and *t* (thermal) may be used to point out the performance under cooling and heating mode, respectively. However, when no ambiguity can take place, in the sequel the use of the prime or of the subscript *t* to indicated EHP performance is avoided.

The upper bound for a heat pump is given by the Carnot efficiency

$$\varepsilon_C' = \frac{T_1}{T_1 - T_2} \tag{3.22}$$

Comparison between (3.22) and (3.14) valid for cooling cycles yields

$$\varepsilon_C' = \varepsilon_C + 1 \tag{3.23}$$

which is justified by the fact that in a heating cycle the input electrical power is added to the useful energy effect, whereas in the cooling cycle it is needed only to increase the refrigerant energy level. The relation (3.23), written for Carnot cycles, holds true also for real cycles [AlF97], which stresses from a thermodynamic point of view the intrinsic high energy efficiency of this device, allowing consistent potential energy and economic savings. For instance, if 40 °C and 0 °C (normally possible temperature figures for condensation and evaporation) are substituted in the relation (3.22), a value of about 8 comes out: theoretically, the overall useful heat results to be even 8 times higher than the high quality power spent to make it available at a more "noble" temperature from a poor quality source. Unfortunately, real *COPs* are typically less than half the theoretical one, but the interest for the potential high-efficiency energy conversion remains the same.

The heat pump efficiency is often presented in datasheets or catalogues, especially from American manufacturers, also as *EER* (Energy Efficiency Ratio) [Kre01], *i.e.*, as the ratio of heating capacity, measured in [Btu per hour], to the electric input rate, measured in [W]. Accordingly, the *EER* has the units of [Btu/Wh], and the corresponding dimensionless *COP* is obtained by dividing the *EER* by the conversion factor 3.413 [Btu/Wh].

The EHP types can be listed according to several features, such as thermal source, heat distribution system, compressor used, rated thermal power, and so forth [Via00]. In general, thermal source and distribution system are the most crucial points for the EHP performance and the plant design.

3.3.6.3. Electric heat pump performance

The Carnot efficiency is inversely proportional to the difference between the high and low temperatures of the energy reservoirs. In a real cycle, the heat transfer is not ideal, and the refrigerant is at a temperature T_2' at the evaporator lower than T_2 of the thermal source; likewise, T_1' at the condenser is higher than T_1 of the ambient to heat; also the compression is not ideal, and other losses play a certain role. All these factors together prompt the real efficiency to be far lower than the theoretical one, according to the expression

$$COP' = \varepsilon(T_1', T_2') \frac{T_1'}{T_2' - T_1'} \tag{3.24}$$

where $\varepsilon(T_1', T_2')$ indicates a generic efficiency, function of the cycle temperatures, which takes into account all the losses present in the real model (a similar model is illustrated in (3.15) for chillers). Average representative values for ε range from 0.3 for smaller units up to 0.65 for large advanced units [Cat01][Dan06].

Figure 3.18 (modified from [Kre01]) shows an example of dependence of heat pump performance on the outdoor conditions (cold source temperature), with comparison among theoretical Carnot efficiency (T_1 [K] = 21 [°C] + 273, T_2 [K] = cold source temperature [°C] + 273), Carnot efficiency with allowance for real temperature difference across the heat exchangers (supposed equal to 10 °C in both condenser and evaporator), and efficiency of a real EHP. The difference between real and theoretical performance is apparent. Moreover, the heat pump is not usually able to run properly under very low cold source temperature conditions, so that in these cases auxiliary heat generator, combustion- or electricity-based, are needed [Kre01][Dan06].

The relation (3.24) contains all the major information needed to characterize heat pump performance: considering in first approximation T_1' (and thus T_1) fixed once fixed the application (for instance, the temperature of an ambient to be heated, or the temperature of water for hot water services), the *COP* is strongly dependent on the lower temperature, which depends on outdoor conditions, and so does the heat pump capacity [Kre01][Dan06]. This is apparent from real data from any heat pump datasheet or catalogue, and is a major concern the energy system designer must take into account when selecting the device.

As far as partial-load characteristics are concerned, usually little information is available from manufacturers, while all the stress is put on the efficiency and capacity drop tracking diminishing outdoor temperatures. However, as most of commercial heat pumps are reversible, or anyway use the same compressor type than chillers, the same considerations carried out on chiller compressors can be resorted to in order to help evaluate partial-load performance of EHPs. In particular, the compressors adopted in heat pumps with capacity lower than 1 MW_t (of interest in this work) are basically screw and reciprocating compressors, as well as scroll compressors. Therefore, taking into account what discussed in Section 3.3.3 for electric chillers, in general and with good approximation the *COP* can be considered fairly constant with respect to heating load (and also cooling load, for reversible units) variations, unless more detailed specific data are available. For instance, [LoC02] presents a model of heat pumps with little efficiency drop at partial load (less than 10% at half load), mostly due to the electrical motor efficiency drop. The hypothesis of considering constant partial-load efficiency is even more justified when, as normally occurs in plants of a certain level, the capacity is dwindled over more units and/or variable-speed drives are adopted.

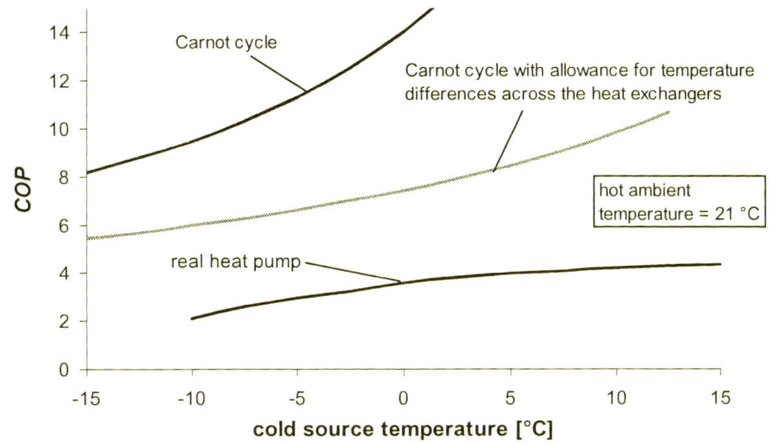

Figure 3.18. Example of *COP* comparison among Carnot cycle, Carnot cycle with heat exchanger penalty, and real heat pumps.

3.3.6.4. The thermal source

Thermal sources play the most fundamental role in heat pump design and selection, affecting performance and useful life, as well as investment, operating and maintenance costs.

The main "cold" sources from which the heat pump can draw heat are, *in primis*, outdoor air or water, from ground source, river, lake, or even the sea [Via00][Kre01][Dan06][KrG07][DiA08]. Moreover, interesting energy saving possibilities can be obtained by cascading heat pumps to other energy processes, for instance by feeding them with heat recovered from industrial processes [Cat01][KeP08], or combining their applications with cooling plant condensers [AlF97][Hav99], as discussed in Section 3.4. Interesting applications in the so-called solar-assisted heat pumps [Dan06][DiA08][KeP08] envisage the presence of solar systems increasing the heat pump input temperature (usually, a solar collector warms up water contained in a storage tank), so as to decrease the required temperature lift and thus increase the energy performance. Geothermal energy [Hep05] can be efficiently exploited as well, and a combination of thermal sources can potentially increase consistently the heat pump *COP* [DiA08]. Some considerations on the heat pump systems using the most frequently available "cold" thermal sources can be summarized as follows:

- *Air-source heat pumps*: air is the most readily available heat source. However, air-source heat pump performance is strongly affected by outdoor conditions, since both heating capacity and *COP* drop when outdoor temperature decreases, as said earlier. In particular, when outdoor temperature decreases, both thermal powers Q_1 and Q_2 of the thermodynamic model decrease. Therefore, in very cold climates, often alternative or integrative systems might have to be adopted.
  Air-source EHPs are normally rated for operation at outdoor temperatures of 6÷7 °C, even though they can normally work (although with often heavy performance drop) at temperatures as low as -15 °C. The same performance and capacity drops occur, in the case of reversible machines, for cooling operation and relatively hot outdoor air (35÷40 °C). The outdoor air-refrigerant heat exchanger normally exploits forced convection on the air side to improve the heat transfer, with difference between the boiling refrigerant and the air typically in the range 6 to 14 °C [Kre01]. In general, the evaporation temperature may represent a severe problem for air-to-refrigerant heat exchangers. Indeed, due to the necessary finite temperature difference across the

condenser, the outdoor air must be at a sufficiently higher temperature than the refrigerant one; furthermore, super-heating is needed at the condenser, so as to avoid the risk to draw liquid drops into the compressor, which requires the air temperature to be even lower. For instance [AlF97], with an evaporating temperature of -2 °C, the refrigerant must be superheated at least to +3 °C at the compressor inlet, which calls for an air temperature of at least (but definitely more than) +5 °C. With lower outdoor temperatures, the compressor may take in liquid drops, and this constitutes another cause of efficiency and capacity drops (not even mentioning possible damages to the compressor) besides the thermodynamic ones from (3.24). In order to avoid this problem typical of direct-expansion refrigerant-to-air heat exchangers, an auxiliary internal exchanger could be used across evaporator and condenser, where heat released to perform under-cooling at the outlet of the condenser is used to perform post-evaporator super-heating (an example is for instance reported in [CaR02]). A practical major problem for air-source machines is the accumulation of frost on the outdoor coil when the temperature drops to about 0 °C, especially in very humid climates. This brings about high worsening of the heat transfer at the evaporator, so that a defrosting process is required, carried out for instance by operating the heat pump, if reversible, in cooling mode for a while. By doing so, the "hot" refrigerant can heat up the outdoor coil [AlF97] [Kre01]. Another typical defrosting technique, the easiest and least efficient, exploits electric resistances. Defrosting creates unavoidable interruptions in the heating provision; this should be (and often is) taken into account in the *COP* data under low-temperature conditions presented by manufacturers [Kre01].

- *Water-source heat pumps*: when using water as a cold energy source, heat pumps can reach performance better than when using air. In fact, water guarantees higher and more constant temperature throughout the year, which improves the thermal transfer and allows higher heat exchanger compactness. For instance, heat pumps using water at 10÷15 °C can reach *COP* of about 4 for normal air conditioning purposes. Several water sources can be used out of urban areas (where often there are problems of water availability, as for water-cooled chillers), such as the sea (water temperature ranges from 10 °C in winter to 25 °C in summer), rivers and lakes (temperatures from 5 to 25 °C), groundwater sources (12 to 18 °C all around the year) [Via00]. However, using these water sources calls for installing filtering systems, as well as for taking special care with corrosion problems (especially for sea water). Moreover, special requirements, aimed at minimising the environmental impact from exploitation of natural sources, need to be met. Water allows better heat transfer at the evaporator than in air-source EHPs, so that a temperature difference of only 5 to 10 °C between refrigerant and water can be adopted, and a better efficiency can be achieved, according to (3.24). Whereas the water-to-refrigerant evaporator sets no serious constraints, the main limitations in using water-source heat pumps stay in the condensation temperature. With R-22, indeed, temperatures not higher than 70 °C can be reached, corresponding to condensation pressure of about 30 bar. However, this pressure can be hard to obtain, depending on the evaporation temperature (and thus pressure) and on the compression ratio that the specific compressor can operate [AlF97]. This kind of technical constraints have been so far a major hurdle to large spread of heat pumps for industrial applications and is stirring research efforts towards new more suitable refrigerants and devices [Cat01][Cat01b].
- *Ground-source heat pumps*: the benefits provided by adopting water as opposed to air as the cold source for heat pumps apply to ground-source (also called *geothermal-source* or *earth-source*) heat pumps as well [KrG07][Ome08]. In this case, thermal

energy is directly extracted from soil or ground water (thus, sometimes both water-source and ground-source terms may be used), whose temperature is nearly constant throughout the year, above all going down to several meters of depth. In addition, for increasing depths not only the ground temperature oscillations over the year decrease, but also the lag between peak surface temperature and peak sub-surface temperature increases [Dan06]. Thereby, a lower temperature lift is required to the compressor, bringing performance even higher than for water-source machines. Normally, an *earthing connection* or *ground loop* transfers heat from the ground to water and then to the refrigerant, while less frequently (mostly due to reliability reasons) refrigerant is directly expanded in the loop [Dan06][KrG07]. Loops may be installed horizontally or vertically, in the ground or in ponds/lakes, be closed or open [KrG07], and represent the major additional costs with respect to air-source heat pumps. Hence, better performance and thus energy saving and smaller operational costs must be traded off with higher initial investment. Nevertheless, in the last years the attention towards ground-source EHPs is increasing consistently due to the higher attention paid to environmental issues, so that this kind of equipment is likely to occupy a more prominent role in the development of high-efficiency distributed energy systems.

- *Recovered heat-source heat pumps*: these heat pump systems can be exploited in industrial processes where heat, that otherwise would be wasted, is available at a temperature relatively low, but still higher than outdoor (industrial wastes, air expelled from heated rooms or productive processes, cooling water, recuperated steam, and so forth [Cat01]). The heat pump can therefore use this heat as thermal source to produce ambient heating or for other purposes [CaR02] (with the limitations mentioned above on the maximum condensation temperature), at higher efficiency than by using colder outdoor air. However, it is important to highlight that waste heat might cause fouling and corrosion problems onto the heat exchanger surfaces, thus shortening the useful life, decreasing the performance, and increasing operation and maintenance costs. Similar applications can refer to heat recovered within buildings, in the case that air hotter than the outdoor one is available for any reason, such as exhaust air from HVAC processes [AlF97][Kre01][Dan06]. In this case, the heat pump essentially transfers and increases thermal power from a place to another within the same building or group of buildings, operating at the same time air conditioning. Also applications in which the "poor" thermal source is represented by heat discharged by absorption chillers [Hav99][BrV05] or other heat recovered from the CGP [AlF97][Cat01] are of rising interest (see Section 3.4).

3.3.6.5. Electric resistance heating

Electric heating may often be coupled (also as a part of the equipment package) to air-source heat pumps to shave the thermal peaks at the lower temperatures. It can also represent a heating resource itself. In fact, electric resistance heating can be used in boilers and furnaces to produce hot water or air for a variety of applications, from small residential furnaces (5 to 15 kW_e) to large boilers for commercial buildings (up to 20 MW_e), with the following upsides [Kre01]:

- lower initial costs with respect to combustion devices and heat pumps;
- almost 100% energy efficiency, roughly constant at partial load;
- no production of exhaust gases *in loco*.

Of course, operating costs counter-balance all these advantages, so that electric heating is mainly used where first costs are primarily important, when fossil fuels are not readily available or where electricity rates are relatively low (northern European countries, for instance). Electric boilers are also normally present in many houses in a number of countries around the world, no matter the electricity rates, because of low installation costs, almost no maintenance needs, and no major safety requirements to comply with.

However, from an efficiency and environmental point of view their use should be carefully evaluated. Indeed, an electric resistance heating system can be modelled, from a thermodynamic standpoint, as a mono-thermal inverse cycle, where the thermal reservoirs are both at the ambient temperature. In this case, no heat is drawn, in practice, from the lower quality source, and all the high-quality energy supplied to the cycle is converted into thermal energy by Joule effect. In other words, electric resistance heating corresponds to a heat pump with *COP* equal to 1. Therefore, apart from the initial investment cost benefits mentioned above, an EHP with efficiency of 3÷4 should rather be installed. Moreover, if one considers the whole life cycle costs of electric systems, also in terms of environmental impact, electric resistance heating might "cost" to the society far more than combustion or heat-pump heating (see for instance [Via00b] for numerical examples referred to Italy).

3.3.7. Engine-driven chillers

Engine-driven chillers (EDCs) [ASH00][RDC03][ZhS03], often reversible machines able to operate also as heat pumps [LaN06b][HeE09], represent interesting multi-energy devices able to supply at the same time cooling (the primary useful effect) and thermal power (recovered) [SuW04][Sun07], while, likewise absorption chillers, displace electricity that typically would be bought or self-produced to feed conventional chillers. Therefore, their application could be profitably exploited to supply coincident cooling and thermal loads, especially in presence of high electricity rates and low gas rates, somehow in a dual fashion with respect to heat pumps. For cooling-only use, instead, engine-driven chillers should be naturally compared with direct-fired absorption chillers.

Although their use is not widespread as EHPs or absorption chillers, EDCs are already well present on the market because of their potentially good economics, especially for industrial applications [RDC03]. In the sequel of this section, basic hints are given on engine-driven chiller main characteristics and applications, most of which are inferable by considering the characteristics of single units that compose them, namely, an ICE and a vapour compression chiller whose compressor is directly connected to the engine shaft.

3.3.7.1. General aspects

EDCs are essentially conventional chillers driven by an engine instead of an electric motor. Indeed, their thermodynamics is based on the vapour compression cycle (Section 3.3.3.1) and the compressors used are the same as for electrical chillers, depending upon the specific application and the size. The only difference is that the mechanical power to drive the compressor is rendered available at the shaft of a usually gas-fired reciprocating engine, likewise for instance a chiller directly coupled to the power shaft of a split-shaft gas turbine [BoK01].

In particular [ASH00], reciprocating compressors are used for cooling applications in a range up to about 700 kW_c, in which often compressor and engine are packaged together [RDC03], while screw compressor chiller sizes range from 350 to 4400 kW_c. Higher capacities not considered in this work typically adopt centrifugal compressors.

Figure 3.19 (elaborated from [ASH00]) shows a typical relation between cooling capacity and engine mechanical size for commercial units.

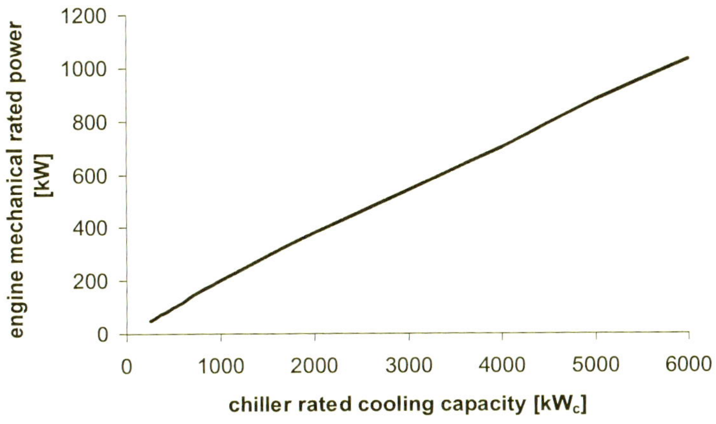

Figure 3.19. Typical relation between cooling load and engine mechanical size.

Reciprocating engines are normally designed to operate optimally between 1200 and 2400 rpm, depending on the size, with the higher speeds for the smaller engines. However, given the size, the speed is kept as low as possible to extend the engine life. Chiller compressors are characterized by an optimal speed as well, so that the connection to the engine shaft is typically carried out through a gearbox in order to optimise the overall performance. The gearbox is subsequently speed-increasing (for screw compressors, optimal speed from 1000 to 4000 rpm) or speed-decreasing (for reciprocating compressors, optimal speed from 1200 to 1800 rpm, with sometimes direct coupling).

The engines used to drive the chiller have the same characteristics as the "bigger" gas and (less frequently) diesel engines. The main upsides and downsides with respect to electric chillers can be summarized as follows:

- High mechanical efficiency, with consequent low operating costs, potentially the lowest ones among any chillers, especially in the presence of favourable fuel rates.
- Possibility of cogenerating heat.
- Possibility of electrical peak shaving in the presence of high electricity rates.
- High levels of noise and vibrations, due to the large number of moving parts and the operation itself of the reciprocating engine, which often calls for being placed in separate buildings and continuous maintenance, also considering the gearbox coupling the engine shaft to the chiller compressor. The high maintenance frequency, in particular, higher than for electric chillers, might represent a major barrier to more widespread adoption, regardless of the potentially good economics [RDC03].
- High emission levels are a further key limiting factor, especially in the presence of stringent local regulatory constraints on air pollution.

3.3.7.2. Engine-driven chiller performance

From an energy point of view, engine-driven chillers can be seen as devices that cogenerate cooling and thermal power; indeed, while driving the chiller compressor, the heat contained in the cooling water and the exhausts can be recovered in the same way as for heat recovery in CHP engine-based systems.

As for chillers, the performance of these devices is characterized by the cooling *COP*, conventionally defined as cooling energy output *R* divided by the fuel energy input *F* (in case calculated on the basis of the fuel *HHV* [ASH00]):

$$COP = \frac{R}{F} \tag{3.25}$$

In order to take into account also the possible heat recovery from jacket coolant and exhaust gases, often manufacturers present performance values according to a modified definition of *COP*, here called ε:

$$\varepsilon = \frac{R+Q}{F} \tag{3.26}$$

in which Q is the total heat recovered. Assessment models based on similar types of indicators can be found in [SuW04][SuG06][Sun07].

Since Q/F is in practice the thermal efficiency of the driving engine, it is simply possible to write

$$\varepsilon = \frac{R+Q}{F} = COP + \eta_t \tag{3.27}$$

where η_t is the thermal efficiency of the combustion engine (as for normal ICEs). Typical *efficiencies* (*COP*, ε and η_t) are given in Table 3.7 (adapted from [ASH00]).

Table 3.7. Typical efficiencies for engine-driven chillers

heat recovery option	ε *at full load*	*COP at full load*	η_t *at full load*
no heat recovery	1.2 ÷ 2.0	1.2 ÷ 2.0	0
jacket water heat recovery	1.5 ÷ 2.25	1.2 ÷ 2.0	0.2 ÷ 0.3
jacket water and exhaust heat recovery	1.7 ÷ 2.4	1.2 ÷ 2.0	0.3 ÷ 0.5

The performance figures indicated in Table 3.7 are defined at nominal capacity and standard conditions for condenser and evaporator temperatures, as for electrically driven chiller. Of course, the performance varies with the external conditions, in particular with condenser and chilled water temperature. Furthermore, the fact that engine-driven chillers can operate at variable speed provides significant advantages over the electricity-fed single-speed chillers when operating at partial loads. Indeed, the *IPLV* (see Section 3.3.3.5) is often higher than the rated *COP*. For instance, for a typical 100 kW_c engine-driven screw compressor chiller the full-load *COP* would be about 1.3 against an *IPLV* of about 1.8, justified on the basis of typical partial-load performance models as the ones shown in Figure 3.20.

3.3.7.3. Heat recovery

Fuel energy is supplied to the engine through combustion, converted into mechanical power to drive the compressor, and eventually discharged as heat to the outside. The heat is basically released in form of cooling water, exhaust gas and radiation. About 70% of the total fuel energy input is converted into heat, most of which can be recovered from the engine exhaust gases and jacket coolant (as well as, in case, lubricating oil, accounting for a much smaller share), to generate steam or hot water for various processes (for instance, service water heating, hot water and steam for restaurant kitchens, space heating, and so on). Typical energy balance for an engine used to drive chillers is shown in Figure 3.21. Of course, unless a thermal storage system is installed, to be cost-effective the cooling and thermal load must be coincident.

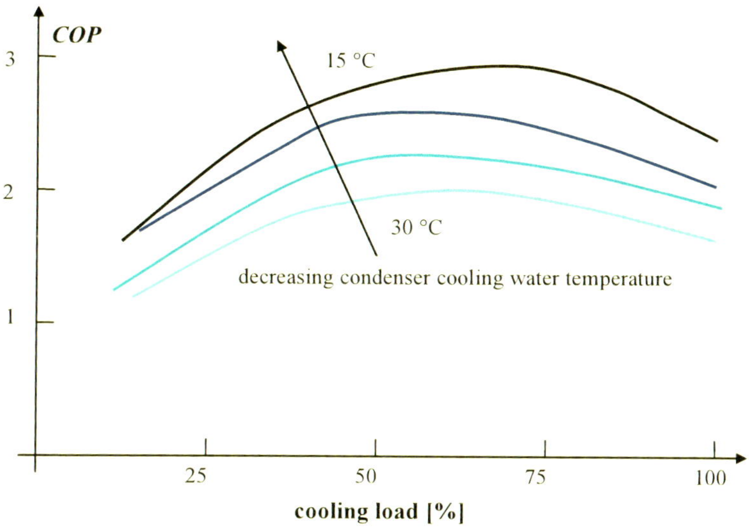

Figure 3.20. Typical partial-load characteristics for an engine-driven screw compressor chiller.

The heat contained in the jacket coolant (and in case in the lubricating oil) is almost completely recoverable and can usually produce hot water at about up to 90 °C; on the contrary, only a portion of the exhausts, whose temperature at the stack is between 450 and 650 °C, can usually be recovered, down to 110 to 175 °C, in order to keep above the condensation thresholds, as already discussed in Section 3.1.2.1. It is interesting to notice that the engine-recovered heat might even fire absorption chillers, in which case the engine would actually operate as a *total cooling energy system*, providing both mechanical energy and thermal energy to be converted into cooling energy.

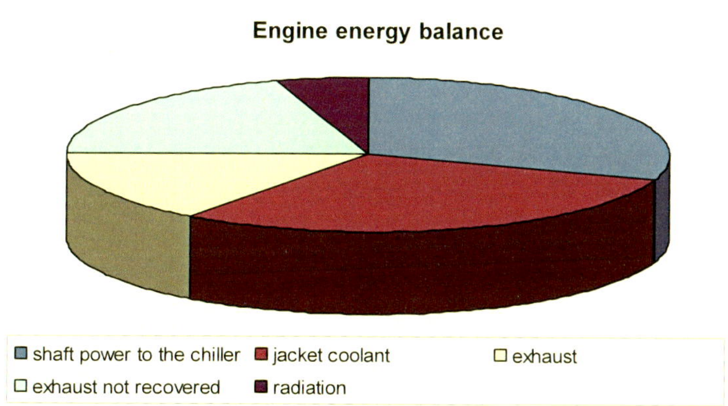

Figure 3.21. Typical energy balance in an engine used to drive a chiller compressor.

3.4. Heat Recovery in Cooling Plants

CHP units lend themselves well to recover thermal power not only for cogeneration purposes properly called, with direct use of the produced heat, but also for cascaded bottoming cycle applications (trigeneration by means of a thermal bottoming cooling plant). Apart from this, in a host of industrial applications thermal power could be recovered but would need to be heated up in order to be reused. In addition, several sources for heat recovering are also available from the condenser coolant of most chiller technologies.

Some characteristics of cooling equipment with possibility of heat recovery and opportunities offered by such a recovery are discussed in this section.

3.4.1. General models for bottoming cycle heat recovery in cooling plants

In general, a bottoming cycle model exploiting recovered heat within cooling plants can be referred to three major categories (Figure 3.22):

1. The heat recovered from a cooling plant condenser is directly used by the user. This might be for instance the case of the heat recovered from a Heat Recovery Condenser (HRC) of a CERG (or also a WARG), or from other source within the air conditioning system [AlF97][Wul99][Kre01]. It has to be underlined that the numerical values of the cooling-mode COP_c, in case of heat recovery at the HRC, might be lower than the corresponding ones without heat recovery [Wul99]. For instance, in order to recover heat at a certain temperature useful for specific purposes, it might as well be required to increase the condensing temperature, with subsequent decrease in the COP_c. Alternatively, the heat might be more profitably brought to the desired temperature by means of an EHP (see below).
2. The heat recovered from a CGP condenser is not at a temperature high enough to be used: for instance, the cooling water of an absorption chiller circulates at about 30 °C in a cooling tower in order to discharge the condenser heat; in general this level of temperature is useless for nearly any application. In this case, an EHP (see for instance [Hav99][BuT03]) or an AHP [AlF97][KeP08] could raise the thermal level up to the required one.
3. The recovered heat, for instance from industrial processes, is not at a temperature high enough to be used for other purposes: also in this case, an electrical [Cat01] or an absorption heat pump [AlF97][KeP08] could be used to boost the thermal level up to the desired one.

Devices such as EHPs and AHPs can play a fundamental role to optimise the energy economy within an energy system, owing to the benefits of upgrading the thermal quality of the input flow with relatively low energy cost, and thus allowing efficient recovery of low-level thermal power that otherwise would be wasted. Details on this aspect from a DMG standpoint are provided in the next chapters.

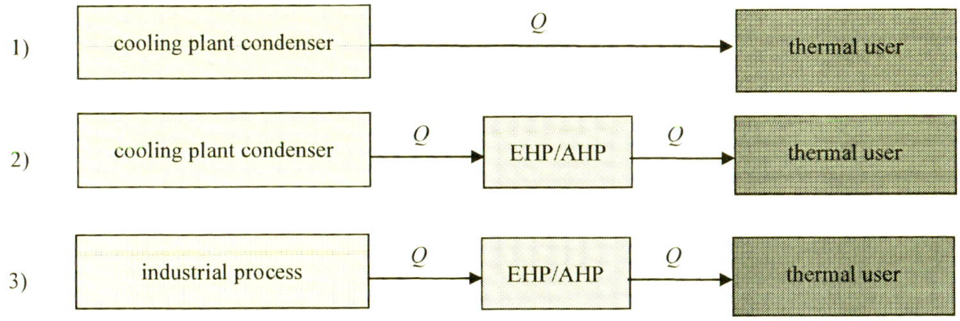

Figure 3.22. Models for bottoming cycle heat recovery within cooling plants.

3.4.2. The EHP for heat recovery bottoming cycles

Following the considerations carried out in the previous sections, the importance of EHPs (and, to a lesser extent, AHPs) within DMG energy systems is related not only to their potentially very high effectiveness, but also to the great flexibility and versatility that allow them to be used in a wide range of heat recovery applications. In this outlook, an example of application of EHP to increase the thermal generation within an MG plant is provided in Section 7.5.

Whilst EHPs can be readily used for HVAC purposes, technology limitations have to be considered yet for their full exploitation for relatively higher temperature (industrial) uses. For instance, the maximum temperature reachable by the heated fluid (hot water, for instance) is related to the maximum temperature reachable by the refrigerant. In the case of R-22 (typical reference, still used in many EHPs) this temperature is about 70 °C in order not to overcome pressure limits. Moreover, also the compressor has to be able to perform high pressure ratio, which are limited to about 3 for scroll chillers, 5 for screw chillers, and 8 for reciprocating chillers. The heat pumps available on the market can typically operate well up to 60÷70 °C, with temperature drops of 30÷40 °C [AIF97][Cat01], so that they are suitable for heating and hot water production.

Design and adoption of EHPs for industrial heat recovery applications with far higher temperatures to be reached are under study in several countries (for Italy, see for instance [Cat01][Cat01b]).

Chapter 4

DISTRIBUTED MULTI-GENERATION PLANNING

The previous chapters have been dedicated to illustrate typical multi-generation plant components and structures. The relevant contents have formed the background to address system-based analyses of various kinds.

This chapter presents a comprehensive approach to multi-generation system characterization and planning. This approach is formulated in terms of the so-called *lambda analysis*, consisting of a unified framework to study multi-generation systems, which extends the classical models based on the analysis of the heat-to-power cogeneration ratio in cogeneration plants. In particular, the representation of the energy interactions within the multi-generation plant is summarized in terms of transformations of an array of original energy or cogeneration ratio values into an equivalent set of values, mathematically expressed by means of specifically defined lambda transforms. The conceptual scheme presented provides effective characterization and modelling of the production side, the demand side, and their interactions in multi-generation systems.

The details of the approach presented are illustrated with reference to alternative schemes and equipment available on the market for setting up multi-generation plants. For each alternative, represented by the corresponding equipment models, the expressions of the relevant lambda transforms are presented.

Numerical applications are provided for a multi-generation system with electrical, thermal, and cooling power production. The results highlight the potential of the lambda analysis framework and of the associated lambda transforms as an effective tool to assist the energy system planner.

Availability of such a synthetic and powerful tool is of utmost importance in order to effectively cope with the increasing complexity of future electro-energetic systems, in which efficiency enhancement will strongly depend on the integration of the equipment for local combined production of manifold energy vectors. These aspects are illustrated in the next chapters.

4.1. PLANNING ISSUES WITHIN THE MULTI-GENERATION FRAMEWORK

Planning a multi-generation plant is a challenging task, due to the variety of alternatives available to set up the energy system. The adoption of different equipment can be considered at a planning stage within a comparative analysis that takes into account the user's needs and the economic conditions. Starting from the requested load patterns and from availability of various pieces of equipment with different energy performance, the economic profitability of a DMG solution also depends on the site-specific characteristics in terms of energy prices and accessibility to energy networks and markets. In particular, the possibility of feeding different heat/cooling generation equipment with energy vectors produced *in situ* or coming from external networks, as well as of selling the produced energy vectors to external networks, represent options that can be effectively investigated at the planning stage and that can change the results of an energy-based assessment [ChM06][ChM06b][Man06].

Let us first address some concepts referred to cogeneration – the simplest case of multi-generation. In general, the first step towards the planning of an energy system is the analysis of the load patterns. One of the traditional approaches is based on the analysis of the *load duration curve* [WiS00] for both electrical and thermal power [EDU01][CaP03]. In addition, since the duration curve analysis does not allow for considering simultaneity of the loads, the *cogeneration ratio* (heat to power ratio [Hor97]) is used as an auxiliary indicator. Load duration curve analysis, along with cogeneration ratio analysis, are usually sufficient to deem, at a first evaluation, which characteristics a prime mover is supposed to boast in order to satisfy the plant needs. However, often these analyses are run under rated or average conditions [Hor97][VoD03]. In this light, this chapter introduces a more general approach that takes into account off-design conditions and partial-load operation. We have dubbed this approach *cogeneration lambda analysis*, being the Greek letter λ used in many texts (for instance, [Hor97]) to indicate the cogeneration ratio.

Electricity and heat are the types of energy produced in a cogeneration plant, with the corresponding duration curves and indicators. However, from a general viewpoint, a CHP plant could sometimes be seen as an MG plant in which the thermal energy may be produced at different enthalpy levels, in response to specific user's requests. For instance, a small-scale ICE can produce typically hot water at about 80 °C and steam at about 10 bar [WiS00][BoK01][EDU01][Dan06][EPAww]. In this case, more duration curves might be adopted for characterizing both thermal needs and thermal generation. Indeed, sometimes the term *trigeneration* is used to point out two different physical energy carriers used for thermal power (for instance, hot water and steam) besides electricity. Thus, when also cooling power is generated (*e.g.*, chilled water at 7 °C for air conditioning purposes), one could use the term "*quad-generation*". A relevant application is for instance discussed in [ChM08b].

However, for the sake of simplicity the analysis carried out in this chapter refers to a unique thermal demand curve, ruling out the details of a second-stage further level of planning, involving for instance the production of steam besides "classical" hot water cogenerated by small-scale systems. This approach has the upside of dealing with thermal demand and production likewise electrical demand and production, allowing both simplified and unified characterization and evaluation techniques. In any case, the tools presented can be easily extended to entail more general options within the DMG framework.

When planning a multi-generation system that includes cooling power besides heat and electricity (trigeneration), the duration curve for the cooling load, as well as the interactions among different types of energy, have to be further considered. In fact, it is possible to interpret the effect of different heat/cooling generation equipment as change in the "equivalent" electrical and/or thermal load seen from the CHP unit, thus affecting the selection and control of the cogeneration prime mover itself. For instance, in classical trigeneration absorption chillers are fed by heat from the cogenerator in order to produce cooling power. The final outcome is an "aggregate thermal consumption", sum of the "pure" heat load and the heat needed to generate cooling [CaP03].

This order of concepts leads to formulate the *multi-generation lambda analysis* presented here, as a generalization of the *cogeneration lambda analysis* [Man06] to multiple energy vectors. In particular, as a special and important sub-case, the description of this methodology can be effectively exemplified with reference to a generalized trigeneration system in which electricity, heat and cooling power can be produced according to different combined equipment and plant schemes. Thus, in the sequel the term *trigeneration lambda analysis* is adopted, although the approach can be readily extended to any other kind of energy vector produced. In mathematical terms, the proposed approach leads to the definition of the *lambda transforms* introduced in Section 4.3.5.

The lambda analysis and the lambda transforms are able to synthetically describe all the different energy flow interactions within a combined MG energy systems and with the external networks (such as electricity, gas, district heating and district cooling). On the other side, this approach is orientated towards the implementation of numerical codes for time-domain simulations that enable to take into account further issues such as control strategies (Section 2.2.2) and economic interactions with the energy markets and the energy networks [ChM06][ChM06b][Man06] (see also Section 8.2) within a DMG framework.

4.2. Characterization and Planning of a Cogeneration Plant

Various techniques for CHP characterization and planning are well consolidated and widely used in most studies. In particular, the load duration curve analysis is one of the favourite tools for first approximation considerations. The studies based on the CHP load duration curves and on the cogeneration ratio represent useful standpoints to start with in order to carry out the lambda analysis in MG systems. Thus, the concepts illustrated in the following subsections form the basis for the lambda analysis and the generalization to the lambda transforms defined in a multi-vector space.

4.2.1. Load duration curve analysis

The *load duration curve analysis* [WiS00][EDU01][CaP03] provides a first snapshot of the load levels and, accompanied by equipment efficiency curves and control strategies, can also give information on the overall fuel consumption. In particular, on the basis of the (thermal or electrical) duration curves, it is possible to select the prime movers and estimate the duration of their operation under a certain control strategy. Some examples are provided in Section 4.7. It has to be pointed out that this kind of analysis is approximated and represents only a first step in the planning procedure. However, it can give hints to the planner about the prime mover sizing and control, as well as on the selection of the auxiliary heat generator group, before performing further analyses, for instance based on time-domain simulations [WiS00][Man06].

Conceptually, there are some limits to the load duration curve approach the planner should be aware of [WiS00][Man06]:

- In a comparative analysis there is no explicit allowance for considering the efficiency (thermal or electrical) of the prime mover; this downside might be overcome by plotting, once decided the control strategy, the fuel thermal consumption duration curves for the prime mover (passing through the actual efficiency point by point).
- There is no possibility to include operating schedule issues and dynamic constraints, which basically depend upon the time evolution of the load pattern, whereas they

might strongly influence the selection of the prime mover and the fuel consumption to some extent.

- There is no allowance for including the "simultaneity" of thermal and electrical load in the analysis. This is the main downside of this type of analysis when applied to cogeneration plants. In fact, depending on the plant sizing and on the control strategy adopted, simultaneity of the load might be a key issue to enable energy saving and economic profits.
- As there is no allowance for including "time-domain" considerations within the duration curves, certain "side" conditions are hard to be taken into account, *e.g.*, the potential drop of power and efficiency depending on outdoor temperature (in particular for MTs [Loz00][AmH04][Hwa04][ZaP06], as also discussed in Section 3.1.3.2).

From the above considerations, it emerges how although load duration curves are an important tool for cogeneration planning analysis, they may be not sufficient in the bulk of practical cases. Other auxiliary tools are thus needed, such as the *lambda analysis* presented here, as well as time-domain simulations.

4.2.2. The cogeneration ratio for generation and load

The planning of a CHP plant has to encompass several and manifold aspects such as, *in primis*, taking into proper account the load patterns and the characteristics of different technologies of prime movers. The subsequent outcome is the need, at the planning stage, for having at disposal simple but sound tools and indicators able to synthetically characterize the plant loads and assist the planner in properly selecting and sizing the prime mover as well as, in case, deciding the control strategy. The approach followed here, although essentially based on the classical approach to cogeneration plant analysis, is aimed at pointing out the characterization of a cogeneration plant on a *time-domain* basis, as opposed to the classical approach carried out on a *rated-value* or *average-value* basis. This is aimed at avoiding some common pitfalls that might mislead the CHP evaluation [Man06]. In particular, considering only nominal values in the analysis might lead to overlook the potentially different energy system characteristics when operating under off-design conditions. In this sense, off-design operation could occur frequently, above all if load-following control strategies are exploited. Since the equipment off-design performance can change consistently with respect to the full-load performance [EDU01][Man06][EPAww], the actual energy evaluation, as well as the economic one, could differ even dramatically from the evaluation based upon rated characteristics. In addition, specific time-domain aspects cannot be adequately accounted for through classical tools [WiS00][VoD03][Man06], and as such need to be tackled with different approaches.

In order to characterize the user's final needs, a CHP plant is often described by the well-known *cogeneration ratio* λ [Hor97], that is, the heat-to-electricity ratio. In particular, it is possible to define a *demand-related* cogeneration ratio λ_d as the ratio of the thermal to the electrical power demand (user's side):

$$\lambda_d = \frac{Q_d}{W_d} \tag{4.1}$$

Similarly, it is possible to introduce the *production-related* cogeneration ratio λ_y as the ratio of the thermal to the electrical power developed by the prime mover (generation side), as already mentioned in Section 3.1.1:

$$\lambda_y = \frac{Q_y}{W_y} \tag{4.2}$$

As general indications on some numerical practical aspects [WiS00][BoK01][EDU01] [Dan06][EPAww], ICEs feature typical rated cogeneration ratios from 1 to about 1.5, with the lower figure for larger machines (due to the higher electrical efficiency and subsequent lower thermal efficiency). MTs, instead, are characterized by higher cogeneration ratios with respect to ICEs, due to lower electrical efficiency and higher thermal one; typical values are between 1.5 and 3 (see also Figure 3.1).

In principle, a perfectly planned CHP plant should provide the "matching" $\lambda_d = \lambda_y$ ("*matched plant*" [Hor97]), and the problem for the planner could be seen as best matching a prime mover to the load, considering the families of machines available in the capacity range to supply the requested thermal and/or electrical demand. In this ideal case, the plant satisfies the power needs without resorting to auxiliary means (namely, auxiliary boilers for thermal production and electricity exchanges with the power grid). Unfortunately, all real cases are made up of "*unmatched" plants*.

References such as [Hor97] provide an insightful description of the thermodynamic aspects related to the load/plant coupling by using the parameters λ_d and λ_y. However, these considerations are mostly limited to rated or average power values.

4.2.3. "Unmatched" plant and energy interaction modelling

Before discussing a suitable approach to time-domain characterization of the energy system, few considerations are needed on the definition of both demand and production cogeneration ratios. Indeed, the definitions given above include the thermal and electrical user's demand, on the one hand, and the thermal and electrical prime mover production, on the other hand. This would be sufficient only in the case of *matched* plant, but when the prime mover production and the user's load are not matched, there are four basic possibilities (or combinations of them), provided that the plant is grid-connected:

1. If the cogenerated electricity is in excess with respect to the user's electrical demand, the excess electricity is sold to the grid; of course, for stand-alone plants, or when no selling to the EDS is allowed, the plant must operate in electrical load-following mode.
2. If the cogenerator fails to completely cover the electrical load, the needed additional electricity is bought from the grid; again, if the plant cannot be backed up by the EDS this situation has to be avoided in order not to bring about a plant shut-down.
3. If the cogenerator produces heat exceeding the user's thermal request, the exceeding heat may be in general discarded.
4. If the cogenerator fails to completely cover the user's thermal demand, the plant is generally backed up by heat produced in the CHG.

Let us now focus on how to model these plant configurations in terms of cogeneration ratio or, better, in terms of *generalized production system* and *generalized load system*. For this purpose, and from an extended point of view, the prime mover plus the CHG plus the energy networks the plant could interact with (such as EDS and DH network) can be all together seen as a *CHP generalized production system* (Figure 4.1) with respect to the prime mover-only, featuring a *generalized production cogeneration ratio* λ_y^{CHP} expressed as

$$\lambda_y^{\mathrm{CHP}} = \frac{Q_y^{\mathrm{CHP}}}{W_y^{\mathrm{CHP}}} = \frac{Q_y + Q_{\mathrm{CHP}}^{\mathrm{CHG}} + Q_{\mathrm{CHP}}^{\mathrm{DH}}}{W_y + W_{\mathrm{CHP}}^{\mathrm{EDS}}} \tag{4.3}$$

where Q_y and W_y are the energy amounts produced by the prime mover, and $Q_{\mathrm{CHP}}^{\mathrm{CHG}}$ and $W_{\mathrm{CHP}}^{\mathrm{EDS}}$ are the energy back-ups to the CHP system from respectively the heat generator group and the grid. Back-up heat may in case also come from a DH network ($Q_{\mathrm{CHP}}^{\mathrm{DH}}$). Thus, when somehow the prime mover production fails the user's demand ($\lambda_y \neq \lambda_d$) the back-up systems (EDS and/or CHG/DH) has to balance out the energy spreads, so as to obtain $\lambda_y^{\mathrm{CHP}} = \lambda_d$. However, this reasoning is sound for production *failings*. When the production *exceeds* the user's demand, the exceeding heat may be discharged off or sold to a DH network [Hor97][Dan06][Man06], exactly as the exceeding electricity is sold to the grid. Therefore, these possibilities have to be taken into account in the formulation of the planning problem considering the presence of a "*generalized load system*", corresponding, namely, to the user plus the EDS plus the DH network. Of course, rather than *loads* or *requests*, the energy shares sold to the EDS or for DH should be seen as *opportunities*.

From a modelling viewpoint, it is possible to include the heat and electricity to be sold to external networks within the user's load model (Figure 4.2), resulting in the *generalized demand cogeneration ratio* λ_d^{CHP}, expressed as

$$\lambda_d^{\mathrm{CHP}} = \frac{Q_d + Q^{\mathrm{DH}}}{W_d + W^{\mathrm{EDS}}} \tag{4.4}$$

In this way, the final match between production and load should be such that $\lambda_y^{\mathrm{CHP}} = \lambda_d^{\mathrm{CHP}}$. In general, there are two possible approaches, namely, to include the exceeding energy amounts into the user's demand figures, as above, or to take it into account separately. Depending on the case, either approach can bring advantages over the other. The latter approach, particularly effective at a first stage of analysis, is followed in the sequel. This is done essentially because of two reasons:

- the focus is meant to be on the user's loads, as shown in detail in Section 4.3 on multi-generation plant characterization, where the major purpose is to analyze the impact of the cooling/heat load over the rest of the plant;
- from the economic point of view, the energy produced to feed the user is worth differently of the one potentially sold outside: keeping the two "shares" of produced energy vectors separate helps the planner better understand the economic potential of the plant [ChM06][ChM06b].

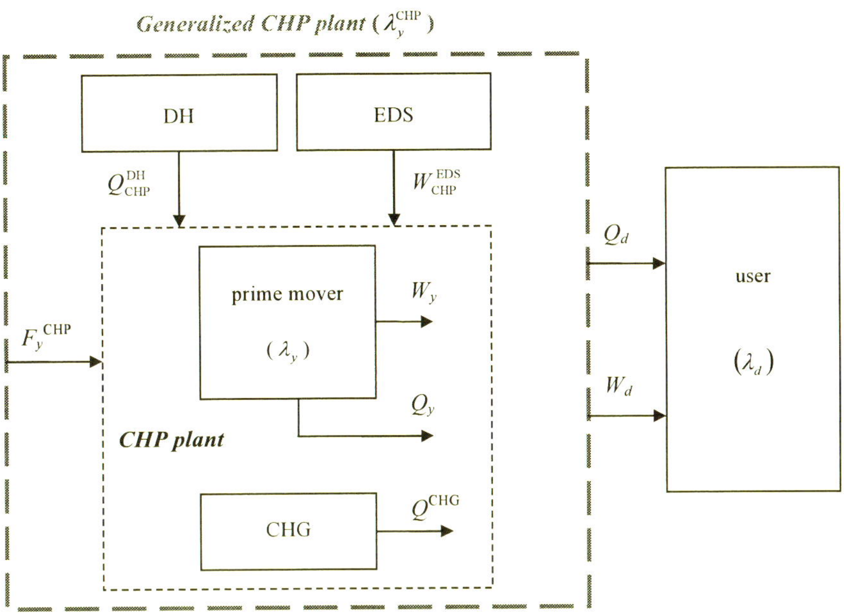

Figure 4.1. Generalized CHP production system model.

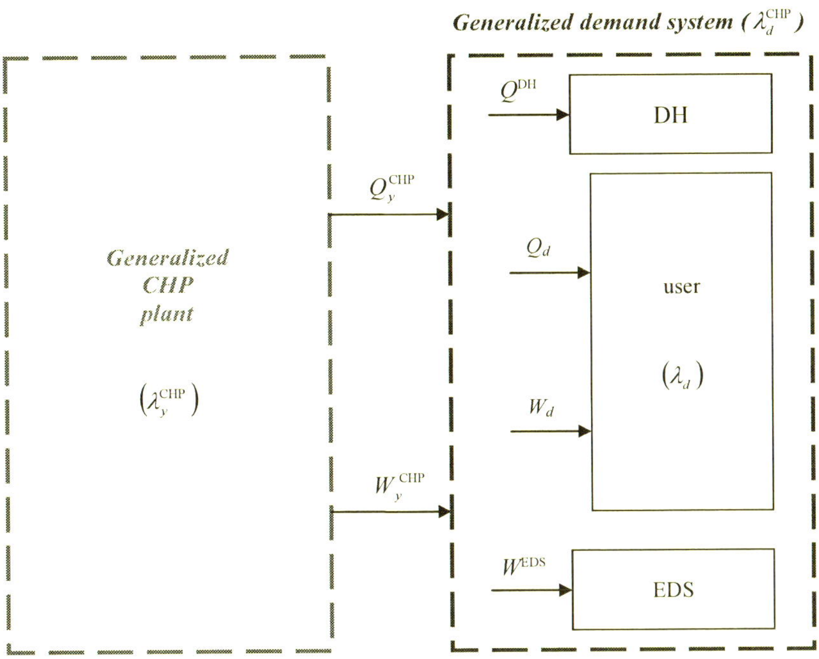

Figure 4.2. Generalized CHP plant and demand system model.

To conclude the reasoning, a last question on the energy bought from DH or EDS arises: why are heat from district or electricity from the grid considered *within* the generalized production model and λ_y^{CHP} when bought, whereas when they may be sold they are treated differently? This could look "asymmetric" or "incoherent", since also electricity and heat bought from outside are worth (from the economic and energy standpoints) differently from the ones produced inside. The difference, however, stands in the final purpose: selling energy outside is *allowed* when the production *exceeds* the user's demand; likewise, buying energy from outside is *mandatory* when the production *fails* the user's demand. The approach followed is then fully consistent with the standpoint of "*serving the user as the first goal*".

4.2.4. Time-domain load characterization of a cogeneration plant

Let's now focus on the user's loads and the prime mover production. The definition of cogeneration ratio is completely general and can refer to any operating condition. Thus, apart from trivial cases, it is apparent that the thermal and electrical load may largely vary throughout a year (because of seasonal weather effects, for instance) or even hourly within a single day (because of the day-night cycle, at least) so as to often make it hard to simply consider rated values or to even define the rated value for λ_d, say $\hat{\lambda}_d$. Indeed, in most cases it is not easy to figure out an "average" load, whereas the peak load may be considered as rated load for planning purposes. On the other hand, it is tougher to choose a representative cogeneration ratio as *the* rated one. The best approach is therefore to plot the electrical and thermal loads on an hourly basis, together with the λ_d characteristics, so as to have a hint on what the interaction with the prime mover might be, also taking into account the time-domain cogeneration characteristics of the CHP plant, as shown below. An exemplificative numerical case study is provided in Section 4.7.

4.2.5. Time-domain production characterization of a cogeneration plant

A cogenerative prime mover is more naturally characterized by its thermal and electrical rated capacities, so that it is easier to define a rated value $\hat{\lambda}_y$. Thus, the prime mover can be selected on the basis of the electrical or thermal load to cover or follow, and of the relevant cogeneration ratio, also depending on the control strategy adopted. However, in general the CHP system is not able to follow the demand-related cogeneration ratio, since λ_d changes continuously over time. In practice, according to the classical approach to CHP planning, the equipment may be chosen in such a way that the rated cogeneration ratio is close to the demand-related cogeneration ratio values occurring in the largest part of the operating conditions. When the difference between $\hat{\lambda}_y$ and λ_d is large, it is necessary to provide energy integration by means of additional heating equipment, heat from DH networks, or electricity import from the EDS.

However, in the bulk of cases this approach based *only* upon rated values of the prime mover cogeneration ratio can result poor and the actual plant performance may not correspond to the forecast one. Indeed, the reasoning developed so far is somehow incomplete, essentially due to the fact that, in load-following operation, λ_y is not constant. In fact, both thermal and electrical efficiency vary in a different fashion at partial loads, corresponding to variations of the electricity and heat cogenerated, and eventually of the cogeneration ratio. Therefore, the approach undertaken has to be improved by suitably modelling changes in λ_y during operation at partial loads, so moving on from a completely *static* point of view to a more *dynamic* one [Man06]. For instance, once chosen the control strategy, by comparing λ_y and λ_d hour by hour the planner may obtain a time-domain picture of the interaction between prime mover and load characteristics.

An effective way to describe the prime mover cogeneration characteristics is provided by exploiting its *lambda characteristics* [Man06], describing how λ_y changes during the actual operation and how it is suitable to someway follow the changes in λ_d.

An example of partial-load performance characteristics for a gas-fuelled ICE has been shown in Figure 3.4. With reference to the same characteristics, Figure 4.3 shows the "*functioning line*" [Hor97], that is, thermal output against electrical output, for different levels of heat recovery (considering separately steam and hot water production). For the same prime mover, the lambda characteristics, again based on the efficiencies in Figure 3.4, are shown in Figure 4.4.

With reference to Figure 4.3 and Figure 4.4, if only steam were profitably recovered, then the cogeneration ratio would be only about 0.5 over the modulation interval (50%÷100% of the electrical output). If only the heat (from intercooler, oil coolant and jacket coolant) available in the hot water circuit were used, the cogeneration ratio would be about 0.6 at full load and would increase slightly at partial load because of the larger heat available for recovery due to the electrical efficiency reduction (prompting higher thermal discharge in the thermodynamic cycle). Finally, if heat were recovered from both sources, the cogeneration ratio would be equal to about 1.1 at full load and would increase slightly at partial load. Between the upper level (full-source recovery) and the lower level (steam-only recovery) all the states are feasible by wasting off thermal energy. The thermal output levels shown by the functioning line in Figure 4.3 reflect this possibility of exploiting the different heat recovery techniques to modulate the prime mover heat production. It is therefore clear that the relevant lambda and functioning line characteristics emerge to be related to a region of the relevant space, more than to simple curves. For practical cases, the actual user's requests (steam and/or hot water) will drive the modulation of the production characteristics. In addition, the profitability of producing additional hot water instead of steam (or *vice versa*) should also be evaluated given the specific case.

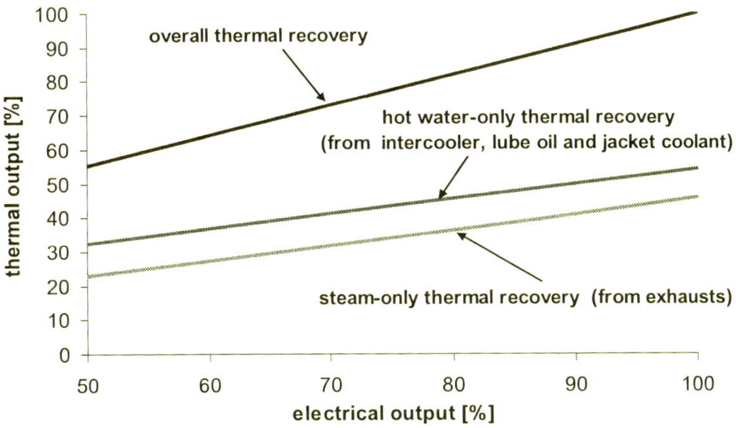

Figure 4.3. Partial-load "*functioning line*" characteristics for a 836-kW$_e$ ICE.

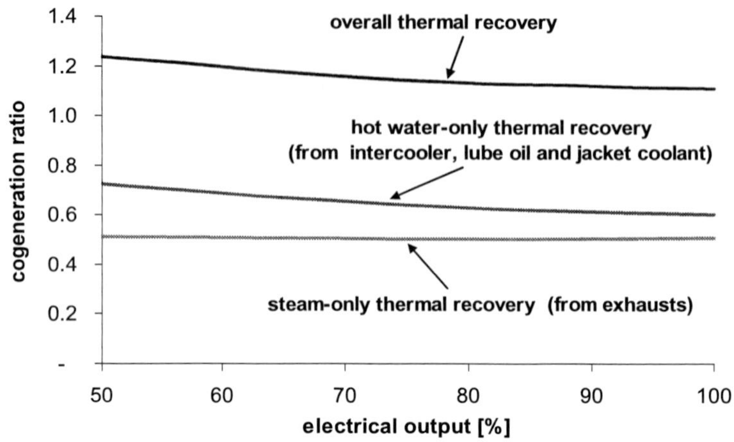

Figure 4.4. "*Lambda characteristics*" of a 836-kW$_e$ ICE for different levels of thermal recovery.

All the aspects discussed need to be analyzed already at the planning stage and, eventually, it must be taken into account that some possible controls would be *dissipative*, and as such to avoid when possible.

Besides the "intrinsic" modification of λ_y as a consequence of modifications of thermal and electrical efficiency at partial loads, it is also possible to modify *on purpose* the characteristics of the CHP system itself by means of several techniques, aimed at operating on the plant components in such a way to change either or both the thermal and electrical efficiencies [Man06].

A peculiar possibility to exploit in MTs would include switching the recuperator on and off [Hor97], or adopting a *step recuperator* [BaC99], in order to change the thermal output and the cogeneration ratio of the prime mover, a very suitable option in the presence of wide range of variation of thermal loads (especially seasonal loads) [Hor97]. Post-combustion [Loz00][EDU01][BrV05] is also possible for both microturbines and ICEs, to raise the thermal output by exploiting additional fuel (and in case also additional air, for ICEs) injected into a HRSG. Finally, modifications of the conditions at which heat is produced may *strongly* impact on the thermal efficiency: in general, the higher the temperature of the heat to provide, the lower the amount of this heat [Han04][Man06][CoA07], as it can be inferred from manufacturers' data (see also Fig. 3.5).

Whatever is the technique used, it has to be highlighted that the modulation of the thermal load and the variation of the cogeneration ratio might occur at the cost of overall efficiency loss while diminishing the recuperated thermal power and/or the electrical one. All the options presented have to be included in the formulation of the planning problem.

4.3. Characterization and Planning of a Multi-Generation Plant

The lambda analysis approach to an MG system can be effectively illustrated with specific reference to CCHP plants. In particular, the first step is to characterize the impact of the cooling power generation on the equivalent load seen from the CHP production side (*trigeneration "lambda" analysis*), considering different technologies for the CGP, according to a broader approach to trigeneration [ChM06][ChM06b][Man06]. Then, the analysis can be generalized to multi-generation systems (for examples of this kind of systems, see for instance [LiN06]) in which an AGP is present, for production of both heat and cooling power. This novel point of view, illustrated firstly for CHCP systems and easily extendable to multi-generation, calls for revisiting the classical approach adopted for CHP system characterization. In particular, the modelling of the impact of the AGP equipment in the interaction with the CHP side leads to the *multi-generation lambda analysis*, whose details are given in the sequel.

4.3.1. The effect of cooling power generation: the trigeneration lambda analysis

Let us now focus on a classical trigeneration plant, with a cogeneration plant coupled to a CGP in which only cooling power is produced, with no heat recovery. The cooling power production impacts somehow on what the cogeneration side "sees" as the load. In fact, by inspection of Figure 4.5 it is immediate to notice that the cooling power could represent an additional load to the CHP unit, which could impact or not on either or both the thermal and electrical load, by "coupling" the cooling need to the other needs of the user.

The first and straight effect on the characterization of the trigeneration CHCP system is that the load duration curves of electrical and thermal demand (as seen from the prime mover) change as well, so bringing possible modifications to the selection of the trigeneration system prime mover with respect to the cogeneration-only case. For instance, the reference [CaP03] reports examples of planning approaches for classical trigeneration as a consequence of the so-called *aggregate thermal consumption*.

The second straight consequence on the characterization of the trigeneration plant is that the cooling plant impact on the absolute thermal and/or electrical loads seen from the CHP unit implies to reconsider and revise the definition of demand-related cogeneration ratio. Indeed, the definition of λ_d is aimed at comparison with the cogeneration ratio prime mover to simplify its selection. In this case, then, the demand needs as seen from the cogeneration side of the plant should take into account the *overall* electricity and thermal demand. In fact, once given the user's *base* thermal load Q_d (*e.g.*, for heating purposes) and the user's *base* electrical load W_d, the presence of different types of cooling equipment to produce the cooling power R_d can change the *overall* electricity or heat demand and subsequently the value of the demand cogeneration ratio.

From this point of view, in CHCP plant planning the selection of the prime mover looking at λ_d is more complex than for CHP plants. The problem can be tackled by performing what we call *trigeneration lambda analysis*, *i.e.*, the analysis of the cooling power production impact over the thermal and electrical demand (as seen from the cogeneration side) and thus over the prime mover selection, sizing and control.

4.3.2. Cooling power generation effect on the cogeneration ratio

The new issues brought in by the presence of the cooling side can be handled by introducing the *trigeneration demand-related cogeneration ratio*, λ_d^z in which the thermal and electrical load include the corresponding share due to cooling demand:

$$\lambda_d^z = \frac{Q_d^z}{W_d^z} = \frac{Q_d + Q_R}{W_d + W_R} \tag{4.5}$$

where the thermal and/or electrical loads (with respect to the cogeneration plant) Q_R and W_R, needed to produce the cooling power R_d, are added to the base Q_d and W_d loads. The terms Q_d^z and W_d^z then represent the overall thermal and electrical trigeneration loads. In the same way as for CHP plants, the possibility of selling electricity to the grid or heat to a DH network might be incorporated within the definition of λ_d^z. However, for the sake of illustration we will focus on the user's loads.

Figure 4.5 and Figure 4.6 compare the two standpoints of a *classical* demand-related cogeneration ratio and *trigeneration* demand-related cogeneration ratio. In the former case, the trigeneration load is seen from the "trigeneration plant" side. In the latter case, the trigeneration load is seen as a whole from the cogeneration side, and as such the demand cogeneration ratio has to include the cooling load (with respect to the CHP side) Q_R and/or W_R necessary to produce the cooling effect R_d.

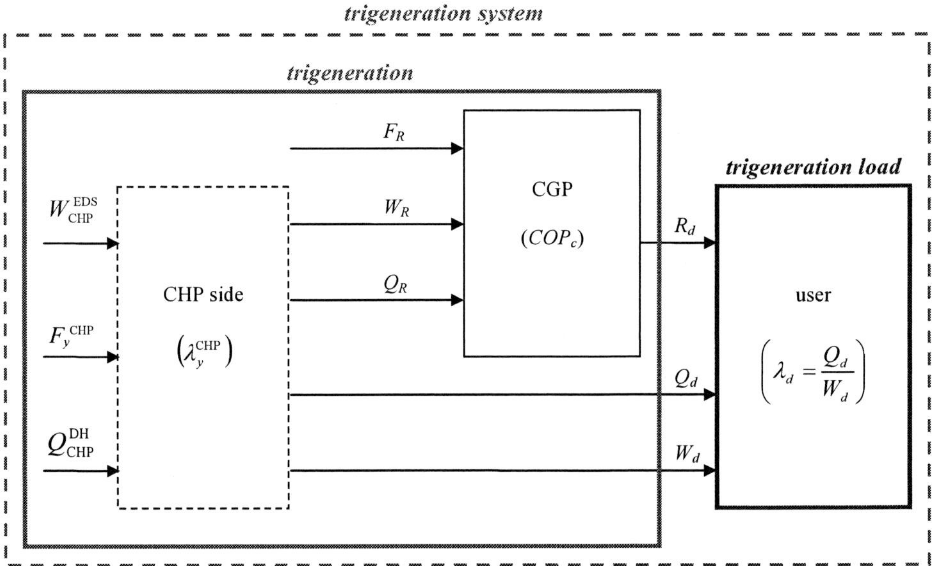

Figure 4.5. General trigeneration plant model with *classical* demand-related cogeneration ratio λ_d.

As for cogeneration-only plants, λ_y^{CHP} refers to the overall CHP-side cogeneration ratio, including the heat produced by the CHG or bought from a DH network and the electricity bought from the EDS.

As a follow-up of the reasoning, the trigeneration system planning is based on the search for the best "match" between the λ_y of the prime mover and the overall trigeneration demand-related cogeneration ratio λ_d^z. When this is not possible, the prime mover cogeneration production is backed up by heat produced in boilers, electricity drawn from the

EDS, and, in case, heat from a DH network (thus λ_y turns into λ_y^{CHP}), so as to reach the final match $\lambda_y^{\mathrm{CHP}} = \lambda_d^z$.

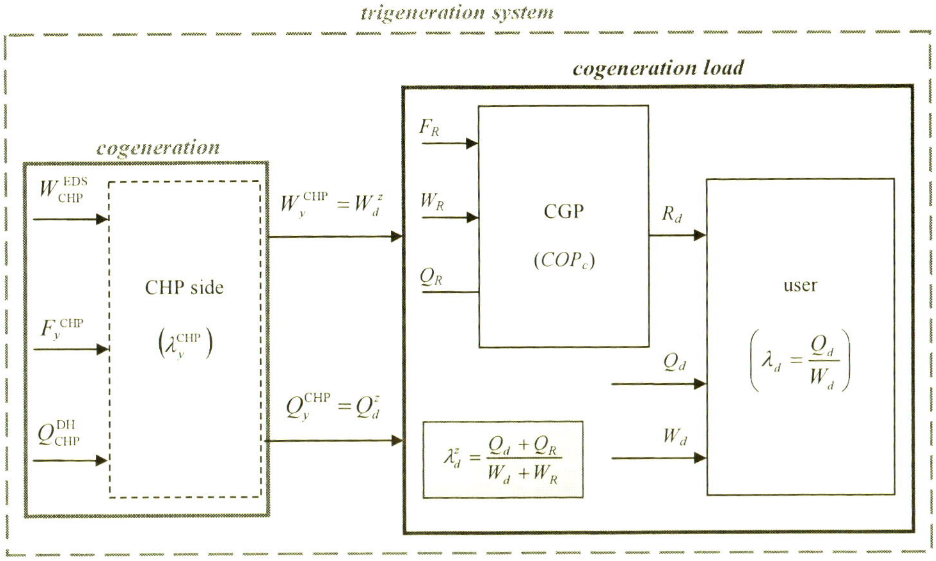

Figure 4.6. General trigeneration plant model with *trigeneration* demand-related cogeneration ratio λ_d^z.

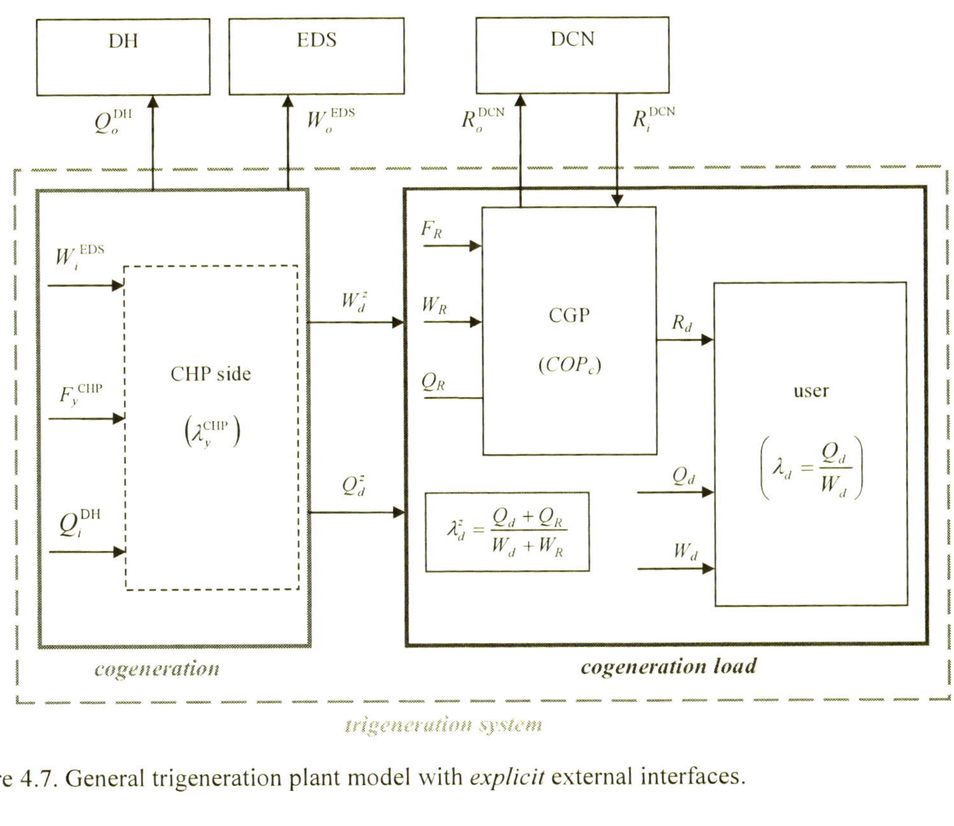

Figure 4.7. General trigeneration plant model with *explicit* external interfaces.

At this stage, while formulating the planning problem, also the opportunity to sell electricity or heat to the outside should be considered *separately* (basically for economic reasons). As such, the CHP system with its λ_y^{CHP} would in practice see as possible demand interfaces the user and the electricity/district heating networks. On the contrary, the possibility of selling/purchasing cooling power to/from a DCN should preferably be taken into account explicitly within the definition of cooling power R, as it impacts directly on the thermal and electrical load seen from the CHP system at the interface with the user.

In order to exemplify this approach, Figure 4.7 shows a schematic model of the trigeneration system including the input/output interface with the DCN and the output interfaces with the EDS and the DH networks; the input interface with EDS and DH is, instead, already taken into account within the definition of CHP side, as discussed above. In terms of notation, the subscripts i (input) and o (output) for the relevant entries are used to point out the energy system incoming and outgoing flows, respectively.

4.3.3. Cooling power generation effect on the load duration curve analysis

Planning trigeneration systems is more complex than cogeneration systems. The approach to follow reflects and is part itself of the trigeneration lambda analysis. Indeed, the various trigeneration alternatives, impacting differently as a thermal or electrical load, lead to change, sometimes consistently, the thermal and/or electrical load duration curves. For instance, the reference [CaP03] provides interesting insights on CHCP system planning, starting from the impact on the thermal load duration curve due to adoption of a WARG coupled to a CHP unit.

In order to properly evaluate the impact of the cooling load, the off-design characteristics of the CGP equipment involved should be possibly implemented. In particular, seasonal effects, which can be relevant when analyzing cooling generation equipment, should be adequately accounted for by extending the time interval of observation to at least one year.

The load duration analysis carried out following these lines can help the planner get a sound view on the changes in the load absolute levels due to the different cooling equipment used. The CHP side planning is then arranged consequently, by receiving from the analysis useful hints for the prime mover sizing and for suitable control strategies. In fact, for every different CGP considered, and thus for every different additional electrical and/or thermal load over the CHP unit, the analysis is developed as for the cogeneration-only case, by comparing different prime mover technologies and sizes. Numerical examples of trigeneration load duration curve analysis are provided in Section 4.7.

4.3.4. Heat/cooling power production effect in the AGP: the multi-generation lambda analysis

In order to extend the *lambda analysis* developed for trigeneration systems to multi-generation systems, it is interesting to evaluate the impact of having at disposal different means to produce cooling and heating power in an AGP. Depending on the equipment used, the resulting effect from adopting specific equipment for heat generation or heat recovery can be twofold. On the one hand, the equivalent load seen from the cogenerator changes due to the heat produced in the AGP. On the other hand, the characteristics of the cogenerator itself could be changed, for instance by exploiting the "heat multiplier" effect of an electric or absorption/adsorption heat pump [Hor97][Meu02][Man06][KeP08]. From the standpoint of the lambda analysis, this is accounted for by changing the equivalent cogeneration ratio seen by the CHP side and of the CHP side itself, as for the trigeneration case, considering also the effect of heat generation besides cooling generation. In particular, the equivalent thermal load seen by the CHP side can be reduced by adopting for instance heat recovery within the AGP, but at the same time the cogenerator could see the electrical load increased, for instance because part of the thermal power is produced by an EHP (see Section 4.5 for details).

The result in terms of demand-side cogeneration ratio is that the simple λ_d turns into λ_d^p, where the superscript p points out poly-generation (taken as a synonym of multi-generation). Similarly, the production-side cogeneration ratio λ_y^{CHP} turns into a multi-generation production-related cogeneration ratio λ_y^p, considering the effects of possible additional means to change the produced electricity and heat, such as heat pumps. Therefore, according to the *lambda analysis*, the aim of the planner is now to study the match between the generalized multi-generation production-related cogeneration ratio λ_y^p and the multi-generation demand-related cogeneration ratio λ_d^p.

Exhaustive details on lambda analyses for trigeneration and multi-generation systems, for the cases most frequently encountered (plant schemes and heat/cooling generation technologies), are provided in Section 4.4.

4.3.5. The lambda transforms

From the considerations carried out so far, the multi-generation planning problem from the standpoint of lambda analysis is outlined in terms of transformation of the relevant "simple" electrical, thermal and cooling loads into "equivalent" electrical and thermal loads. The same holds true for the production side, where the cogeneration characteristics of the prime mover are changed by the CHG, external networks (DH and EDS), and equipment in the AGP (in particular, heat pumps).

In mathematical terms, the effect of heating/cooling generation on the interaction between CHP generation side and demand side can be effectively described by means of two types of transforms, that we have called *lambda transforms*. The lambda transforms can be applied on the one hand to the demand side, and on the other hand to the production side. These transforms operate on a set of original energy vectors (for instance referred to a single hour of operation) by transforming them to equivalent energy vectors that embed the interactions among the various components in the MG system. The description of the characteristics of the lambda transforms is provided here with reference to the trigeneration case. The analysis can be easily extended to other types of energy vectors (for instance a manifold level of heat production/utilization).

The first type of lambda transforms operates on the single energy vectors (*e.g.*, heat, electricity and cooling power), providing in general the transformation from the vector space $\Re^h$ of the original variables to the vector space $\Re^k$ of the equivalent variables, with $h \geq k$. Focusing on a trigeneration case, considering the demand side, the *lambda transform* $\Lambda_d^{WQR,WQ}$ ($\Re^3 \rightarrow \Re^2$) transforms the electrical, thermal and cooling loads into the equivalent thermal and electrical loads seen from the CHP side. From the production side, the *lambda transform* $\Lambda_y^{WQ,WQ}$ ($\Re^2 \rightarrow \Re^2$) transforms the electrical and thermal prime mover production into the equivalent thermal and electrical production after adopting additional generation means such as CHG, heat pumps, and so forth.

The second type of lambda transforms operates on the cogeneration ratio, providing the $\Re \rightarrow \Re$ transformation from the "classical" cogeneration ratio to the equivalent cogeneration ratio. Thus, considering the demand side, the *lambda transform* $\overline{\Lambda}_d$ transforms the "simple" demand-related cogeneration ratio λ_d into the trigeneration equivalent cogeneration ratio λ_d^z seen from the CHP side. From the production side, the *lambda transform* $\overline{\Lambda}_y$ transforms the "simple" cogeneration ratio λ_y, characteristic of the prime mover generation, into the trigeneration equivalent cogeneration ratio λ_y^z.

Practical applications of the lambda transforms, referred to technologies mostly already available on the market, are provided in the following sections.

4.4. Performance Indicators for Multi-Generation Equipment

According to the structure outlined in Section 4.3 for trigeneration analyses, it has been pointed out how, depending on the equipment used in the CGP, the cooling load seen from the CHP side represents a further electrical or thermal load to be characterized from a time-domain point of view. In turn, production of heating power in the AGP, as well as heat recovery from chillers, can be seen as an equivalent change in the load characteristics or in the production characteristics for a multi-generation plant. All these issues are addressed by means of the lambda transforms, which can be expressed through the efficiency characteristics of the relevant equipment, as shown in Section 4.5.

In this perspective, this section describes the efficiency indicators and models referred to equipment for cogeneration, trigeneration and MG systems. Further details are then provided in the case study presented in Section 4.7 and in the following chapters.

4.4.1. Input-output black-box modelling approach

The lambda analysis for multi-generation equipment is time domain-oriented, potentially accounting for the off-design equipment characteristics to be implemented in time-domain simulation codes, in line with modern energy system planning techniques. From this standpoint, an effective approach to describe characteristics and performance of the energy system equipment is to adopt *input-output black-boxes* [Man06] describing every piece of equipment by means of the relevant efficiency indicators. In fact, regardless of the physical content of the box representing the equipment, the rationale of the lambda analysis is simply to *track back*, starting from the user's needs, the input energy requested by each machine, adding it up to the overall electrical or thermal load seen from the CHP side. For instance, the cooling power, if produced by a CERG, represents an additional electrical load to the CHP side, and the amount of this load is calculated through the cooling machine *input-output performance indicator* (*i.e.*, its *COP*, Section 3.3.3.1). In addition, once known the overall electricity production and efficiency characteristics of the prime mover, its fuel input could be tracked back as well. In this way, starting from the desired output, it is straightforward to calculate the input energy consumption for each component and the plant (that can be in turn seen as a black box with respect to the external networks [Man06]) as a whole. Moreover, this approach allows evaluation of the energy system performance in *all* the *actual* operating points, provided that *off-design* models of the components are available and implemented.

4.4.2. Efficiency indicators for black-box models

The formulation of efficiency indicators as output-to-input energy ratios makes a black-box approach of straightforward application for energy and economic analyses [Man06]. The main indicators for the equipment mostly used in MG systems are reported in Table 4.1 for cogeneration systems (prime mover and CHG), and in Table 4.2 and Table 4.3 for AGP equipment respectively in the bottoming configuration and in the separate configuration (Section 2.1.1). The subscripts used highlight the "final goal" of the specific entry; in particular, *y* points out "cogeneration" production, *c* general cooling production, *t* indicates general thermal production or recovery; *Q*, *W*, and *R* refer to the final goal of producing heat, electricity, and cooling power, respectively.

Table 4.1. Equipment and performance indicators for cogeneration plants

equipment	*indicator*	*output*	*input*	*equation*
CHP prime mover	$\eta_W = W_y / F_y$	electrical power	fuel thermal power	(4.6)
	$\eta_Q = Q_y / F_y$	thermal power	fuel thermal power	(4.7)
	$EUF = \eta_W + \eta_Q$	thermal + electrical power	fuel thermal power	(4.8)
CHG	$\eta_t = Q / F_Q$	thermal power	fuel thermal power	(4.9)

Table 4.2. Equipment and performance indicators for bottoming heat/cooling generation

equipment	*operation mode*	*indicator*	*output*	*input*	*equation*
CERG	cooling	$COP_c = R / W_R$	cooling power (heat extracted from the cold sink)	electrical power	(4.10)
	heat recovery	$\varepsilon_t = Q / W_R$	recovered thermal power	electrical power	(4.11)
EHP	heating	$COP_t = Q / W_Q$	thermal power released to the hot sink	electrical power	(4.12)
	cooling	$COP_c = R / W_R$	cooling power (heat extracted from the cold sink)	electrical power	(4.13)
WARG	cooling	$COP_c = R / Q_R$	cooling power (heat extracted from the cold sink)	thermal power	(4.14)
	heating	$COP_t = Q / Q_Q$	thermal power released to the hot sink	thermal power	(4.15)
	heat recovery	$\varepsilon_t = Q / Q_R$	recovered thermal power	thermal power	(4.16)
WAHP	heating	$COP_t = Q / Q_Q$	thermal power released to the hot sink	thermal power	(4.17)

Table 4.3. Equipment and performance indicators for separate heat/cooling generation

equipment	*operation mode*	*indicator*	*output*	*input*	*equation*
GARG	cooling	$COP_c = R/F_R$	cooling power (heat extracted from the cold sink)	fuel thermal power	(4.18)
	heating	$COP_c = Q/F_Q$	thermal power released to the hot sink	fuel thermal power	(4.19)
	heat recovery	$\varepsilon_t = Q/F_R$	recovered thermal power	fuel thermal power	(4.20)
GAHP	heating	$COP_t = Q/F_Q$	thermal power released to the hot sink	fuel thermal power	(4.21)
EDC	cooling	$COP_c = R/F_R$	cooling power (heat extracted from the cold sink)	fuel thermal power	(4.22)
	heat recovery	$\varepsilon_t = Q/F_R$	recovered thermal power	fuel thermal power	(4.23)
EDHP	heating	$COP_t = Q/F_Q$	thermal power released to the hot sink	fuel thermal power	(4.24)

Concerning the various pieces of equipment, more specifically it can be observed that:

- *Cogeneration prime movers*: the performance of a CHP prime mover can be described (as already mentioned in Section 3.3.1) through the electrical efficiency η_W (4.6), electrical output to fuel thermal input ratio, and the thermal efficiency η_Q (4.7), thermal output to fuel thermal input ratio. In addition, as an indicator of overall cogeneration performance the *EUF* (4.8) is also often used, giving a first idea of the actual exploitation of the fuel thermal content. The numerical values of (4.6) and (4.7) depend upon the technology, the loading level, the outdoor conditions, the heat recovery system, and the application, as illustrated throughout Section 3.1. For instance, a complete partial-load model for a gas-fed CHP ICE has been shown in Figure 3.4; another model, referred to the engine used for the case study in Section 4.7, is reported in Table 4.7.
- *Combustion heat generators*: the performance of CHG units is normally characterized by the thermal efficiency η_t (4.9), thermal output to fuel thermal input ratio (Section 3.2.2). Also in this case, a typical efficiency model is reported in Table 4.7, while details on off-design models are provided in Section 3.2.3.
- *AGP plant equipment*: the *COP* is the common way to describe the performance for the various equipment that can set up an AGP. However, the *COP* can be defined in different ways, depending on the specific machine and on the relevant input and output energy vectors, as widely discussed in Section 3.3. For instance, the desired output can be cooling power for electric and engine-driven chillers as well as heat pumps and absorption/adsorption machines operating under cooling mode. An alternative output could be heating power for electric heat pumps as well as engine-driven and absorption machines operating under heating mode. The input can be electrical power for electric chillers and electric heat pumps, thermal power in the form of steam or hot water for indirect-fired absorption/adsorption machines, fuel (or sometimes exhaust gases) for direct-fired absorption machines and engine driven

equipment (Table 4.2 and Table 4.3). Typical models for a number of components have been discussed in Section 3.3. Some examples are reported in Table 4.7. In addition, heat can be sometimes recovered from chillers (Section 3.4), which can be modelled for instance through a *thermal recovery effectiveness* ε_t, conventionally defined as ratio of the recovered thermal power to the input power (electrical power for CERG, thermal power for a WARG, fuel thermal power for a GARG and an EDC) (Table 4.2 and Table 4.3).

4.5. HEAT/COOLING GENERATION IMPACT ON THE COGENERATION SIDE: EXPRESSIONS FOR THE LAMBDA TRANSFORMS

After synthesizing typical performance indicators for multi-generation equipment, it is now possible to analyze in detail how different cooling and heat generation components in an AGP can change in a different fashion the *equivalent* load seen from the CHP side. In addition, it is possible to show how the thermal power generated or recovered in the AGP can be seen as a modification of the CHP cogeneration characteristics (besides the techniques illustrated in Section 4.2 for the prime mover). This is the core of the lambda analysis, resulting in the expressions representing the Λ_d-transforms (Section 4.3.5) reported in Table 4.4 and which exploits the performance indicators introduced above. The entries in Table 4.4 are organized so that all the expressions relevant to the lambda analysis are clearly indicated. In particular, F^p is the overall fuel thermal input to the MG system, W_d, Q_d and R_d are the user's energy demand, F_y^{CHP}, W_y^{CHP} and Q_y^{CHP} are respectively the fuel thermal input (including in case the amount for the CHG) and electricity and heat output from the CHP system (including the shares from CHG, DH and EDS). In addition, according to the rationale behind the lambda analysis, W_y^{CHP} and Q_y^{CHP} are case by case equal to the relevant expressions for the trigeneration Λ_d-transformed W_d^p and the trigeneration Λ_d-transformed Q_d^p, respectively. This section ends with showing the alternative point of view of transforming the electricity and heat generated from the CHP side into equivalent electricity and heat seen from the demand side. This is carried out by means of the Λ_y-transforms (Section 4.3.5) and illustrated for cogeneration prime movers coupled to heat pumps.

For the sake of completeness, the equivalent and relevant expressions for the $\overline{\Lambda}_d$-transforms introduced in Section 4.3.5, containing the equivalent expressions of the demand-side cogeneration ratio λ_d^p in the various cases analyzed, are summarized in Table 4.5.

The considerations carried out in this section are supported by MG plant schemes representing various equipment and interactions. Each piece of equipment is characterized by the relevant energy efficiency indicator, according to the black-box model approach presented above. In addition, although normally present and considered in the general theory, for the sake of simplicity the plant models shown here do not envisage CHG production, nor purchase from EDS or DH networks.

Table 4.4. Relevant energy balances and expressions of the Λ_d-transforms for different heat/cooling generation equipment in the AGP

equipment and operation mode		*alternative expressions for the energy balances and Λ_d-transforms*					
		F_m		$W_y^{CHP} = W_d^P$		$Q_y^{CHP} = Q_d^P$	
		form 1	form 2	form 1	form 2	form 1	form 2
separate cooling/heat generation							
GARG or EDC	cooling	$F_y^{CHP} + F_R$	$F_y^{CHP} + \frac{R_d}{COP_c}$	W_d	---	Q_d	---
	heat recovery	$F_y^{CHP} + F_R$	$F_y^{CHP} + \frac{R_d}{COP_c}$	W_d	---	$Q_d - Q^{AGP}$	$Q_d - \frac{\varepsilon_t}{COP_c} \cdot R_d$
GAHP or EDHP	heating	$F_y^{CHP} + F_Q$	---	W_d	---	$Q_d - Q^{AGP}$	---
bottoming cooling generation							
WARG	cooling	F_y^{CHP}	---	W_d	---	$Q_d + Q_R$	$Q_d + \frac{R_d}{COP_c}$
	heat recovery	F_y^{CHP}	---	W_d	---	$Q_d + Q_R - Q^{AGP}$	$Q_d + \frac{R_d}{COP_c}(1-\varepsilon_t)$
CERG or EHP	cooling	F_y^{CHP}	---	$W_d + W_R$	$W_d + \frac{R_d}{COP_c}$	Q_d	---
	heat recovery	F_y^{CHP}	---	$W_d + W_R$	$W_d + \frac{R_d}{COP_c}$	$Q_d - Q^{AGP}$	$Q_d - \frac{\varepsilon_t}{COP_c} \cdot R_d$

Table 4.4. Continued

equipment and operation mode		*alternative expressions for the energy balances and Λ_d-transforms*					
		F_m		$W_y^{\mathrm{CHP}} = W_d^p$		$Q_y^{\mathrm{CHP}} = Q_d^p$	
		form 1	form 2	form 1	form 2	form 1	form 2
bottoming heat generation							
EHP	heating	F_y^{CHP}	---	$W_d + W_Q^{\mathrm{AGP}}$	---	$Q_d - Q^{\mathrm{AGP}}$	---
WARG or WAHP	heating	F_y^{CHP}	---	W_d	---	$Q_d + Q_Q^{\mathrm{AGP}} - Q^{\mathrm{AGP}}$	$Q_d + (1 - COP_t) \cdot Q_Q$

Table 4.5. Equipment and relevant expressions for the $\overline{\Lambda}_d$-transforms for different heat/cooling generation equipment in the AGP

equipment	*operation mode*	*alternative expressions for* λ_d^p	
		form 1	form 2
CERG	cooling	$\frac{Q_d}{W_d+W_R}$	$\frac{Q_d}{W_d+\frac{R_d}{COP_c}}$
	cooling with heat recovery	$\frac{Q_d-Q}{W_d+W_R}$	$\frac{Q_d-\frac{\varepsilon_t}{COP_c}R_d}{W_d+\frac{R_d}{COP_c}}$
EHP	cooling	$\frac{Q_d}{W_d+W_R}$	$\frac{Q_d}{W_d+\frac{R_d}{COP_c}}$
	heating	$\frac{Q_d-Q}{W_d+W_Q}$	$\frac{Q_d-Q}{W_d+\frac{Q}{COP_t}}$
WARG	cooling	$\frac{Q_d+Q_R}{W_d}$	$\frac{Q_d+\frac{R_d}{COP_c}}{W_d}$
	heating	$\frac{Q_d+Q_Q-Q}{W_d}$	$\frac{Q_d+Q(\frac{1}{COP_t}-1)}{W_d}$
	cooling with heat recovery	$\frac{Q_d+Q_R-Q}{W_d}$	$\frac{Q_d+R_d\left(\frac{1-\varepsilon_t}{COP_c}\right)}{W_d}$

Table 4.5. Continued

equipment	*operation mode*	*alternative expressions for λ_d^p* form 1	form 2
WAHP [a]	heating	$\frac{Q_Q}{W_d}$	$\frac{\frac{Q_d}{COP_t}}{W_d}$
GARG or EDC	cooling	$\frac{Q_d}{W_d}$	---
	heating	$\frac{Q_d - Q}{W_d}$	$\frac{Q_d - COP_t \cdot F_Q}{W_d}$
	cooling with heat recovery	$\frac{Q_d - Q}{W_d}$	$\frac{Q_d - \frac{\varepsilon_t}{COP_c} R_d}{W_d}$
GAHP or EDHP	heating	$\frac{Q_d - Q}{W_d}$	$\frac{Q_d - COP_t \cdot F_Q}{W_d}$

[a] The expressions indicated are used if all the cogenerated heat feeds the absorption heat pump, otherwise the expression for λ_d^p is the same as for the WARG under heating mode.

4.5.1. Separate cooling/heat generation

If a separate CGP (Section 2.1.1) is used to produce cooling power by means of a GARG or an EDC, the cooling power does not impact on the other loads. In fact, gas is (typically) used to supply directly the chillers, so that the input energy for cooling is completely independent of the cogeneration side operation; therefore, there is no additional electrical or thermal load. The cooling energy generation is substantially "*decoupled*" from cogeneration.

In terms of energy balances, with reference to Figure 4.8, the relevant expressions for a GARG or an EDC under cooling mode are reported in Table 4.4. In particular, F_R and COP_c are respectively the fuel thermal input to the refrigerator and its cooling-mode coefficient of performance.

As a direct consequence of the separate CGP selection, the prime mover can be sized as in the classical cogeneration case (apart from possible thermal recovery from the cooling side, as shown in the sequel).

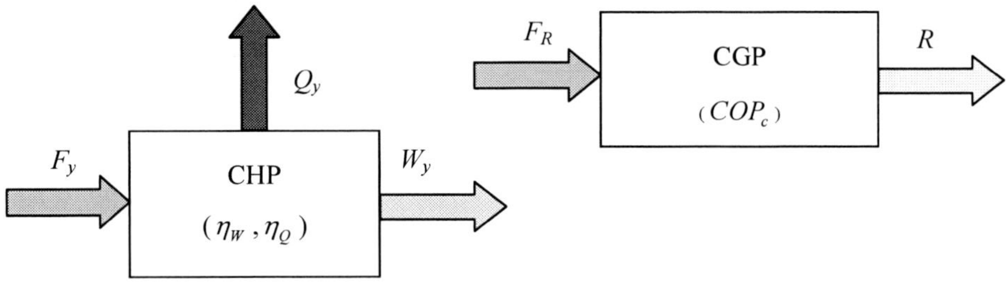

Figure 4.8. Black-box model for separate generation of cooling power.

Differently from the case of separate cooling production, the equipment used to produce thermal power in a separate AGP affects the production side (Figure 4.9). In particular, if an EDHP or a GAHP are adopted, they produce an amount of thermal power equal to

$$Q^{AGP} = COP_t \cdot F_Q^{AGP} \tag{4.25}$$

where Q^{AGP} and F_Q^{AGP} are respectively the heat produced in the AGP and the fuel thermal input to produce that heat. From the relevant energy balances (Table 4.4) it can be seen that the cogenerator is "underloaded" by a certain amount of heat production to supply the thermal demand, at the cost of additional fuel input to the GAHP or to the EDHP.

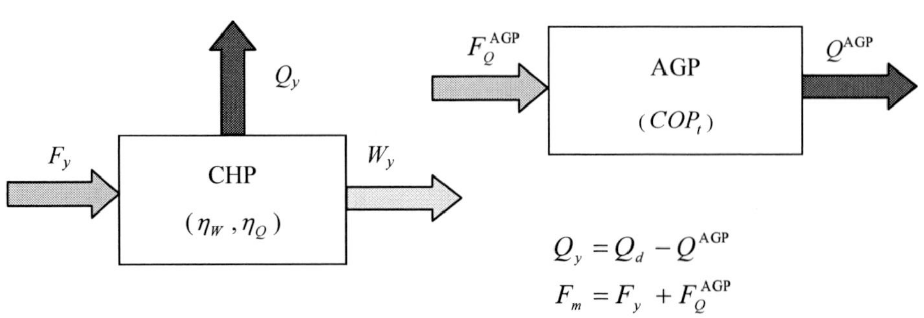

Figure 4.9. Black-box model for separate generation of thermal power.

4.5.2. Bottoming cooling generation

Using a bottoming CGP (Section 2.1.1) for cooling production, cascaded to the CHP production, the impact on the CHP plant depends on the equipment adopted.

If the CGP contains a WARG (Figure 4.10), the input energy vector for cooling is heat, so that the *thermal load* of the CHP unit is increased proportionally to the cooling power required by a rate depending on the COP_c of the WARG. The cooling side is therefore *thermally-coupled* to the cogeneration side. This case represents the classical literature reference for trigeneration plants [WuW06][TRIww]. The prime mover characteristics have to take into account the increase of thermal load due to the cooling load, according to the relations in Table 4.4. In these relations, in particular, the change in Q_d^p, *i.e.*, the change in the equivalent trigeneration thermal load seen by the CHP side, has to be pointed out. The same model would also apply to an adsorption chiller.

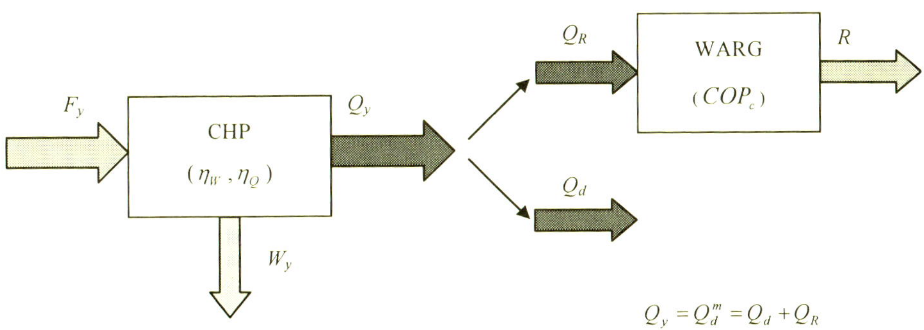

Figure 4.10. Black-box model for thermal bottoming generation of cooling power.

If a CERG is used as refrigerator group (Figure 4.11), the input to the refrigerator is electricity. Therefore, the cooling load raises the CHP electrical load proportionally to the cooling need by a rate depending on the COP_c of the CERG. The cooling side is *electrically-coupled* to the cogeneration side. The prime mover characteristics have to take into account the rise in electrical load due to the cooling load, according to the relevant expressions in Table 4.4. In this case, the equivalent trigeneration electrical load W_d^p seen by the CHP side undergoes a modification with respect to the base W_d.

An EHP is usually a reversible machine. When it works under refrigeration mode, it has the same impact onto the electrical load as for the CERG, namely, the need for cooling power generation increases the electrical load by a rate depending on the refrigeration-mode COP_c.

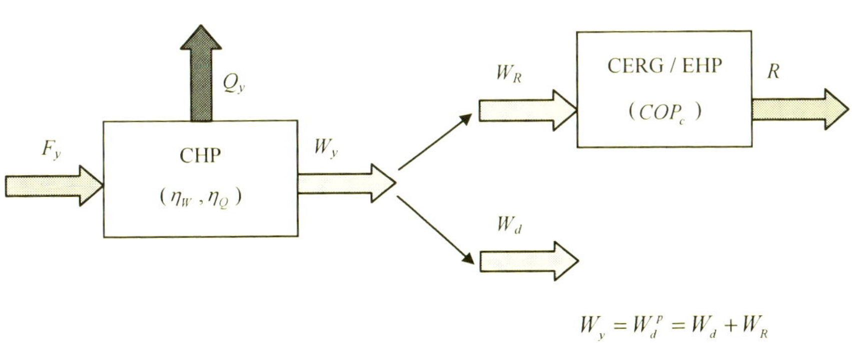

Figure 4.11. Black-box model for electrical bottoming generation of cooling power.

4.5.3. Bottoming heat generation

An EHP operating under heating mode (Figure 4.12) is fed by electricity and produces heat in a proportional fashion by a rate equal to its heating-mode COP_t. In this way, the net result is to "electrically" load the prime mover to produce thermal power, so that the heat generation in the AGP is, to some extent, *electrically-coupled* to cogeneration.

The energy balance of this process can be expressed as

$$W_Q^{\mathrm{EHP}} = \frac{Q^{\mathrm{EHP}}}{COP_t} \tag{4.26}$$

corresponding to the energy flows in the plant (excluding for simplicity the presence of simultaneous cooling load) indicated in Table 4.4. Therefore, an EHP changes both the electrical and thermal equivalent loads seen by the cogeneration side, resulting in the equivalent MG electrical load W_d^p and the equivalent MG thermal load Q_d^p.

The amount of heat produced by the EHP depends on the control strategy adopted and, as far as the machine performance is concerned, on the thermodynamic quality of the requested heat and on the cold sink temperature. In fact, the heat pump is more suitable to supply heat at temperatures below 60÷70 °C in order to obtain good COP_t. Similarly, if the cold sink temperature is too low, it is not profitable to run an EHP due to its decrease in thermal capacity and efficiency (Section 3.3.6.3). On the contrary, the cogenerated heat can be supplied in the form of hot water, superheated water or steam, also depending on the technology used (Section 3.1). In addition, the amount of thermal energy Q^{EHP} to produce in the heat pump could depend on economic factors, such as the profitability of producing "heat" from an electrical source rather than from a thermal one, depending for instance on tariff or market conditions [OsV97][Luc01][Ber02][Mar04].

A WARG under heating mode or a WAHP (Figure 4.13) results in changing only the thermal load profile by a rate equal to the COP_t, leading to the expression

$$Q^{\mathrm{AGP}} = COP_t \cdot Q_Q^{\mathrm{AGP}} \tag{4.27}$$

and to the related expressions in Table 4.4. In particular, Q_Q^{AGP} is the thermal input to the WAHP (or the WARG under heating mode), Q^{AGP} is the thermal output from the WAHP (or WARG under heating mode), and Q_d^p is the equivalent multi-generation thermal load seen by the CHP side. According to (4.27), if the COP_t of the absorption machine is greater than unity [Meu02][Dan06], the equivalent effect is to decrease the thermal load to be produced at the CHP side.

The amount of thermal heat feeding the absorption unit depends again on the control strategy and in case on the thermodynamic quality of the required heat, while the performance of absorption machines is relatively less dependent on the outdoor conditions than for electric heat [DOE04][Dan06][LiN06][KaK07]. In particular, if all the cogenerated heat Q_y is used to feed the WAHP (or the WARG under heating mode), the expression (27) turns into

$$Q_y = Q_d^p = \frac{Q_d}{COP_t} \tag{4.28}$$

in which it is again possible to point out the effect of decreasing the equivalent thermal load as seen from the cogenerator if COP_t is higher than unity.

The same model would also apply to adsorption.

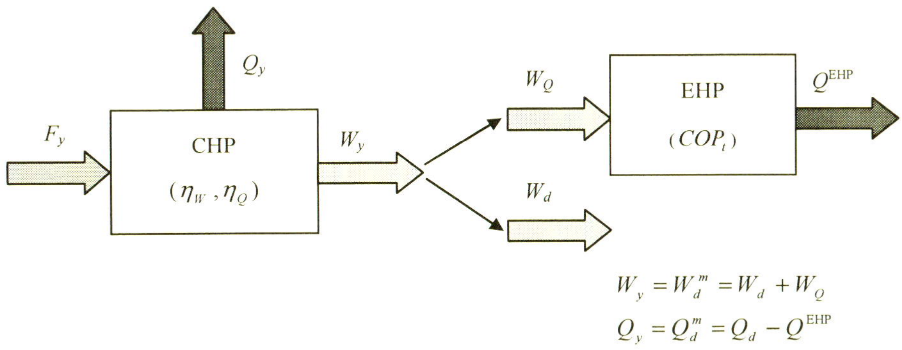

Figure 4.12. Black-box model for electrical bottoming generation of thermal power.

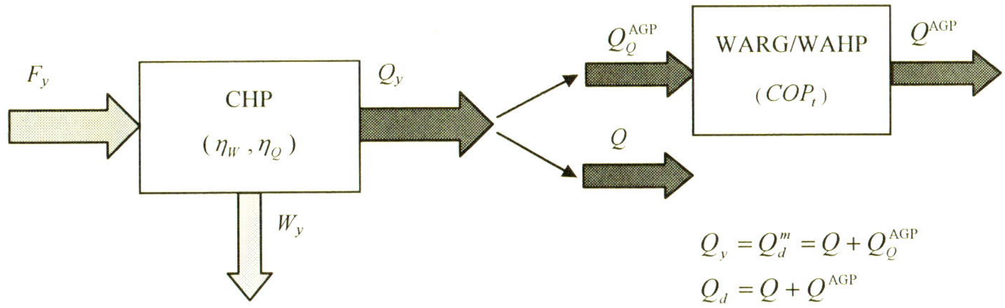

Figure 4.13. Black-box model for thermal bottoming generation of thermal power.

4.5.4. The heat recovery from chillers in the AGP

Heat recovery from that produces cooling power is another means to get the effect of decreasing the equivalent thermal load seen by the cogenerator. In addition, this occurs while producing cooling power, which in turns impacts on the equivalent electrical or thermal load. Therefore, according to the equipment used, there might be different effects:

- if heat is recovered in a CERG by means of a HRC [Wul99] (Section 3.4), there is a "tri-coupling" of the loads [Man06], since electricity is used to produce cooling power and, through heat recovery, also to produce thermal power; the relevant energy flows are shown in Table 4.4, in which, in particular, Q^{AGP} is the recovered heat, COP_c is the chiller cooling-mode coefficient of performance, and ε_t is the heat recovery efficiency (4.11);
- if a GARG (or an EDC) is operated under heat-recovery mode, the equivalent effect is to decrease the thermal load, so that:

$$Q_d^p = Q_d - \varepsilon_t \cdot F_R^{\mathrm{AGP}} = Q_d - \frac{\varepsilon_t}{COP_c} \cdot R_d \qquad (4.29)$$

- finally, similar expressions (Table 4.4) hold true if heat is recovered from the condenser of a WARG.

4.5.5. An alternative point of view: transformation of the prime mover characteristics and Λ_y-transforms

As generally discussed in Section 4.3, it might also be possible to entail the equivalent effects brought by the AGP within the production side rather than within the demand side. In terms of lambda analysis, this corresponds to exploit the Λ_y-transforms introduced in Section 4.3.5 from the generation side rather than from the user's side. In particular, this alternative approach allows interpretation of the impact of additional heat generation in terms of change of the prime mover cogeneration characteristics (in analogy to what discussed in Section 4.2.5) or of the characteristics of the CHP side in general.

For instance, considering an EHP, seen from the CHP side its adoption corresponds to reduce the thermal request and to increase the electrical one. Now, let us consider the EHP within the CHP side, *i.e.*, at the interface with the user: in this case, because of the same reasons as for the impact on the demand side, the electrical heat pump *increases* the overall CHP-side cogeneration ratio λ_y^{CHP} (Figure 4.14, in which for the sake of simplicity there is no CHG production nor electricity bought in from the EDS). In this respect, likewise a GT or MT plant that may use fired HRSG to increase the thermal output, the EHP can somehow be seen as an "*electric post-combustor*".

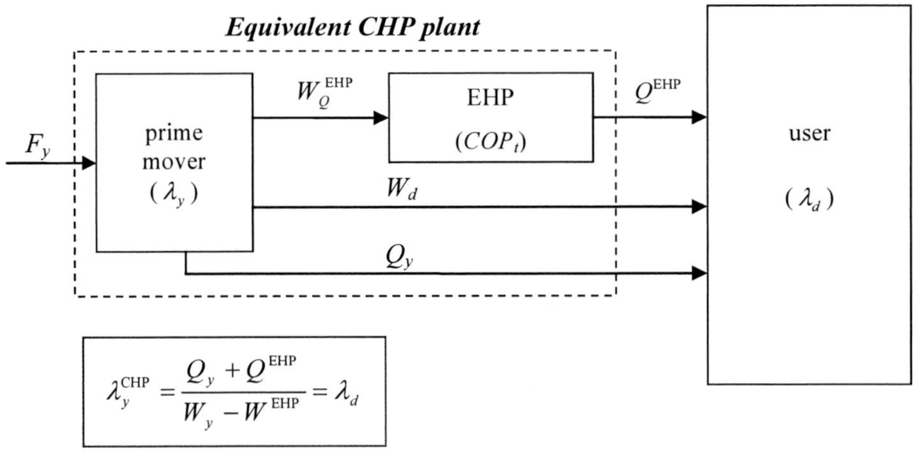

Figure 4.14. Multi-generation scheme with electric heat pump as an "electric post-combustor".

In this case, the relevant energy flow expressions are:

$$Q_y^{\mathrm{CHP}} = Q_y + Q^{\mathrm{EHP}} = Q_d \qquad (4.30)$$

$$W_y^{\mathrm{CHP}} = W_d - W_Q^{\mathrm{EHP}} = W_d - \frac{Q^{\mathrm{EHP}}}{COP_t} \qquad (4.31)$$

$$F_y^{\mathrm{CHP}} = F_y \tag{4.32}$$

Similarly, the "heat multiplier" characteristics of a WAHP (as well as of a WARG in heating mode) could be seen as a means to increase the thermal production of the cogenerator, so that the WAHP could be interpreted as a "*thermal post-combustor*" (Figure 4.15). In this case, the corresponding energy balances are:

$$Q_y^{\mathrm{CHP}} = Q_y - Q_Q^{\mathrm{WAHP}} + Q^{\mathrm{WAHP}} = Q_y + Q^{\mathrm{WAHP}}\left(1 - \frac{1}{COP_t}\right) = Q_d \tag{4.33}$$

$$W_y^{\mathrm{CHP}} = W_y = W_d \tag{4.34}$$

$$F_y^{\mathrm{CHP}} = F_y \tag{4.35}$$

Again, if all the cogenerated heat is used to feed the WAHP, the expression (4.33) becomes

$$Q_y^{\mathrm{CHP}} = COP_t \cdot Q_y = Q_d \tag{4.36}$$

making it apparent that the absorption/adsorption heat pump exhibits "post-combustion" characteristics if COP_t is greater than unity.

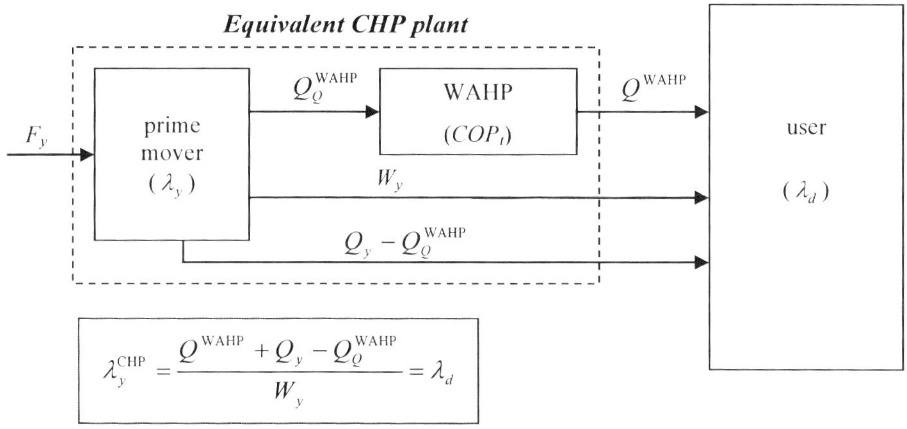

Figure 4.15. Multi-generation scheme with absorption heat pump as a "thermal post-combustor".

In a similar way, the heat recovered from a chiller in the AGP could be encompassed into the overall heat generated Q_y^{CHP}, so as to represent another means to change the cogeneration characteristics.

As a further comment, it has to be noticed that the heat potentially produced by the cogenerator, by heat pumps, and by chiller heat-recovery condensers, can be at different enthalpy levels, so as to yield actual multi-generation in the plant.

In analogy with the $\overline{\Lambda}_d$ *-transforms* reported in Table 4.5, it is also possible to express the lambda transforms applied to the production side in terms of production side cogeneration ratio, which is carried out by means of the $\overline{\Lambda}_y$ *-transforms* (Section 4.3.5). For this purpose, Figure 4.14 and Figure 4.15 report the relevant expressions for the resulting λ_y^{CHP} in the specific case considered.

4.6. The Lambda Analysis as a Planning Tool

The effects on MG system planning brought by adopting different alternatives for the AGP can be effectively exploited at a planning stage. In this respect, the lambda analysis can actually be seen as an additional planning tool.

4.6.1. The multi-generation energy system planning process

In the light of the concepts relevant to the lambda transforms illustrated above, and developing the considerations carried out in Section 4.3, the steps needed while planning a multi-generation energy system, with specific reference to the production of electricity, heat and cooling power, can be summarized as follows:

- select a possible set of AGP equipment suitable to supply the cooling load and/or the thermal load, also on the basis of the units available on the market in the needed range of power and (in case) of a first rough economic analysis;
- apply the relevant lambda transforms to carry out the multi-generation *lambda analysis* (including the multi-generation load duration curve analysis) in order to characterize the user's needs from the CHP side perspective;
- select some of the cogenerative prime movers available on the market within the relevant power and cogeneration ratio ranges;
- analyze the potential of different prime movers to follow the load variations and the user's multi-generation lambda variation (described by the multi-generation λ_d^p after applying the relevant lambda transform) according to different control strategies;
- evaluate, at first approximation, the technical/economic performance of the different multi-generation systems considered as alternatives;
- run time-domain simulations to support the preliminary technical and economic analyses.

Often this process is not so clear-cut, and iterations may be needed before going on with the next step. For instance, the selection of the prime mover and/or of the control strategy may change after evaluating the overall performance of the system. Similarly, different options for the AGP can be considered after analyzing the prime mover alternatives. A numerical case study illustrating some of the concepts pointed out here is presented in Section 4.7.

4.6.2. AGP selection resorting to the lambda analysis

In the previous sections of this chapter we have explored how different pieces of equipment for cooling/heat generation, with specific input/output characteristics, have a different impact on the loads seen from the CHP side. In particular, production of cooling/heat power in the AGP can be interpreted as means to change λ_d into λ_d^p through the lambda transforms. In the same way, it is sometimes possible to modify the production by resorting to different prime mover configurations through electric or thermal heat pumps operating as post-combustors, or by recovering heat from chillers in the AGP.

Therefore, to a certain extent the multi-generation planner has the *opportunity* to exploit the AGP equipment as a further variable in the selection, seeking a better match between load and production characteristics. Somehow, this opportunity is similar to the possibility of modifying the prime mover cogeneration characteristics as described in Section 4.2.5. Indeed, once given the manifold load (*e.g.*, electricity, heat and cooling), the planner can carry out a first selection of prime movers, with size and characteristics suitable to supply the load, on the basis of the demand-related cogeneration ratio. Then, the AGP can be selected by taking into account the impact it may have over the cogeneration ratio in order to make λ_y (or the generalized λ_y^{CHP}) and λ_d^p better match. The more the load-production characteristics match, the better the plant is properly selected.

For instance, were the "basic" λ_y/λ_d ratio too high (without accounting for cooling generation/load), the use of a WARG could "charge" the cooling load onto the thermal one, with the effect of obtaining a reduced actual λ_y/λ_d^p ratio. On the other hand, an excessively low λ_y/λ_d ratio could be tackled by adopting a CERG (whose cooling load production impacts on the electrical demand load) or a reversible EHP (with both cooling and thermal loads impacting on the electrical one, when the heat pump works respectively under cooling mode and heating mode). Similarly, if a good generation-load matching were possible without considering the cooling load, a GARG or an EDC could be a viable alternative, by decoupling the cogeneration of electricity and heat from the cooling production. In addition, the generation of heat in the AGP, directly produced (with AGP equipment working as heat pump) or recovered from chillers, could be exploited as a further variable, thus enhancing the energy system flexibility. Finally, combinations of different cooling/heat generation equipment could be considered in some cases as viable options, bringing interesting energy saving and economic potential to the MG system [Hav99][Meu03][MiL03][MiL03b] [BuT03][Man06].

4.6.3. Suitability of multi-generation solutions to different load configurations

According to the considerations carried out so far, it is apparent how the possibility of exploiting different cooling and/or heat production alternatives for multi-generation in the AGP represents a high-potential opportunity. In fact, the energy recovery intrinsic to energy system cascading and equipment combination may provide better performance in terms of energy saving. At the same time, the different input/output characteristics of the equipment may provide alternative solutions to cover different plant needs in the most profitable way, also from an economic point of view [ChM06][ChM06b][Man06].

Some general hints on possible applications of the lambda analysis, according to the considerations drawn so far, to different types of cooling/heat generators are summarized in Table 4.6, in which equivalent multi-generation cogeneration ratios and equivalent multi-generation loads are taken into account. In particular, it is possible to synthetically notice how the main types of actions refer to situations with thermal load higher or far higher than the electrical one. This is the typical situation in the bulk of small-scale applications (residential blocks, *in primis*, along with commercial malls, offices, hospitals, schools, hotels, restaurants, and so forth [MaT02][CaP03][ZiP06][NeD06][TRIww]), as well as larger ones (for instance, airports [CaP06b][CaS06]). Specific trigeneration applications within the lambda analysis framework are illustrated in the next section.

4.6.4. Suitability of specific trigeneration solutions to load configurations

4.6.4.1. CHP-WARG/WAHP scheme

The scheme of a CHP plant coupled to an absorption machine driven by cogenerated heat and set to cooling mode is shown in Figure 4.16. This represents the classical trigeneration case as reported in the literature and in most of the current implementations. The same model would also apply to coupling with adsorption chillers.

In Figure 4.16, that recalls Figure 4.10, Q_d is the thermal power demanded by the user, Q_R the thermal power that fires the absorption chiller to produce the cooling power R, W_y is the produced electricity; η_W and η_Q are respectively the electrical and thermal efficiency of the CHP unit (the heat produced in the CHG is not considered, for the sake of simplicity). The fuel input is represented as F_z, using the subscript z to explicitly indicate trigeneration. Borrowing the terminology adopted in [GeA07], α_R is the so-called *dispatch factor*. In this specific case, α_R represents a *heat-to-cooling* dispatch factor [Man06], that is, the per unit share of the overall thermal power Q_y used to produce $Q_R = \alpha_R Q_y$. The dispatch factor is also used as a variable in various assessment models, as described in Chapters 6÷8. Utilization of dispatch factors as a modelling tool for general DMG systems is discussed in Section 5.1.2.

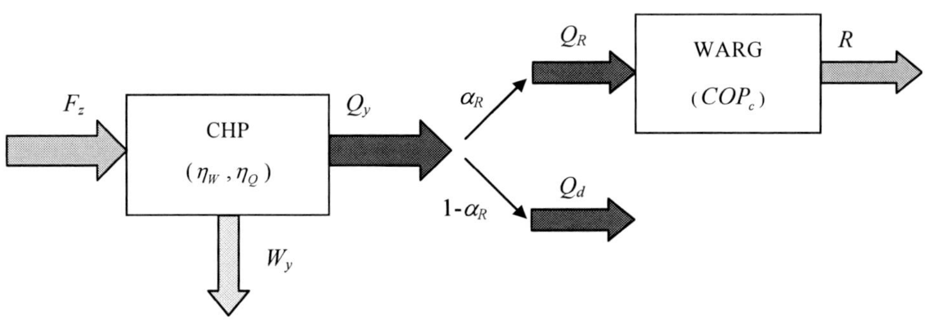

Figure 4.16. CHP-WARG trigeneration system scheme.

The application of a WARG is *in primis* related to seasonal thermal loads. Indeed, the fact that in most applications thermal and cooling loads are inversely correlated allows a flatter use of the cogenerator throughout the year in a heat-following operation mode. In particular, heat-and-electricity cogeneration can be performed in the wintertime (also exploiting, in case, the possibility of using the absorption machine in heating mode), and cooling-and-electricity cogeneration in the summertime.

When the load is purely "trigenerative", in the sense that heat, electricity and cooling are, to a certain extent, needed all the year around, the WARG suits well in applications in which the demand-related cogeneration ratio (4.1) is low with respect to the typical cogeneration ratio (4.2) of the prime movers available in that size [Man06]. In addition, the possibility of recovering the discharged heat, directly or through an EHP driven in turn by cogenerated electricity, so as to satisfy part of the thermal need, may represent a viable manner for "tri-coupling" the loads. In this case, a "trigenerative" load with consistent seasonal effect might be effectively matched: in the wintertime, when the cooling need is lower, the chiller plant can operate partialised (or "modularized") producing an amount of recovered heat for the thermal demand; in the summertime, the chiller plant could work full-capacity and in case the recovered heat might be used to save hot-water production from the auxiliary CHG. This kind of use depends on the temperature of the rejected heat and the user's request (needing in case a heat pump). However, it has to be pointed out that in general it is not possible to save money by not installing a cooling tower, which is needed in any case when the thermal needs are lower.

Table 4.6. Summary of the multi-generation lambda analysis main results

AGP equipment	*operating mode*	*suitable cases*	*effect on* W_d^p	*effect on* Q_d^p	*effect on* λ_d^p
CERG	cooling	$\lambda_y < \lambda_d$	increases	---	decreases
	cooling with heat recovery	$\lambda_y << \lambda_d$	increases	decreases	decreases
EHP	heating	$\lambda_y << \lambda_d$	increases	decreases	decreases
	cooling	$\lambda_y < \lambda_d$	increases	---	decreases
WARG	cooling	$\lambda_y > \lambda_d$	---	increases	increases
	cooling with heat recovery	$\lambda_y \approx \lambda_d$	---	stationary	stationary
	heating	$\lambda_y < \lambda_d$	---	decreases	decreases
WAHP	heating	$\lambda_y < \lambda_d$	---	decreases	decreases
GARG	cooling	$\lambda_y \approx \lambda_d$	---	---	---
	cooling with heat recovery	$\lambda_y < \lambda_d$	---	decreases	decreases
GAHP	heating	$\lambda_y < \lambda_d$	---	decreases	decreases
EDC	cooling	$\lambda_y \approx \lambda_d$	---	---	---
	heating	$\lambda_y << \lambda_d$	---	decreases	decreases
	cooling with heat recovery	$\lambda_y < \lambda_d$	---	decreases	decreases
EDHP	heating	$\lambda_y << \lambda_d$	---	decreases	decreases

Finally, a possible scheme with CHP-WAHP can be adopted with high thermal loads, exploiting the "heat multiplier" effect of the absorption (or also adsorption) heat pump (Figure 4.15), and when the thermal power is needed at relatively low temperatures (*e.g.*, winter air conditioning, hot water services).

4.6.4.2. CHP-GARG/GAHP and CHP-EDC/EDHP schemes

As far as load pattern analysis is concerned, parallel cooling plants for cooling production (with GARGs and EDCs) might be used when the threefold load pattern is such that there is an optimised match between cogeneration production from the prime mover and the electrical and thermal loads, so that there can be separated and "dedicated" cooling production (Table 4.6). From the economic point of view, the profitability of this configuration is strongly related to the energy price scenario: in general, if low gas rates are available, then this might be a viable solution; if the gas rates are high, profitability goes down, especially in presence of simultaneous low electricity rates. In the latter case, a better choice might be to select and size the CHP unit on the basis of the thermal-only and electrical-only loads, and to produce cooling power by means of a CERG fed by low-cost electricity from the grid.

An EDC lends itself well to be exploited within load configurations similar as for the GARG. However, it should be considered that EDCs feature cooling *COP* in general higher than GARGs, so that, given the same amount of fuel input, the cooling capacity an EDC could supply would be higher. Furthermore, a considerable amount of high-quality heat can be recovered from the engine exhaust gases, so making it suitable to be used in plants with demand cogeneration ratio higher than for the GARG. Therefore, EDC applications could be profitably exploited to supply coincident cooling and thermal loads, especially in the presence of high electricity rates and low gas rates, somehow in a dual fashion with respect to electric heat pumps.

In addition, for both types the possibility of operating reversibly as EDHP and GAHP allows exploitation of a wide range of thermal and cooling loads throughout the year [FoD00][DOE04][LaN06][LaN06b]. At the same time, electricity could be produced in a completely independent fashion. For instance, such plants could be adopted when the main loads are heat or cooling, independently supplied by the CGP, and the CHP unit could be operated according to market logics, for instance selling electricity in peak load hours or providing strategic ancillary services when needed.

However, drawbacks of the EDC solution might be higher investment and maintenance costs with respect to other alternatives; furthermore, the presence of environmental constraints upon the plant emissions might render it either unfeasible or too expensive (due to the additional cost of required emission abatement system).

4.6.4.3. CHP-EHP and CHP-CERG schemes

The CERG is the typical cooling production solution. From a load analysis standpoint, its application could suit high demand cogeneration ratio (heat demand much higher than electric demand, for instance by 5 times), as for the WAHP. In fact, in this case additional electrical load to produce cooling power would lower the net cogeneration ratio seen by the cogenerator, so as to better match the typical cogeneration characteristics of ICEs (heat-to-power ratio around 1) and MTs (heat-to-power ratio around 2) (Table 4.6). In addition, if part of the discharged condensing heat is recovered at the condenser, again it is possible to have a sort of "tri-coupling" among loads, suitable for applications such as hospitals, offices, hotels, and so on.

A major upside of electric heat pumps lies in the possibility of being exploited also as a chiller and then in "seasonal" applications, allowing for an optimised "electrical" rating of the prime mover to which it is cascaded. Indeed, the prime mover can be sized on the basis of the overall summertime electrical load (including the electricity needed to feed the EHP in cooling mode). Then, when chilled water is not needed, the EHP keeps on working to provide thermal energy (provided that there is a higher heat need in wintertime). Also in this case, in general there is a wide range of thermal loads that can be supplied, by exploiting the "heat multiplier" effect of the heat pump plus the cogenerated heat.

Figure 4.17 shows a cogeneration prime mover coupled to a reversible EHP. In particular, following the general concepts introduced above for the CHP-WARG scheme, when the EHP operates under heating mode $\alpha = \alpha_Q$ represents the *electricity-to-heat* dispatch factor, that is, the per unit share of electricity cogenerated that goes to feed the EHP to produce heat (in this case, the scheme in Figure 4.17 coincides with the one in Figure 4.12); on the other hand, under cooling mode $\alpha = \alpha_R$ is the *electricity-to-cooling* dispatch factor, that is, the per unit share of electricity cogenerated that goes to feed the EHP to produce cooling (in this case, the scheme in Figure 4.17 coincides with the one in Figure 4.11) [Man06].

The economic profitability of the overall system is more complicated to evaluate, and requires a dedicated analysis case by case [ChM06][ChM06b][Man06]. In particular, it is interesting to evaluate whether it is more effective to deploy a higher CHP thermal efficiency, as opposed to a higher electrical one associated to a combined EHP to produce additional heat, in all those cases (which represent the major share) when the heat request is far higher (*e.g.*, double) than the electricity one. A practical application, for instance, might lie in the comparison between an ICE coupled to an EHP and an MT. However, from an economic point of view low electricity rates generally encourage the use of EHPs, as well as of electric chillers, for both cogeneration and trigeneration applications. Also, the specific cost of electro-thermal technologies is in general lower than for absorption/adsorption or engine-driven technologies.

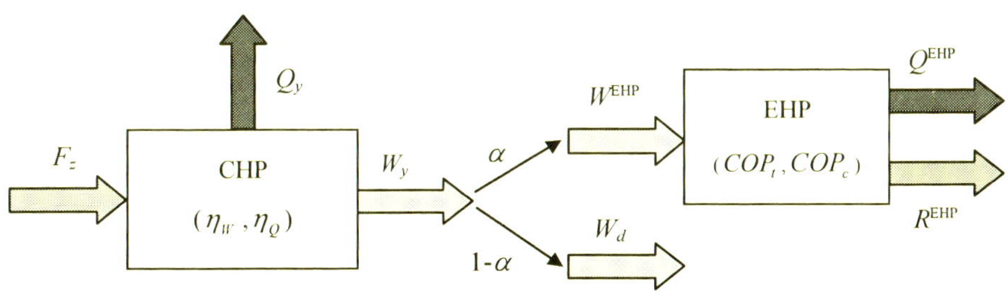

Figure 4.17. CHP-EHP reversible trigeneration system scheme.

4.7. Case Study Application

4.7.1. Description of the trigeneration user

A practical case study application referred to a CHCP system is presented in this section. The aim is to point out the quantitative impact of different cooling generation equipment on the demand-related cogeneration ratio ($\overline{\Lambda}_d$-transforms) and to highlight the *net* equivalent loads seen at the interface with the CHP side (Λ_d-transforms).

A threefold load pattern (heat, electricity and cooling power) is assigned as input. In particular, the hourly thermal load is given for three reference days, related to the winter, mid season and summer seasons, respectively. The daily electrical load is assigned through the hourly pattern, that is reproduced without variations throughout the year. The thermal load pattern (hot water at 80 °C) and the electrical load pattern are shown in Figure 4.18. In addition, Figure 4.19 shows the hourly cooling load pattern (chilled water at 7 °C) for the same three seasons of the year as for the thermal load.

The load models presented in Figure 4.18 and Figure 4.19 reflect a classical approach to the planning of a combined energy system. The thermal loads (heat and cooling) are supposed to follow the night-day activity cycle as well as a seasonal behaviour, essentially related to outdoor conditions and heating loads. For instance, the winter period could correspond to five months, the midseason period to four months and the summer period to three months. The electrical load, instead, is supposed to be "steady" over the year, but of course it varies throughout the day. An application with loads similar to the ones shown could be for instance a hotel in a mild/warm climate.

From graphical inspection, it is clearly difficult to figure out an "average" load. In general, the "rated" load is considered at the maximum peak. On the other hand, it is tougher to choose a representative cogeneration ratio as the rated one, even though the considered load patterns are fairly simplified models. These aspects are in line with the considerations drawn in Section 4.2.

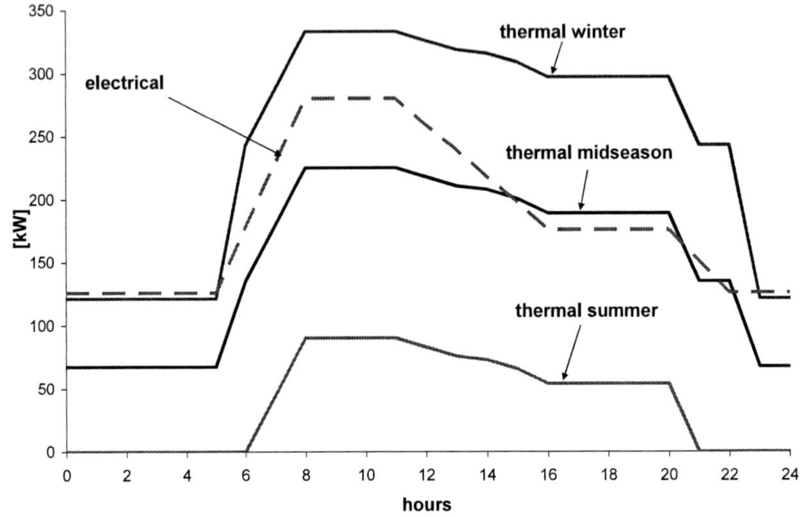

Figure 4.18. Hourly load patterns for the electrical power W_d and the thermal power Q_d with seasonal effects.

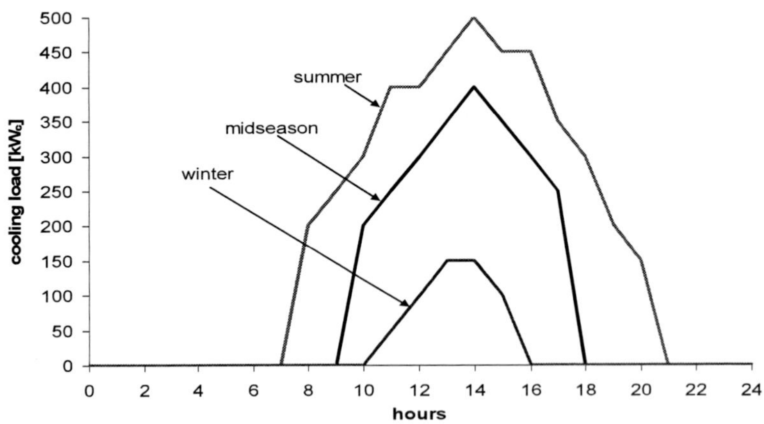

Figure 4.19. Hourly load patterns for the cooling power R_d with seasonal effects.

The duration curves for the thermal, electrical ad cooling load, together with the corresponding "basic" demand-related cogeneration ratio $\lambda_d = Q_d / W_d$ (no cooling generation impact) for the three seasons, are shown in Figure 4.20 and Figure 4.21, respectively.

An example of possible equipment selection to supply the user's trigenerative loads is presented below. The main data and the models for the equipment used are summarized in Table 4.7.

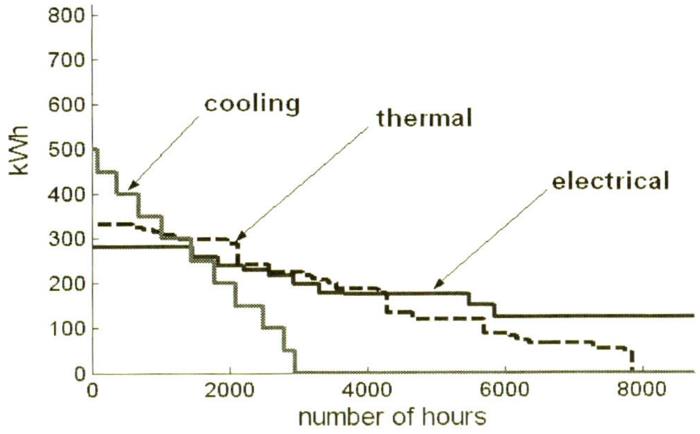

Figure 4.20. Threefold load duration curve for electricity, heat and cooling power.

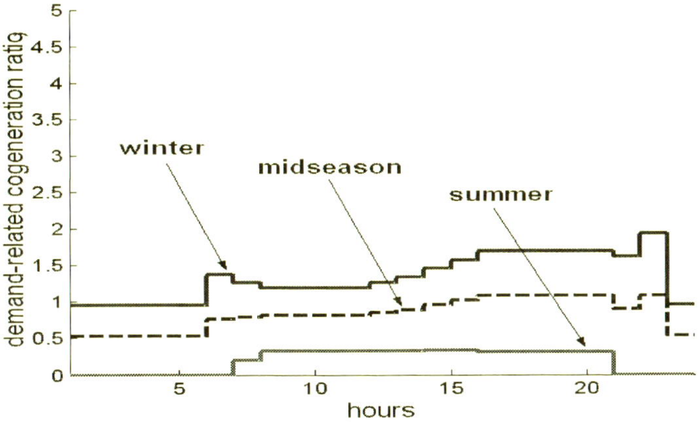

Figure 4.21. "Basic" hourly demand-related cogeneration ratio λ_d.

4.7.2. The lambda analysis applied to the cooling power generation equipment: results of the lambda transforms

Let us consider five cases for the CGP equipment, aimed at showing the impact of the different cooling generation characteristics on the electrical and thermal loads, that is, the results of the lambda transforms upon the duration curves and the demand-related cogeneration ratios. In particular, the impact of the cooling load is calculated by means of the expressions found in Section 4.5 for the relevant lambda transforms, and considering off-design *COP* characteristics for all the equipment, according to models drawn from data from equipment available in the market (Table 4.7). The results obtained from the application of the Λ_d-transforms (Table 4.4) and the $\overline{\Lambda}_d$ -transforms (Table 4.5) are shown in Figure 4.22 for the five cases considered, described below.

4.7.2.1. Case 1: GARG

Let us consider that the CGP is composed of a GARG. In particular, the chilling group is made up of two direct-fired double-effect 281-kW$_c$ units, directly fed by gas from a GDS. The rated *COP* is about 1.05, and the off-design model takes into account partial-load operation change (Table 4.7) (see also Section 3.3.4.3). The two units are supposed to operate so as to equally share the load. Since a GARG "decouples" the cooling side from the cogeneration side (heating and electrical loads remain unchanged by the presence of the cooling load), the duration curves and the user's lambda shown in Figure 4.22 are exactly the same as in Figure 4.20 and Figure 4.21 ($\lambda_d^z = \lambda_d$), respectively. Thus, the cogenerator sizing is independent of the cooling production. In addition, from the cooling load duration curve it is possible to estimate the gas consumption due to the GARG, as if it were a prime mover, according to what discussed in Section 4.2.1.

4.7.2.2. Case 2: CERG

In this case, the CGP plant is assumed to be composed of a CERG, made up of two 270-kW$_c$ electric chillers in parallel, fed by the CHP system and/or the EDS. The rated *COP* is about 3, and the off-design model takes into account *COP* variations with seasonal average variations of average outdoor temperature, whereas the performance is considered at first approximation constant at partial load (Table 4.7), as also discussed in Section 3.3.3.5. Again, the two units are supposed to be controlled so as to equally share the load.

The trigeneration electrical and thermal load duration curves (after applying the relevant Λ_d-transform) and the demand-related cogeneration ratio λ_d^z (after applying the relevant $\overline{\Lambda}_d$ -transform) for this solution are again shown in Figure 4.22. The cooling load goes to impact on the electrical load, raising it up in the morning hours of the day, when the chillers are mostly on, thus lowering the λ_d^z. By direct comparison with *Case 1*, it is apparent that the electrical load increases in a time span equal to about one half of the yearly load duration.

4.7.2.3. Case 3: WARG

The classical concept of trigeneration leads to consider for the CGP the cooling production through a WARG, namely, one 515-kW$_c$ machine, fed by hot water at 80 °C produced in cogeneration when the CHP system is switched on, otherwise by the CHG. In this case, a single-effect unit is used, with rated *COP* equal to about 0.65. The off-design model takes into account partial-load variations and seasonal outdoor temperature-based variations of the *COP* (Table 4.7).

The lambda-transformed trigeneration load duration curves and the relevant λ_d^z are shown in Figure 4.22. The thermal load undergoes a significant increase because of the cooling load impact, due to the relatively "low" *COP* (about 0.65 at rated conditions) and the relatively "high" cooling load. The curve profiles are completely modified, if compared to the other cooling machines considered.

4.7.2.4. Case 4: WARG (base-load) + CERG (modulation)

In this case, the CGP is composed of two different pieces of equipment, namely, one 260-kW$_c$ unit for the WARG and one 270-kW$_c$ unit for the CERG. The off-design characteristics of the equipment are set as in *Case 2* and *Case 3*. Being there two types of cooling equipment, a better defined control strategy for the cooling plant needs to be set up. Thus, in this case the

WARG is used to cover the base-cooling load (so as to better exploit the exceeding cogenerated heat), whereas the CERG is controlled to follow the peak cooling load (modulation). This actually represents a case that can be encountered quite often in existing plants, whose performance is thoroughly analysed from a technical point of view, for instance in [MiL03][MiL03b].

The duration curves and the user's trigeneration lambda for this case are shown in Figure 4.22. The lambda transforms for the WARG and the CERG (Table 4.4) are applied to the duration curves and the cogeneration ratio accounting for the considered control strategy. The results, as expected, represent something in the middle between *Case 2* and *Case 3*. In practice, the thermal curve is somehow "flattened" and "regularized" with respect to the case of thermal-only equipment (WARG, *Case 3*); at the same time, also the electrical load is modified with respect to the case with cooling production through electrical-only equipment (CERG, *Case 2*). This intermediate situation could enable to better exploit the prime mover to cover either or both of the aggregated electrical and aggregated thermal loads.

4.7.2.5. Case 5: CERG (base-load) + WARG (modulation)

The same cooling machines as in *Case 4* are used. However, in this case the cooling-side control strategy is such that the CERG follows the base-cooling load, whereas the WARG is used for the cooling load modulation. The duration curves and user's trigeneration lambda are shown in Figure 4.22, after applying the relevant lambda transforms as in *Case 4*, but with "opposite" control strategy. The results, as expected, represent again something in the middle between *Case 2* and *Case 3*. Indeed, as in *Case 4*, the effects of the additional thermal and electrical loads due to the cooling generation are mitigated with respect to the WARG-only and CERG-only cases, respectively, as apparent from visual inspection of Figure 4.22.

4.7.3. Discussion on the prime mover selection

The five cases illustrated above provide in practice the results of the lambda analysis as a means to characterize the user's need by applying the relevant lambda transforms (Figure 4.22). Following the steps indicated in Section 4.6, it is further possible to attempt to work out a possible choice for the cogenerator and its performance, also according to different control strategies. In this respect, for the sake of simplicity, and as in most applications, in the following considerations the possibility of incorporating the heat production in a district heating network is not included, whereas connection to the electrical grid is explicitly taken into account.

Hence, in general the thermal demand can be satisfied in part by the heat produced in cogeneration and in part, if necessary, by a CHG. For instance, the CHG can be composed, in *Case 1* and *Case 2*, of two 200 kW_t boilers in parallel with rated efficiency η_t = 0.9, which ensure also adequate back-up reserve (suitable in applications such as hotels). The partial-load model for this equipment is shown in Table 4.7. However, if WARG units are used in the CGP (*Case 3*, *Case 4* and *Case 5*), the number of boilers is adequately increased in order to guarantee complete back-up for the cooling production as well (Table 4.7). At this point, looking up to the different cases, it could be possible to search for a suitable prime mover from commercial catalogues. In this outlook, the approach illustrated is consistent with classical methods based on load duration curve analyses [WiS00], as well as with newer techniques, such as the one based on aggregated thermal loads suggested for planning WARG-based trigeneration systems in [CaP03][PiC08], on whose basis the cogenerator selection could be readily carried out.

Table 4.7. Equipment data and models used in the case study application

equipment	*number of units*	*unit rated capacity*	*unit rated efficiency*	*off-design models*
prime mover (gas-fuelled ICE)	1	330 kW$_e$	$\eta_W = 0.36$ $\eta_Q = 0.43$	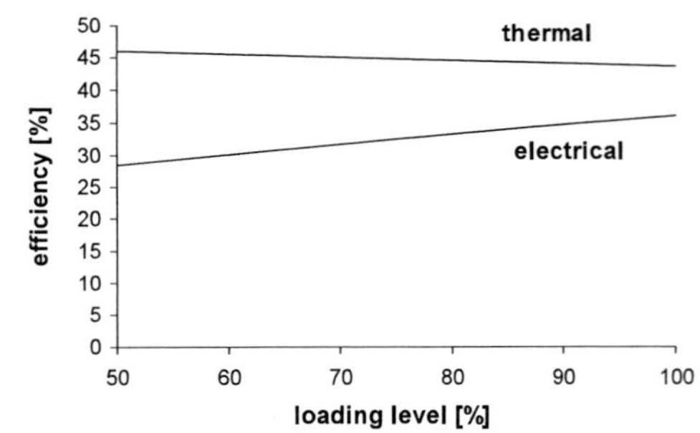
CHG (gas-fuelled boilers)	2 [a], 3 [c] or 4 [b]	200 kW$_t$	$\eta_t = 0.9$	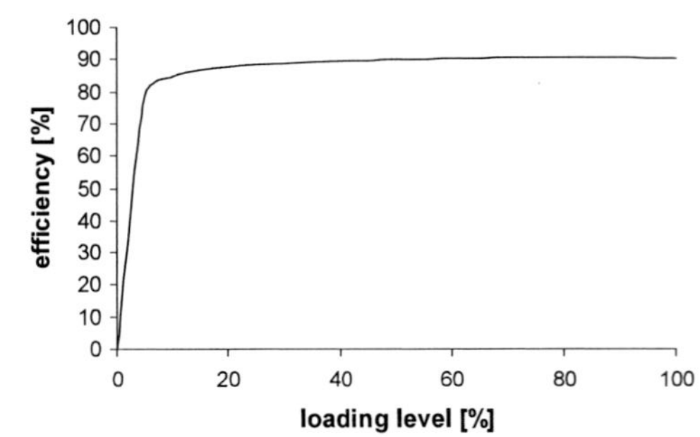

Table 4.7. Continued

equipment	*number of units*	*unit rated capacity*	*unit rated efficiency*	*off-design models*
CERG	2 [a] or 1 [c]	270 kW_c	$COP = 3.1$	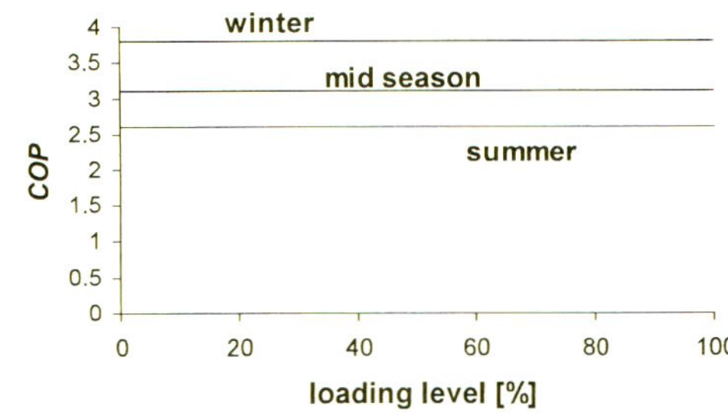
WARG	1	515 kW_c [b] or 260 kW_c [c]	$COP = 0.65$	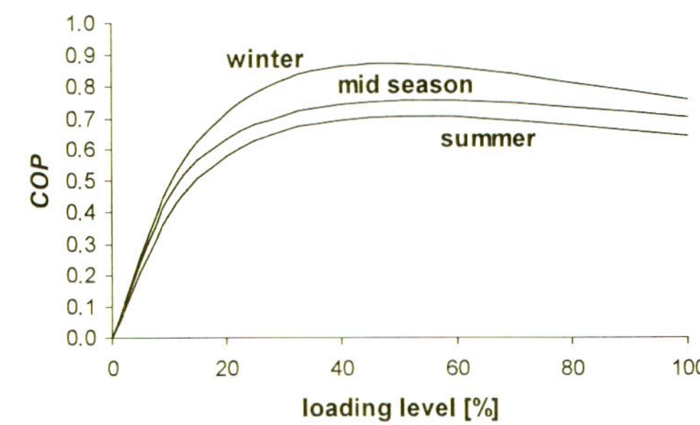

Table 4.7. Continued

equipment	*number of units*	*unit rated capacity*	*unit rated efficiency*	*off-design models*
`GARG	2	281 kW_c	$COP = 1.05$	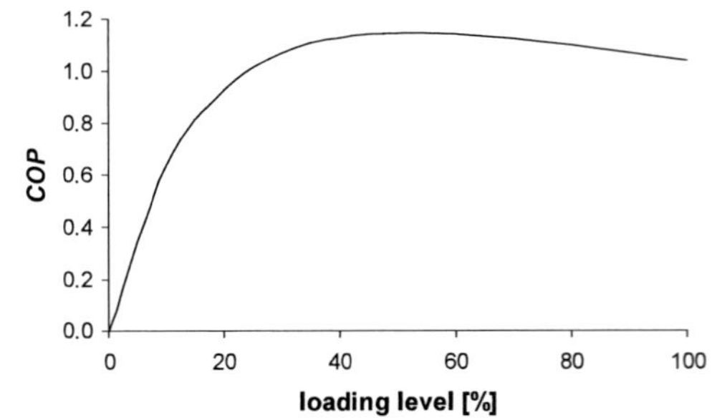

[a] *Case 2;* [b] *Case 3;* [c] *Case 4* and *Case 5.*

An alternative approach, shown in the sequel, consists of selecting a certain prime mover roughly suitable for all the different CGP cases, and to work out the potential plant performance for the different cases. For instance, let's consider a 330 kW_e gas-fed ICE, whose technology guarantees high electric efficiency also at partial loads and good dynamic performance. Its size is a good trade-off for matching the electrical and thermal peak demand among the alternatives commercially available. The partial-load characteristics of the selected ICE are shown in Table 4.7. In particular, the prime mover λ_y is about 1.2 at the rated power, and increases at partial loads (see also Section 4.2.5).

In general, at first approximation the following considerations can be carried out, case by case, about the possible prime mover behaviour, the suitable control strategy, and the estimated plant performance:

- GARG (*Case 1*): in this case, being $\lambda_y > \lambda_d$ in the summertime and in the midseason, the engine would mostly find itself operating, if run under electrical load-following mode, wasting part of the thermal power cogenerated. The alternative would be to adopt a heat-following strategy, which would imply electricity sale to the grid, to be further evaluated.

- CERG (*Case 2*): changing the cooling equipment to a CERG would bring about an additional electrical load, as discussed above. Therefore, holding the same engine, there would be an increase in the number of operation hours (under electrical load-following mode), and the engine would also work closer to its rated conditions (at higher electrical efficiency) for longer time, thanks to the higher electrical load. This would mean in general overall performance better than with a GARG, although still with thermal waste.

- WARG (*Case 3*): a hot-water absorption machine could allow for a reduction of the thermal waste, thanks to the increase in the thermal load. However, in this case, the control strategy could rather be turned into thermal load-following and the electricity overproduced could be again sold to the grid. In case, it might also be viable to increase the size of the engine in the light of a thermal, rather than electrical, load-following strategy, provided that it is profitable to sell the overproduced electricity. Of course, a longer time interval with partial-load operation should be taken into account, as well as the risk of switching off the engine more often and for longer time, if subject to constraints on the minimum load level.

- WARG (base load) + CERG (modulation) (*Case 4*) and CERG (base load) + WARG (modulation) (*Case 5*): in these combined solutions, the overall performance should improve, regardless of the adoption of electrical or thermal load-following strategy. In fact, both the equivalent load duration and lambda curves are flattened with respect to the cases with cooling production only based on electrically-fed or thermally-fed equipment, thus allowing better exploitation of the prime mover sizing. In fact, balancing somehow the thermal load and the electrical load (which is carried out through combination of cooling generation equipment) enables to supply the load more effectively by avoiding waste of cogenerated thermal power and thus improving the prime mover operation for longer time.

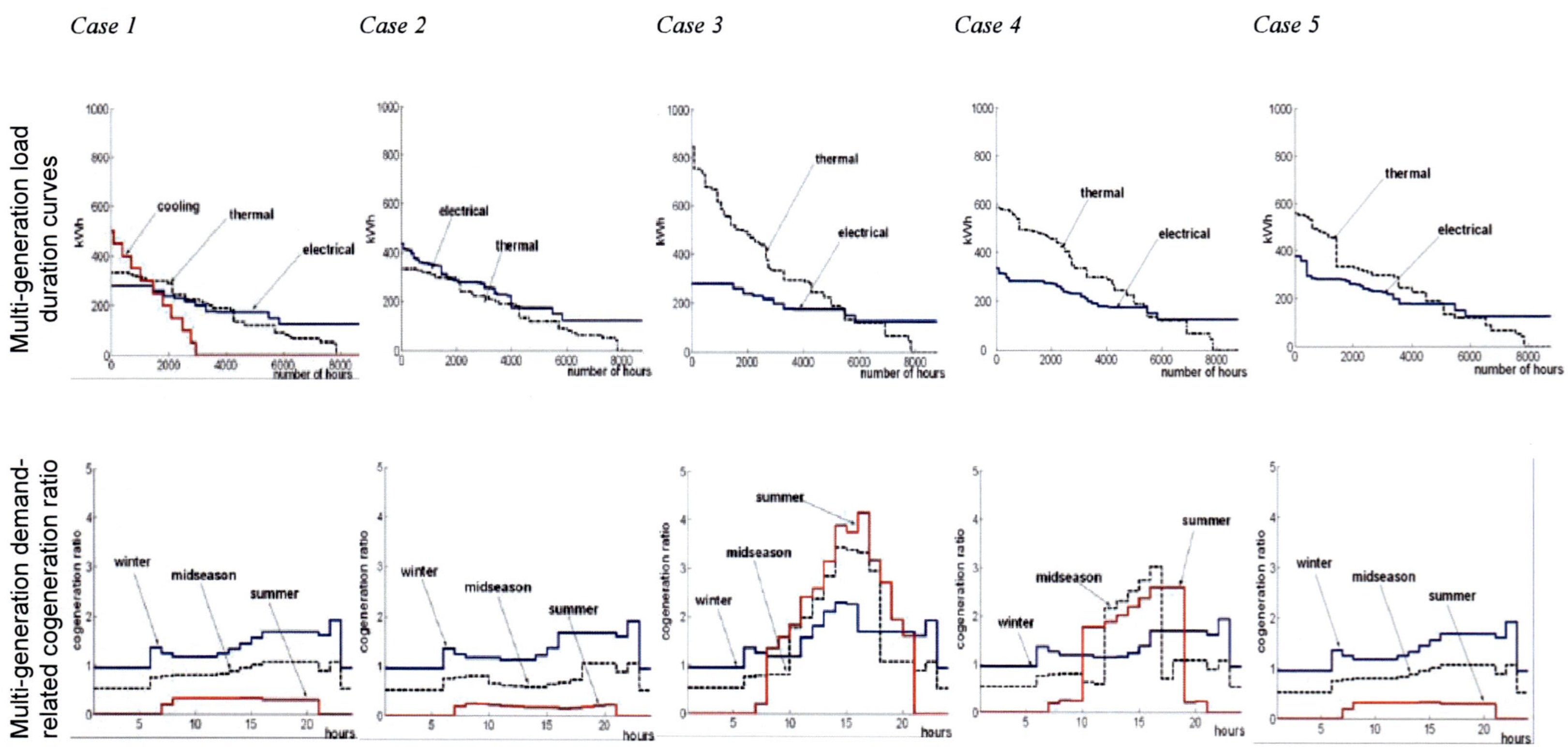

Figure 4.22. Multi-generation load duration curves and cogeneration ratios after applying the relevant lambda transforms for the case study application.

Other alternatives could take into account the possibility of splitting the prime mover into more units, for example a cluster [BaC99][NeD06][LaN06b] of 100-kW_e microturbines. Simultaneous operation of the MTs, with the possibility of exploiting them in different partial-load conditions, could enable a better "match" between the overall λ_y^{CHP} (MT cluster + CHG) and λ_d^z. Of course, λ_y^{CHP} would not depend only upon the machine characteristics any longer, but it would also be related to the number of MTs used and on their control strategy. These options should be analyzed in further details through time-domain simulations.

4.8. Remarks on Multi-Generation Planning

The diffusion of systems for combined production of manifold energy vectors (such as electricity, heat and cooling power) on a small scale is representing a new challenge for energy planners and operators. At the same time, the potential opportunities provided by such systems calls for adequate planning techniques.

In this chapter, a novel and comprehensive approach to multi-generation system characterization and planning, with specific reference to cogeneration and trigeneration, has been presented. This approach, called *lambda analysis*, is formulated in mathematical terms by introducing the so-called *lambda transforms*, that is, transformations of the relevant loads into equivalent loads, with the objective of facilitating the selection and sizing of the equipment involved in the analysis.

More specifically, the *cogeneration lambda analysis* focuses on characterizing the cogeneration loads and the cogeneration production taking into account time-domain aspects (for instance, seasonal, daily, and hourly load variations), as well as off-design aspects (for instance, variation of the prime mover cogeneration characteristics at partial load), thus revisiting from a modern standpoint "classical" tools such as the cogeneration ratio indicator widely adopted for cogeneration studies. In addition, the interactions with energy networks (and, in case, related markets) such as the electrical grid, district heating and district cooling networks, whose presence is intrinsic in the concept of "distributed" energy system, are adequately modelled in the novel approach formulated.

The *trigeneration lambda analysis* is a further extension including the new aspects brought about by the need for cooling power generation. In particular, if different alternative pieces of equipment are considered within the CGP, each of these pieces can impact differently on the thermal and/or electrical load, so as to make it necessary to build new models for the energy system characterization.

In a more general case, the CGP turns into an AGP that also envisages adopting equipment for heat generation, so that the interaction between the production side and the demand side (both including flow exchanges with energy networks) becomes even more intertwined, and can be adequately described within the so-called *multi-generation lambda analysis*.

After the introduction of the theoretical framework, the relevant formulas to apply the lambda transforms have been presented for several cooling and heat generation equipment and plant schemes. In addition, different MG cases have been analyzed, pointing out the suitability of specific cases to particular load configurations, and thus the possibility of exploiting the multi-generation lambda analysis as a planning tool. From a more practical standpoint, some numerical applications exemplificative of the concepts presented have been provided in the case study application, in order to highlight, also from a graphical point of view, the use and the results of the lambda transforms.

The lambda analysis, considered as a first step in the energy system characterization and planning procedure, provides a useful representation of the energy flow interactions within the plant and with the external energy networks in a synthetic and at the same time comprehensive way. As such, it represents a powerful tool to be exploited at a planning stage and has the upside of taking into account aspects such as time-domain variations of the loads and off-design characteristics of the equipment, paving the way to more detailed analyses, based for instance upon time-domain simulations, also facing further issues related to energy saving and economic aspects [ChM06][ChM06b][Man06].

One of the most promising directions for extending the approach presented consists of including within the lambda approach the possibility of shifting the electrical, thermal and cooling loads by means of *energy storage* devices [Has98][Has98b][Wul99][ASH00][Kre01][WuW06] [Dan06], evaluating the related impact in terms of energy efficiency and economic profitability.

The next chapters represent a natural follow up of the planning models discussed here. In particular, once potential alternatives have been explored, suitable methodologies for energy, (and then for environmental and economic) assessment are needed. These aspects are dealt with in the sequel.

Chapter 5

ENERGY PERFORMANCE ASSESSMENT: RATIONALES AND INDICATORS

This chapter introduces and discusses a number of models, methodologies and indicators to assess the energy performance of multi-generation systems. The models are outlined under a unified theoretical framework based on the black-box approach introduced above. This framework is developed starting from the classical approaches to combined heat and power evaluation, and enables to take into account the distributed nature of small-scale plants by exploiting energy chain models. The benefits provided by distributed multi-generation systems can be acknowledged by incorporating transmission and distribution losses in the energy efficiency assessment. The adoption of specific metrics for energy performance evaluation of different multi-generation plant structures is illustrated and discussed. These metrics lead to the definition of various types of indicators, defined in both absolute and incremental terms, based on the plant component efficiencies and on the primary energy rate concept. When the aim of the energy performance assessment is the comparison between combined production and separate production of multiple energy vectors to supply the same outputs, the key issue of setting up the reference values for separate production is addressed in detail. The indicators developed are applied in the successive chapters to various multi-generation cases.

5.1. GENERALITIES ON THE METHODOLOGY ADOPTED FOR DMG ENERGY ASSESSMENT

5.1.1. DMG energy system assessment approaches

Multi-generation systems can be composed of different types of units, fed by different types of energy, and producing different energy vectors with various characteristics at both the cogeneration side (producing electricity as well as heat in various forms, such as hot water, superheated water, and/or steam) and the AGP side (producing, in particular, cooling power and/or heat, both potentially at different enthalpy levels). The presence of manifold forms of energy calls for developing adequate indicators, tools and techniques to synthetically and effectively evaluate the overall energy system performance and to set up sound comparisons among different solutions.

The approach presented for energy performance assessment is based on the first law of thermodynamics and is aimed at providing indications at a high-level energy planning stage and within an energy *system* vision. In particular, the interactions among the plant components are addressed by formulating specific energy chain models [Lun08]. Although models integrating the use of the second law of thermodynamics and exergy concepts [Bej97][Boe97][Ros99][DiR07][LiZ07] are recalled for the sake of completeness, the application of related theories (such as thermo-economics [VaL97][Gog05] or exergo-economics [Sci05]) is not addressed here. For more detailed analyses, for instance aimed at designing the specific MG equipment, further tools based on second law models might be needed. Relevant applications of these concepts to multi-generation can be for instance found in [TeB04][RoL05][CaP06c][CaP07][VeC08][DeW08].

The formulation of the unified assessment methodology introduced below is based on *reference energy models* for the separate production (SP) of the relevant energy vectors (particularly, electricity, heat and cooling power). These reference models and the corresponding performance characteristics thus represent the benchmark for evaluating cogeneration, trigeneration and multi-generation systems. The complexity of the interactions among the different types of energy in multi-generation systems is handled through black-box modelling of the single plant components and of the overall plant. Hence, by extending the range of classical performance indicators available for CHP systems, a consistent set of indices is introduced with the aim of evaluating the primary energy input needed for producing the required energy outputs. Relevant arrays of (absolute and incremental) indicators, whose number of components corresponds to the output energy vectors involved, are introduced to formalize mathematically the comprehensive approach illustrated. This further leads to formulate synthetic indicators (in particular, for energy saving assessment) for the overall plant seen as a black box, which can be in turn extended to evaluate DMG systems interacting through energy networks. The structure of the proposed approach is flexible enough to be applicable to any kind of system that produces different energy vectors (*e.g.*, chilled water, steam, etc.), and to directly derive the indicators to be used for trigeneration and classical cogeneration system assessment as particular cases. Given the relevance of CCHP systems, special focus is set on trigeneration applications.

5.1.2. DMG energy system model with a black-box-based approach

As already discussed in Section 4.4, a convenient system representation of MG plants can be built by considering each component as a *black box*, describing its characteristics by means of input-output relations. This representation has a major upside in limiting the number of variables to be handled in the analysis of plant structures with different numbers and types of interconnected components. At the same time, the contents of any box can be specified with different levels of detail, independently of the other boxes. For instance, a generic component can be modelled through constant efficiency values or more elaborated partial-load efficiency characteristics. Furthermore, the black boxes representing the various pieces of equipment can be aggregated to build black boxes at a macro-level, for instance to represent the overall cogeneration side, the cooling side, and eventually the whole plant [Man06][ChM09b]. This is consistent with the need for endeavouring to develop solutions leading to efficient use of energy, and is practically effective in terms of minimizing the input energy needed for obtaining a given energy output for the specific application.

The performance assessment models that are developed below for multi-generation and in particular trigeneration systems rely indeed on the rationale of black-box aggregation. Under this approach, in particular, indicators of various types (*e.g.*, energy, environmental, economic) can be formulated in function of design parameters (for instance, the energy efficiencies of the equipment). In this way, it is possible to carry out sensitivity studies over the variables and parameters involved, as well as to draw preliminary conclusions on the comparative evaluations of different alternatives.

The energy flow interactions within the plant components must be adequately described accordingly. This is done by means of *dispatch factors* (sometimes called *splitting ratios* [ShN02][GrF07]), already introduced in practical bottoming configuration examples in Section 4.6.4. More specifically, the dispatch factors represent the share of a given energy vector that splits up to feed two or more components (or plants), and account for the plant topology as well as for loading levels/control strategies [Man06][ChM09b]. Hence, starting from the user's manifold energy needs (plant outputs), DMG modelling based upon black boxes and dispatch factors enables to track back the energy inputs to the whole system (for instance, gas from a GDS, electricity from an EDS, heat from a DH network, and so on). In this way, once known the output, it is readily possible to calculate the input through the relevant efficiency chains identified by cascading the various black boxes. For instance, a WARG (thermal bottoming equipment) produces cooling power from a thermal input, which in turn can be produced by the CHP unit, in turn fed by gas, the original plant input. In this way, it is possible, known the desired output, to calculate, point by point, the input energy consumption for every piece of equipment and for the plant as a whole. The same also applies for instance to parallel-type cooling generation equipment (*e.g.*, GARGs or EDCs), for which from the cooling demand it is straightforward to calculate the necessary fuel input that, in turn, sums up to the cogenerator fuel input in order to yield the overall plant fuel input. In particular, by aggregation of black-box components through dispatch factor-based connectivity rules, it is possible to synthesize the whole plant as an overall black box [Man06][ChM09b].

Among other upsides, this approach is easy to be implemented in simulation codes, and potentially allows a thorough evaluation of the performance of every piece of equipment in every actual operating point, provided that *partial-load* or, more generally, *off-design* models are available.

In a similar fashion, it is possible to model the interactions among different plants, each of which is seen as a black box, through dispatch factors. The presence of intermediate energy networks can also be modelled through input-output energy efficiencies.

Automatic construction of black boxes representing aggregated components for system studies is formulated and described in [ChM09b]. Detailed examples of dispatch factor modelling and relevant energy system assessment applications are given in Chapters 6÷8.

5.2. Unified Approach to Single and Multiple Energy Vector Production

5.2.1. Output-to-input first-law performance indicators for equipment and networks

Overall energy systems as opposed to single components may as well be ideally characterized by synthetic indicators that provide a comprehensive view on their performance. Such indicators should be able to represent the various types of energy involved in the thermodynamic transformations, as well as the effectiveness of the components in operating such transformations. Energy-based output-to-input performance indicators are among the most used ones within the energy sector.

First-law energy balances provide the bases for general and simple description of efficiency for thermodynamic transformations, useful for any kind of equipment. The corresponding *first law efficiency* is defined as *useful energy result* (output) to *energy "cost"* (input) ratio, where the *useful energy result* is the difference between the *energy "cost"* and the relevant *energy losses* occurring in the transformation (see also Section 4.4.2). The variables used in the definition of first-law efficiencies and the performance indicators for some of the most used pieces of equipment for the MG systems considered in this work are shown in Table 4.1, Table 4.2 and Table 4.3. For all cases, a black-box model approach is adopted in order to carry out synthetic system analyses, as detailed in the sequel. Further classical equipment indicators, for instance referring to the effectiveness of heat exchangers, may be found in [Bej97][Hor97][Man06].

In order to account for the losses occurring in the presence of networks, relevant output-to-input *network efficiencies* can be defined as the ratio of the energy delivered at the final point of use to the energy injected in the network. However, the loss physical mechanisms change with the specific energy vector, and numerical calculations may be in general not straightforward.

5.2.2. Evaluation of different types of energy: the need for a common metric

On the above considerations, a basic principle for energy performance assessment of a given component or plant relies on determining ratios of the useful outputs to the inputs introduced to obtain those outputs. Once fixed the performance evaluation tool, this can be applied to comparatively evaluate a whole class of similar devices. For instance, the *COP* can be effectively used to evaluate the whole class of vapour-compression electric chillers (Section 3.3.3), as it is a measure of the performance of the cycle on its own. Yet, the evaluation becomes more complicated for systems that, although producing the same output, are driven by different types of energy. For instance, for absorption/adsorption chillers the *COP* is generally evaluated as output to input ratio, but their driving energy (waste heat, for instance) is well different from the driving energy of electric chillers. Assessment of systems that, besides being driven by different types of energy inputs, also produce different types of energy outputs (as in a multi-generation plant) introduces further complications. Therefore, an approach merely based on the component energy performance may often be insufficient from the energy system standpoint, so that adequate techniques to avoid the related drawbacks must be sought.

When considering electricity-only generation, as for conventional power plants, in alternative to the electrical efficiency (4.6) it is common to adopt the Heat Rate (*HR*) indicator [Hor97], defined as

$$HR = \frac{F_W}{W} \tag{5.1}$$

that is, the inverse of the electrical efficiency. The subscript W in F_W stresses that the fuel power is used for electricity-only production, differently for instance from a cogeneration plant, in which the same fuel is exploited for both electricity and heat production.

The *HR* is a classical tool for power plant appraisal, giving directly the primary energy fuel "cost" to produce a given amount of electricity. Similarly, if a generic energy vector is produced and involved in thermodynamic transformations, a suitable approach to its assessment can be to refer the final outputs to the relevant *original* fuel energy input. Hence, when comparing different types of equipment with possibly different input energy vectors (for instance, absorption chillers and electric chillers) for production of a same energy vector (for instance, cooling in the form of chilled water), a formulation based on a "heat rate"-type approach yields a comparative indication of how much primary resource is being spent. This approach can also provide a common metric to address the effectiveness of production of different forms of energy. In particular, for system energy-based analyses as the ones carried out here, such an approach allows the assessment of the overall energy efficiency (or of the overall resource consumption) of a multi-generation system with respect to conventional separate production means. Various applications related to this general concept are available in the literature (see for instance [Hav99][KoW04][SuW04][KoW05][SuG06]), including the energy chain model [Lun08] outlined below.

A major upside of the heat rate-based approach is the possibility of extending the energy models, analyses and results to environmental impact aspects [ChM08][ChM08b][MaC08][MaC08b], as briefly discussed in Section 8.4. On these premises, a general *unified approach* for the evaluation of MG systems with various types of equipment and input and output energy vectors in terms of primary energy is formulated in the sequel.

A limitation of the heat rate-based methodology lies in the fact that it does not capture the thermodynamic value of the different energy vectors, nor the effects of irreversibilities and of thermodynamic quality degradation occurring in the transformations. These issues may be addressed by means of exergy-based theories [Bej97][DiR07], in which exergy and not primary energy is used as reference common metric. However, even with such more sophisticated tools, several issues, also in terms of agreed nomenclature [LiZ07], are currently not sorted out yet, and for multi-generation systems different theories may lead to different results [GaN99][WaL07]. Nevertheless, as already mentioned, the energy models illustrated below represent a suitable standpoint for the purpose of the analyses carried out here.

5.2.3. Energy chain model for DMG systems

According to what discussed earlier, for fuel-based energy systems the quantity of primary thermal energy contained in the fuel and needed to produce the different outputs represents a suitable comparative evaluation metric, leading to the so-called *energy chain* approach [Lun08]. This approach estimates the overall energy system efficiency in the different transformations that occur from one form of energy into another while undergoing the relevant processes. In particular, the introduction of a unique reference, namely, primary energy, for alternative energy systems allows comparative unbiased assessment.

Among the general benefits from multiple energy vector production (Chapter 1), utilization of DMG systems close to the users avoids electricity flows from centralized power plants through transmission and distribution networks towards the final user. Transport losses associated to these flows are consequently avoided. The same reasoning could be also applied for instance to compare small-scale DMG systems to large centralized heat generation stations with heat networks, in which case thermal and pumping hydraulic losses would be avoided. Transmission and distribution (TD) losses can be taken into account within the energy chain model through the definition of transmission and distribution efficiency. In this respect, Figure 5.1 shows a typical example of electrical energy chain from fuels to final users.

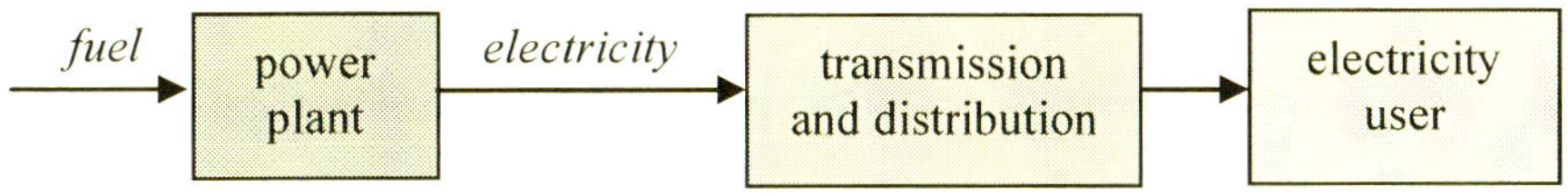

Figure 5.1. Electrical energy chain model.

5.2.4. Single energy vector assessment for trigeneration cases: the Primary Energy Rate (*PER*) indicator and the Absolute Trigeneration Heat Rate (*ATHR*) array

In line with the heat rate-type approaches discussed, the energy efficiency of complex energy cycles can be evaluated on the basis of the *Primary Energy Rate* (*PER*), ratio of the primary energy needed to the required energy output [ZiR93].

It is easy to show that, focusing on the energy vectors relevant to CCHP systems and with respect to the most adopted production means, the typical *PER* indicators for *separate production* of electricity, heat and cooling can be defined as follows:

- for electricity production (from a prime mover):

$$PER_e = \frac{1}{\eta_e} \tag{5.2}$$

- for thermal production (in a combustion heat generator):

$$PER_t = \frac{1}{\eta_t} \tag{5.3}$$

- for cooling production (in an electrically-driven compression chiller):

$$PER_c = \frac{1}{\eta_e \cdot COP_c} = PER_e \frac{1}{COP_c} \tag{5.4}$$

where η_t, η_e and COP_c are, respectively, the thermal efficiency of the heat generator, the electrical efficiency of the electricity generation system and the efficiency of the vapour compression chiller.

Apparently, the PER_e indicator corresponds to the *HR* defined in (5.1). Thus, with analogous nomenclature, it is possible to dub the different *PERs* for production of electricity, heat and cooling; this leads to introduce the *electrical heat rate* (*EHR*) (*i.e.*, the *HR*, renamed here for consistency of notation), the *thermal heat rate* (*THR*), and the *cooling heat rate* (*CHR*), respectively. Clearly, the definitions given for thermal and cooling heat rates are relevant to the specific equipment considered, as discussed below, while the *EHR* corresponds in general to the inverse of an electrical efficiency.

The indicators introduced define an *array* of heat rates, referred to the production of three different final forms of energy, here called *Absolute Trigeneration Heat Rate* (*ATHR*) (where the term "absolute" is used to make a distinction with respect to the "incremental" one introduced in Section 5.5.5):

$$\overline{ATHR} = \begin{pmatrix} EHR \\ THR \\ CHR \end{pmatrix} \tag{5.5}$$

The *EHR*, *THR* and *CHR* indicators, opportunely evaluated, represent the energy consumption references for comparing any other generation system that produces the same energy vectors. In the following paragraphs, the *THR* and *CHR* are presented and discussed with respect to different generation technologies and plants.

It can be noticed that the definition given for the *CHR* reflects the rationale of the energy chain model, as the final energy output obtained by means of electric chillers is associated to the original input (fuel) through an intermediate energy chain stage (electricity generation). In this respect, it is apparent how further *PER* definitions could be derived with respect to the energy output *delivered to the user*. In fact, the above definitions refer only to generation equipment inputs/outputs, whereas in network studies also the intermediate stages occurring at the energy transportation level should be considered. In this way, the primary energy input needed to produce the energy amount delivered to the user would also explicitly take into account energy losses, thus improving the overall assessment. For instance, in the case of electricity produced in centralized power plants, and with reference to the energy chain model in Figure 5.1, it is possible to redefine the PER_e indicator as

$$EHR = \frac{1}{\eta_e \cdot \eta_{\mathrm{TD}}} \tag{5.6}$$

where η_{TD} accounts for the transmission and distribution electrical losses. Similar models could apply to indicators for other energy vectors, including for instance the presence of DH networks, DCNs, hydrogen networks, and so on. For the sake of clarity, in the sequel no formal distinction is considered for formulas with and without network losses. The effect of losses may be taken into account by modifying the generation efficiency numerical value according to the relevant network efficiency value.

In (5.5), the $\overline{ATHR}$ is basically defined in order to group together the relevant set of SP benchmarks. In Section 5.5.3, a further definition is given with respect to specifically introduce equivalent efficiencies for a CCHP system.

5.2.5. The Thermal Heat Rate (*THR*) in thermal-only production

Thermal-only production in any heat generator may be evaluated by means of the *Thermal Heat Rate* (*THR*), generally defined as

$$THR = \frac{F_Q}{Q} \tag{5.7}$$

where Q is the desired thermal output and F_Q is the primary thermal fuel input needed to obtain it. Specific *THR* expressions can be derived for CHG, EHP, GAHP, and EDHP, in terms of the relevant performance indicators given in Table 4.1, Table 4.2 and Table 4.3. These expressions and the corresponding energy chain models are reported in Figure 5.2.

In particular, for EDHP and GAHP (grouped together under the acronym DFHP – Direct-Fired Heat Pump) the energy chain model is the same, and is in turn the same as for CHG, with fuel input and heat output. Indeed, when these different systems are assessed through black-box models, the equivalent input/output effects as seen from the outside are exactly the same, regardless of the physical processes occurring within the box. On the other hand, for EHPs the useful effect is represented by the thermal energy Q released to the hot sink at the cost of the electric energy W_Q introduced into the cycle, so that its efficiency under heating mode is defined by means of its heating *COP* (4.12). As W_Q is produced by means of an electric prime mover or the electrical grid, the *THR* can be calculated as

$$THR^{\mathrm{EHP}} = \frac{F_Q}{W_Q}\frac{W_Q}{Q} = \frac{1}{\eta_e}\frac{1}{COP_t} = EHR\frac{1}{COP_t} \qquad (5.8)$$

where η_e is the electrical efficiency of the EHP electrical input source (Figure 5.2). In particular, if the electrical source is the grid, the electrical efficiency (or equivalently the *EHR*) numerical values must also include transmission and distribution losses, according to the model (5.6). The subscript Q in F_Q and W_Q stresses that the final output is thermal power. It should also be noticed that in the energy models described the "free" energy obtained from the environment when using heat pumps is not explicitly modelled as an input, since it does not represent an energy cost.

It is also worth recalling that *electric resistance heating* can be seen as a particular case of heat pump with $COP = 1$ (Section 3.3.6.5), so that all the considerations carried out in this section can be applied to this type of heat generation system as well.

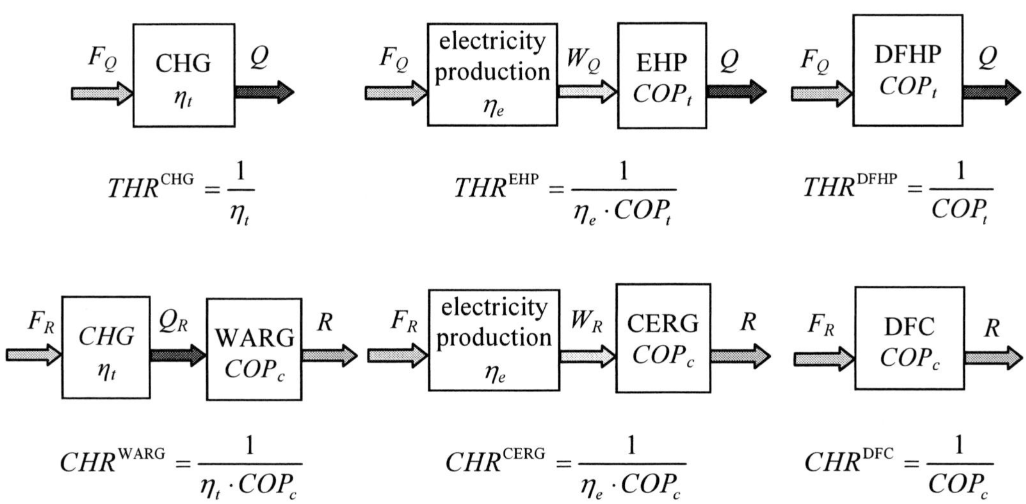

Figure 5.2. *THR* and *CHR* for different types of equipment.

5.2.6. The Cooling Heat Rate (*CHR*) in cooling-only production

For a generic cooling generation devICE, the *Cooling Heat Rate* (*CHR*) can be defined as the ratio of the fuel thermal input F_R to the useful effect R, that is,

$$CHR = \frac{F_R}{R} \qquad (5.9)$$

The *CHR* expressions for CERG, WARG, GARG, and EDC, written in terms of the corresponding performance indicators given in Table 4.2 and Table 4.3, are also reported in Figure 5.2.

For the CERG (or EHP under cooling mode), the model considered for evaluating the efficiency in terms of primary energy is the same as the one introduced above for evaluating the *THR* for an electric heat pump. In particular, now the useful effect is the cooling power *R*, that is, the heat drawn from the ambient to keep cooled at a specified temperature, obtained at the cost of the electrical input W_R. Hence, the CHR^{CERG} is defined through the COP_c ("cooling" mode) in analogy with (5.8) for the heat pump in heating mode, as shown in Figure 5.2.

As discussed in Section 3.3.4, for an absorption (as well as adsorption) machine) the thermodynamic model is *tri-thermal*, with three thermal references instead of two as in the previous cases, namely, the thermal sources from which heat is drawn (useful cooling effect) and to which heat is discarded (not modelled if it is not a useful effect, as it occurs instead in the presence of a HRC – Section 3.4.1), and the thermal input to the device (energy cost). In this case, then, the COP_c is defined as the ratio of the cooling power *R* (useful output) to the thermal power Q_R (energy cost). More specifically, for an indirect-fired WARG (the same model also applies to an adsorption chiller) the thermal energy Q_R (for instance, hot water or steam) that feeds the absorption generator is modelled as being generated in a CHG of efficiency η_t , obtaining the energy chain expression for CHR^{WARG} (Figure 5.2). However, it has to be pointed out that such a model is interesting prevalently from a theoretical point of view, since CHG-WARG coupling is generally not efficient (and in most cases nor economic) in comparison with an electric chiller, for instance. This is further discussed in Section 7.3, where it is shown how the energy efficiency of a WARG-based system greatly benefits from exploiting recovered heat for cooling production. On the other hand, for a direct-fired GARG the thermal energy Q_R that feeds the absorption generator is released when burning the fuel directly supplied to the machine ($Q_R = F_R$). In practice, a direct-fired machine could be described by a CHG-WARG scheme with $\eta_t = 1$, but considering that the numerical value of its COP_c takes already into account the internal efficiency of converting the primary energy contained in the fuel into thermal energy and, in turn, into cooling energy. The EDC model is the same as for the GARG (both types of equipment are grouped together under the acronym DFC – Direct-Fired Chiller), since, as seen from the outside, the fuel thermal input is directly converted into cooling power. Of course, also in this case there is intermediate conversion efficiency (from fuel power into the engine mechanical power that drives the chiller compressor), but the model itself used by manufacturers "hides" the intermediate stage. It has to be recalled that the COP_c to be used in this kind of evaluation is the "cooling-only" *COP*, whereas some manufacturers sometimes indicate a "cogenerative" *COP* that encompasses also the *heat* recovered from the driving engine, as discussed in Section 3.3.7.

5.3. Benchmark Models for Separate Production of Heat, Electricity and Cooling Power

Among the technologies and devices for generating single energy vectors (namely, heat, cooling and electricity, for CCHP applications), there are *standard* technologies spread worldwide. These technologies (and their numerical performance characteristics) represent the conventional *benchmarks* when evaluating other technologies for producing the same energy vectors. Moreover, they represent also the natural reference to adopt when combined MG energy systems are analysed and more complex indicators are used.

A classical example is the definition of conventional *Separate Production* (SP) of electricity and heat as the reference for evaluating the energy saving due to combined production of the same energy vectors. Such models are used in some countries as official regulatory tools for cogeneration plant assessment [CaP05]. Therefore, before introducing the methodologies for multi-generation assessment, the models used to evaluate the *conventional* SP of electricity, heat, and cooling power are presented. Further key discussions on the numerical values to assign to the reference models selected are provided throughout the sequel (see in particular Section 7.1 and Section 8.3.2.5).

Generalization to other energy vectors, if needed, depends on the specific case. A theoretical model for quad-generation systems is for instance shown in [ChM08b].

5.3.1. Conventional reference model for separate production of electricity: equivalent power plant

Electricity is conventionally produced in large power plants. Hence, it is possible to introduce, a black-box model for an *equivalent power plant*, characterized by a reference electrical efficiency η_e^{SP}. The EHR^{SP} for separate production is therefore given by

$$EHR^{SP} = \frac{1}{\eta_e^{SP}} \tag{5.10}$$

The equivalent power plant model could refer to a specific technology, or to the bulk power system in a certain country. In the latter case, when used for comparison to DMG systems, the numerical value of η_e^{SP} should also take into account average electrical transmission and distribution losses that are avoided by distributed systems close to the user, in line with the energy chain model illustrated in Section 5.2.2.

Typical numerical values for thermal prime movers range from about 0.25 for small-scale MTs to above 0.55÷0.6 for large Combined Cycle Gas Turbines (CCGTs), whereas bulk electricity generation efficiency is of course country-dependent (for instance, in Italy the average generation efficiency is about η_e^{SP} = 0.4, including transmission and distribution losses).

5.3.2. Conventional reference model for separate production of heat: equivalent boiler

Conventional production of thermal power typically occurs in combustion heat generators. Hence, it is possible to introduce an *equivalent boiler* black-box characterized by a reference thermal efficiency η_t^{SP}. In terms of primary energy, the conventional THR^{SP} is defined accordingly as:

$$THR^{SP} = \frac{1}{\eta_t^{SP}} \tag{5.11}$$

In most countries, average rated values for heat generator efficiencies are about 0.85 for residential equipment (typically up to 100 kW_t) and 0.9 for industrial equipment.

Consequently, THR^{SP} values may typically range from 1.1 (industrial) to 1.2 (residential), whilst condensing boilers can exhibit THR^{SP} rated values below 1, owing to the efficiency characteristics in Table 3.4.

Although boilers are typically selected as heat production reference in several studies, in some countries also EHPs are widespread as generation equipment. Hence, alternative reference models could also be based on an equivalent EHP, rather than a boiler, through the relevant energy chain illustrated in Section 5.2.5. Comparative assessments between EHPs and boilers are provided in Section 6.1.1.

5.3.3. Conventional reference model for separate production of cooling power: equivalent electric chiller

Unlike for heat and electricity, for which technical literature and regulatory frameworks provide the typical SP reference models discussed above, the selection of cooling production reference is more arbitrary. However, a natural benchmark for cooling production is represented by electricity-fed compression chillers (the most widespread and consolidated technology), whose corresponding reference CHR^{SP} can be defined as

$$CHR^{SP} = \frac{1}{COP^{SP} \cdot \eta_e^{SP}} \tag{5.12}$$

Again, the evaluation is carried out here by considering an energy chain black-box model that exploits the equivalent power plant described above, which is aggregated to a reference *equivalent electric chiller* black-box to set up an overall (fuel-to-cooling) black-box model, with fuel as input and cooling as output. The numerical selection for η_e^{SP} is relevant to the model adopted for electricity generation, as indicated in Section 5.3.1. On the other hand, since the *COP* of a cooling machine is strongly dependent on the operating conditions and in particular temperatures (Section 3.3.3), the numerical reference figures should be considered case by case depending on the application. However, average reference values that could be considered for electric compression chiller *COPs* range from 2 (household application) to 5 (larger centralized chillers).

As for heat production, of course alternative means are available for cooling generation, in particular adopting absorption/adsorption chillers. A performance comparison between general DFC (EDC and GARG) and electric chillers can be found for instance in [MaC08b].

5.4. Performance Evaluation Criteria for CHP Systems

Several techniques have been presented in the technical literature for assessing CHP plants. Classical approaches [Hor97] are based on the analysis of one or more indices aimed at evaluating the technical/thermodynamic effectiveness of combined heat and electricity production. Some of these indices are recalled in this section, focusing on the ones more suitable for generalisation to the multi-generation case. Before discussing them, the model used to describe a cogeneration plant is recalled and presented under the notation of the proposed unified approach.

5.4.1. Model for cogeneration of electricity and heat

In a cogeneration plant the desired outputs W_y and Q_y are both useful energy vectors, and there is a unique fuel input F_y to the plant. In this case, it is not physically possible to distinguish, in the thermal input, the single amounts giving origin to the two outputs. Hence, conventional cogeneration electrical and thermal efficiencies (Section 3.1.1 and Section 4.4.2) are both defined with respect to the *overall* fuel thermal input. In terms of heat rates, considering the cogeneration electrical efficiency η_W and the cogeneration thermal efficiency η_Q, it is possible to define a *cogeneration Electrical Heat Rate* EHR_y and a *cogeneration Thermal Heat Rate* THR_y, in analogy with the ones introduced for separate production of heat and electricity, as

$$EHR_y = \frac{1}{\eta_W} \tag{5.13}$$

$$THR_y = \frac{1}{\eta_Q} \tag{5.14}$$

However, although the definitions given for CHP heat rates here and for CHP efficiencies in Section 4.4.2 correspond to the ones given in Section 5.3 for SP, the interpretation of cogeneration efficiencies must be different from the SP ones. Indeed, the CHP efficiency definitions do not take into account the fact that another energy vector is produced, so that a mere comparison of efficiency (or *EHR* and *THR*) values between CHP and SP means is not adequate. In the light of this simple reasoning, it is apparent that more sophisticated assessment tools need to be developed.

5.4.2. Cogeneration first law efficiency or Energy Utilisation Factor (*EUF*)

In the presence of simultaneous production of the two forms of energies in cogeneration, the application of the first law balance leads to the definition of the *Energy Utilisation Factor* (*EUF*) [PoM82] (or *cogeneration overall efficiency*, or *cogeneration first law efficiency*) introduced in (3.4) and recalled in Section 4.4.2.

The *EUF* provides an indication of the overall fuel energy amount that is turned into *useful* outputs, thus being an indication of the process effectiveness in terms of resource utilization. In addition to the data in Table 3.1, Table 5.1 provides typical ranges for electrical efficiency, thermal efficiency and *EUF* for some cogeneration technologies.

5.4.3. "Value-Weighted" Energy Utilisation Factor (EUF_{vw})

Although the *EUF* is very useful for first-hand efficiency indications, it is not able to address the different thermodynamic (and economic) value of electricity and heat. In order to make up for this shortcoming, adequate weights to the entries in (4.8) may be introduced. Indeed, thermodynamics yields that electrical (or mechanical, more in general) energy is much more difficult to produce and costs more than heat. Therefore, in order to overcome the limits of the *EUF* in considering electricity and heat with the same "value", it is possible to weigh the entries appearing in the definition of the *EUF* with the respective cost/price. In this way, the actual conceptual reference is not given in terms of energy balance any longer, but of economic balance. Hence, if the sale price of electricity and heat are respectively ρ_e [€/kWh$_e$] and ρ_t [€/kWh$_t$], and the purchase price of fuel is ρ_F [€/kWh$_t$], then the *Value-Weighted Energy Utilisation Factor* EUF_{vw} [Hor97] can be defined as

$$EUF_{vw} = \frac{\rho_e \cdot W_y + \rho_t \cdot Q_y}{\rho_F \cdot F_y} = \frac{\rho_e}{\rho_F} \eta_{vw} \tag{5.15}$$

where

$$\eta_{vw} = \eta_W + \frac{\rho_t}{\rho_e} \eta_Q \tag{5.16}$$

is referred to as a 1st law value-weighted equivalent efficiency. This indicator represents an attempt to take into account economics together with thermodynamics. The above expressions can be useful especially at a first comparative evaluation of different equipment, with different values of thermal and electrical efficiency and, in case, also different fuels.

Table 5.1. Typical values of 1st law efficiencies for cogeneration systems

type of cogeneration plant	η_W	η_Q	$EUF = \eta_W + \eta_Q$
Stirling engine	0.10 ÷ 0.20	0.70 ÷ 0.80	0.80 ÷ 0.90
ICE with total recovery	0.25 ÷ 0.45	0.30 ÷ 0.45	0.70 ÷ 0.85
ICE with only-exhaust recovery	0.25 ÷ 0.45	0.15 ÷ 0.22	0.50 ÷ 0.65
microturbine	0.20 ÷ 0.35	0.35 ÷ 0.60	0.70 ÷ 0.85
gas turbine	0.20 ÷ 0.38	0.35 ÷ 0.50	0.70 ÷ 0.85
steam turbine	0.10 ÷ 0.35	0.20 ÷ 0.80	0.75 ÷ 0.85
combined cycle	0.35 ÷ 0.60	0.05 ÷ 0.45	0.60 ÷ 0.80

5.4.4. CHP incremental indicators

If electricity is considered the "main" output of a CHP system, a widespread approach to evaluate the "effective" amount of fuel needed to its production while *also* producing heat is based on the *artificial electrical efficiency* [Hor97][EDU01][Dan06][EPAww]

$$\eta_a = \frac{W_y}{F_y - Q_y / \eta_t^{SP}} \tag{5.17}$$

In the definition of this "equivalent" electrical efficiency, the energy in the fuel supplied to the CHP plant is reduced by the amount that would have been however required to produce the cogenerated heat. This amount of fuel is conventionally calculated by supposing that the CHP heat generation is displacing a reference equivalent boiler of efficiency η_t^{SP}.

The same rationale, suitable to extend the unified heat rate framework introduced for single energy vectors, is used to define the *Electrical Incremental Heat Rate* (*EIHR*), as a generalization of the heat rate (5.1). Simplifying the original reasoning presented in [PoM82] and following [Hor97] (in which the *EIHR* defined below is simply called *IHR*, Incremental Heat Rate), the fuel energy supply F_y is broken down into the sum of the fuel energy that would be supplied to a reference boiler of efficiency η_t^{SP} to produce the same amount of cogenerated heat, and the balance that would go to the production of electricity (at an "*incremental*" heat rate *EIHR*), so as to obtain

$$F_y = \frac{Q_y}{\eta_t^{SP}} + EIHR \cdot W_y \tag{5.18}$$

which yields

$$EIHR = \frac{F_y}{W_y} - \frac{Q_y}{\eta_t^{SP} \cdot W_y} = \frac{1}{\eta_W} - \frac{\eta_Q}{\eta_W \eta_t^{SP}} \tag{5.19}$$

From comparison with (5.17), it is also apparent that

$$EIHR = \frac{1}{\eta_a} \tag{5.20}$$

The general idea behind the definition of the *EIHR* lies in the fact that it is not possible to distinguish between *two* separate amounts of fuel needed for producing electricity W_y and heat Q_y in cogeneration. Therefore, an equivalent model is introduced in order to "discount" the classical electricity-only heat rate (5.13) to account for the heat generation quota. In fact, the expression (5.19) can be also written in terms of primary energy as

$$EIHR = EHR_y - THR^{SP} \cdot \lambda_y \tag{5.21}$$

where EHR_y is the "absolute" cogeneration electrical heat rate (5.13) and λ_y is the production-related cogeneration ratio (4.2). Hence, although EHR_y values in certain CHP systems might be higher than for other non-CHP prime mover, the relevant *EIHR* might be lower, so acknowledging, within the electricity generation performance assessment, the additional production of a further energy vector. Of course, the definition and interpretation of this "actual effectiveness" in producing electricity *while* cogenerating *also* heat is conventional, and other alternative approaches could be considered, for instance based on the exergy theory [Bej97][DiR07]. However, this approach is adopted in some countries as a regulatory tool, although with some possible downsides, as argued in [CaP05].

In analogy with the *EIHR*, it is possible to introduce in a dual fashion a *Thermal Incremental Heat Rate* (*TIHR*) index [Man06][ChM06c]:

$$TIHR = \frac{F_y}{Q_y} - \frac{W_y}{\eta_e^{SP} Q_y} = \frac{1}{\eta_Q} - \frac{\eta_W}{\eta_Q \eta_e^{SP}} \tag{5.22}$$

that can also be written in terms of heat rates as

$$TIHR = THR_y - EHR^{SP} \lambda_y^{-1} \tag{5.23}$$

The *TIHR* allows evaluating the "actual" primary energy cost in order to produce heat in a cogeneration plant, putting aside the fuel amount ideally needed to produce the same amount of electricity in separate production. In practice, the approach defining the *TIHR* privileges the evaluation of the heat production rather than electricity. Again, the definition (5.22) is conventional, but it can be useful in those cases in which the focus of the analysis is on thermal production.

The incremental heat rates presented could be used for economic assessments (for electricity and heat pricing, for instance [Hor97]) or for environmental assessments when fuel or emission allocation models are needed [Luc00][Ros08]. An example of economic application of incremental indicators is provided in Section 8.3.2.

5.4.5. Second law-based models

Although 2nd law-based models are not detailed in this work, it is worth to give a general idea of some possible tools it is possible to encounter in the literature and on their potential use. In this respect, a classical approach to take into account the different thermodynamic value of electricity and heat produced in cogeneration resorts to the results from the exergy theory framework. Among several theories and methodologies proposed to assess MG systems, a simple way to get first-hand useful information on the thermodynamic characteristics of the energy vectors involved in the analysis of CHP systems is to define the *2nd law efficiency* (or *exergy efficiency*) *for cogeneration production* as [Ang98][Loz00]

$$\eta_{II} = \frac{W_y + Q_y \cdot \left(1 - \frac{T_A}{T_u}\right)}{E_y} \tag{5.24}$$

where E_y is the exergy contained in the fuel (at first approximation, for natural gas it can be considered equal to the *LHV*-based primary energy contained in the fuel F_y) [BeT96][Bej97] [Loz00].

If we compare the expression (5.24) with the *EUF* (4.8), the useful heat output is now weighted by the factor $(1-T_A/T_u)$, that is, the efficiency of an ideally reversible cycle (Carnot cycle) operating between the temperatures T_u at which the useful heat is requested and the ambient temperature T_A at which the ideal power cycle would discharge the exhausts. The term $Q_y(1-T_A/T_u)$ is the "*equivalent mechanical power*" usable (in exergy terms) out of the useful thermal power Q_y.

Given two plants with the same *EUF*, η_{II} would "grant" the one with higher electricity production, as well as with higher heat quality level. However, although based on solid thermodynamic bases, this approach might not be suitable in some practical cases, above all for small-scale DG [CaP05], and might present some limitations [FeC98][WaL07].

Another way to assess cogeneration plants by means of 2nd law based theories is to introduce the so called *Rational Criterion* (*RC*) indicator [Hor97], which can be interpreted as

$$RC = \frac{F_{rev}}{F_y} \tag{5.25}$$

where F_y is the actual fuel energy input needed to produce W_y and Q_y, whereas F_{rev} is the fuel that would have been needed to produce the same outputs with an ideally reversible process. At first approximation, *RC* can also be expressed as [Hor97]

$$RC = EUF - \lambda_y \cdot \eta_W \cdot \frac{T_A}{T_u} \tag{5.26}$$

where $\lambda_y = Q_y / W_y$ is again the cogeneration ratio. From (5.26), the *RC* "penalises" the *EUF* as more as the bigger is the heat production and/or the lower is the heat temperature, in line with the exergy analysis for which "low temperature" heat is worth less than electricity. Further details can be found in [Hor97].

Accurate insights on 2nd law-based methodologies and indicators for cogeneration systems can be found in [GaN99][Luc00][WaL07][Ert07][LiZ07].

5.4.6. Fuel Energy Saving Ratio (*FESR*) or Primary Energy Saving (*PES*)

On the above discussions, it is clear how performance evaluation of cogeneration systems is far from being straightforward, above all in the search for properly allocating the cost of resources to electricity and heat. However, as the potential "strength" of a cogeneration plant is represented, *in primis*, by the possibility of saving fuel, one of the most meaningful indicators adopted for CHP assessments is the *fuel energy saving* with respect to the fuel needed for the separate production of the same energy vectors. In this respect, if as usual η_t^{SP} and η_e^{SP} are the reference thermal and electrical efficiency for the SP equivalent models, respectively, the *fuel energy saving* potentially obtained through cogeneration is defined as

$$\Delta F = F^{SP} - F_y = \frac{Q_y}{\eta_t^{SP}} + \frac{W_y}{\eta_e^{SP}} - F_y \tag{5.27}$$

where F^{SP} is the equivalent fuel that would be used in the reference SP generation means to produce the same amount of cogenerated electricity and heat. In terms of heat rates, the expression (5.27) can also be written as

$$\Delta F = THR^{SP} \cdot Q_y + EHR^{SP} \cdot W_y - F_y \tag{5.28}$$

The *Fuel Energy Saving Ratio* (*FESR*) [Hor97] (also indicated as *Primary Energy Saving, PES* in [Hav99] and in the European Directive [EU04b], or as *Quality Index* in [HeB02]) is defined as

$$FESR = \frac{\Delta F}{F^{SP}} = \frac{F^{SP} - F_y}{F^{SP}} = 1 - \frac{F_y}{F^{SP}} \tag{5.29}$$

If Q_y and W_y are completely produced by the prime movers (no heat production from CHG or electricity purchased from the grid), an alternative expression of (5.29) is

$$FESR = \frac{\Delta F}{\frac{Q_y}{\eta_t^{SP}} + \frac{W_y}{\eta_e^{SP}}} = 1 - \frac{F_y}{\frac{Q_y}{\eta_t^{SP}} + \frac{W_y}{\eta_e^{SP}}} = 1 - \frac{1}{\frac{\eta_Q}{\eta_t^{SP}} + \frac{\eta_W}{\eta_e^{SP}}} \tag{5.30}$$

where η_Q and η_W are evaluated case by case as punctual, rated, or average values.

As a matter of terminology, in this work the terms *FESR* and *PES* are indifferently used as synonyms, as both of them are normally used in the current literature.

The expression (5.30) is of straightforward application for evaluating the relative energy saving due to cogeneration with respect to separate generation.

In the presence of shares of electricity and/or heat not cogenerated (for instance, electricity bought in from the EDS and heat produced in the CHG), an alternative definition of the *FESR* would also entail the fuel needed to produce back-up energy [Hor97]:

$$FESR^* = \frac{\Delta F}{F^{SP}} = 1 - \frac{F_y + F^{EDS} + F^{CHG}}{F^{SP}} \tag{5.31}$$

where F^{EDS} and F^{CHG} are respectively the fuel needed to produce the bought-in electricity (calculated according to the relevant chain model that also defines η_e^{SP}) and the auxiliary heat in boilers. A more general approach including the interaction with a district heating network can be extrapolated from the considerations provided in Chapter 4.

The *FESR* is one of the favourite tools applied in many countries to regulate the cogeneration production, thanks to its adequate response and clarity (relatively to other indicators) with respect to energy performance assessment issues [FeC98][CaP05].

5.5. Performance Evaluation Criteria for CCHP Systems

Trigeneration concepts in the technical literature date back to the early 1990s [FoD91][Huf92][ZiR93], and practical applications were not widespread until those days. This explains in part why a suitable framework for trigeneration assessment and policy regulation is still missing, at least with respect to the models available for cogeneration. In this respect, the criteria used to evaluate CHP systems represent a good starting point for discussing performance evaluation criteria for CCHP systems. However, the indices illustrated above need to be revised and generalized, in order to carry out proper assessment of trigeneration plants. In particular, the *FESR* represents a fundamental tool, considering the importance of energy saving issues.

In this section, after illustrating and discussing some literature approaches to CCHP energy evaluation, the *FESR* is opportunely generalised to the trigeneration case. Finally, a comprehensive approach for the evaluation of trigeneration systems, framed within the unified primary energy-based assessment discussed above, is outlined.

5.5.1. The evaluation of cooling power through reference electric chillers

A number of approaches have been discussed in the literature to evaluate the energy efficiency of cooling power generation within CCHP systems (see for instance [Hav99][HeB02][CoG03][MiL03][MiL03b][TeB04]).

An interesting approach is based on assessing the cooling power produced in a combined system with respect to the "classical" separate production (by means of a CERG). Accordingly, for "electricity-and-cooling-only" cogeneration, in [CoG03] the cooling energy produced by an absorption group fed by a cogenerator is referred to the equivalent electricity that would have been needed by using a *benchmark* CERG. Therefore, an overall *trigeneration electric equivalent efficiency* η_e^z is introduced, defined as

$$\eta_e^z = \frac{W_y + W_R^{\mathrm{SP}}}{F_y} \tag{5.32}$$

In (5.32), W_y is the actual electricity produced in cogeneration, F_y is the fuel thermal input to the cogenerator in order to obtain W_y and R_y (cooling power produced in cogeneration through the absorption group), and W_R^{SP} (electricity needed in separate production to produce R_y and displaced by the absorption chiller) is defined as

$$W_R^{\mathrm{SP}} = \frac{R_y}{COP^{\mathrm{SP}}} \tag{5.33}$$

in which COP^{SP} is the coefficient of performance of the reference CERG.

As a matter of nomenclature, the superscript/subscript z is used from now on to indicate "trigeneration" entries.

By referring all the energy vectors generated to electricity, this approach yields useful indications on the performance characteristics of the combined system through an equivalent electrical efficiency. However, it can be readily applied only to electricity-and-cooling cogeneration cases and not for actual trigeneration assessment. Nevertheless, the concept of referring the generated energy vectors to a common basis represents the same standpoint used above for the definition of *PERs*, which leads to the generalized energy saving models introduced in the sequel.

5.5.2. Trigeneration Energy Utilization Factor (*TEUF*)

The simplicity of 1st law-based energy efficiency concepts can be exploited to define a trigeneration efficiency index in analogy to the cogeneration efficiency (4.8). This approach is for instance followed in [MiL03][BrV05][CaT05] for classical trigeneration systems. By extending this approach to a generic trigeneration black-box model, in analogy to the *EUF* the *Trigeneration Energy Utilization Factor* (*TEUF*) [Man06] is defined as

$$TEUF = \frac{W_z + Q_z + R_z}{F_z} = \eta_W^z + \eta_Q^z + \eta_R^z \tag{5.34}$$

In (5.34), W_z, Q_z and R_z represent the *net* electricity, heat, and cooling power *generated* in the CCHP system, respectively, while F_z represents the *overall* fuel input to the system; the entries η_W^z, η_Q^z and η_R^z are equivalent trigeneration efficiencies, defined as

$$\eta_W^z = \frac{W_z}{F_z} \quad \text{(trigeneration electrical efficiency)} \tag{5.35}$$

$$\eta_Q^z = \frac{Q_z}{F_z} \quad \text{(trigeneration thermal efficiency)} \tag{5.36}$$

$$\eta_R^z = \frac{R_z}{F_z} \quad \text{(trigeneration cooling efficiency)} \tag{5.37}$$

in analogy to η_W and η_Q defined in Table 4.1 for cogeneration plants.

The model (5.34) can be applied to any kind of trigeneration plant with generic equipment, and within a generic time window (for instance, on an hourly basis for short-term assessment rather than on annul basis for integral assessment), provided that the involved entries are adequately calculated. In particular, the *output* energy vectors must refer to the *net* outputs from the overall trigeneration systems. Therefore, for instance, in the presence of a CHP-WARG system, the quota of cogenerated heat that is used to feed the chiller must not be taken into account in the entry Q_z, which refers only to the sheer heat output. Similarly, if for instance a CHP-EDC scheme is used, F_z must take into account both fuel energy terms used to supply the CHP *and* the EDC systems. More details on this approach are given in the next sections.

5.5.3. Absolute Trigeneration Heat Rate (*ATHR*) and Overall Trigeneration Heat Rate (*OTHR*)

The presence of equipment with *COP* higher than unity could lead to *TEUF* values higher than unity as well. This represents a drawback in the attempt of devising a suitable efficiency definition, due to loss of the intuitive benchmark according to which the more the indicator is closer to unity, the better is the plant performance.

It is worth mentioning that, for instance, the same drawback applies to the *EUF* in presence of CHP-EHP schemes, due to the fact that the EHP allows for extracting *free* heat from the atmosphere. In this case, the interpretation of the *EUF* as *efficiency* should also be evaluated with care, since a value equal to unity does not mean that the best performance has been reached [Hor97].

The reason for the occurrence of such drawbacks lies in the fact that different forms of energy are considered in the definition (5.34), regardless of their thermodynamic value, in analogy to the discussions for a cogeneration system reported in Section 5.4. This issue could be addressed by means of a 2nd law-based approach [RoL05][DiR07] or, alternatively, through a primary energy-based approach in line with the above considerations. Such an approach for various types of trigeneration systems is for instance used in [Hav99][KoW04].

Consistently with the notation introduced and with the definition provided in (5.5), it is possible to define the trigeneration heat rates EHR_z, THR_z and CHR_z, which in turn form the components of the $\overline{ATHR_z}$ *array* defined as:

$$\overline{ATHR_z} = \begin{pmatrix} EHR_z \\ THR_z \\ CHR_z \end{pmatrix} = \begin{pmatrix} \frac{F_z}{W_z} \\ \frac{F_z}{Q_z} \\ \frac{F_z}{R_z} \end{pmatrix} = \begin{pmatrix} \frac{1}{\eta_W^z} \\ \frac{1}{\eta_Q^z} \\ \frac{1}{\eta_R^z} \end{pmatrix} \tag{5.38}$$

An equivalent *PER* for the overall trigeneration black-box can be defined as the *Overall Trigeneration Heat Rate* (*OTHR*) [Man06]:

$$OTHR_z = \frac{F_z}{W_z + Q_z + R_z} = \frac{1}{TEUF} \tag{5.39}$$

The $OTHR_z$ can provide comparative information on the overall fuel utilisation efficiency for different alternatives, avoiding the misuse of the *TEUF* as an efficiency indicator and without resorting to equivalent models for the separate production case. Nevertheless, it is possible to introduce an $OTHR^{SP}$ also for conventional separate production, expressed in function of the component of the $\overline{ATHR_z}$ (5.38) calculated from the relevant conventional SP models:

$$OTHR^{SP} = \frac{EHR^{SP} \cdot W_z + THR^{SP} \cdot Q_z + CHR^{SP} \cdot R_z}{W_z + Q_z + R_z} \tag{5.40}$$

From the above definitions, it is straightforward to derive the condition for which primary energy saving is obtained from the CCHP system with respect to the conventional SP models:

$$OTHR_z < OTHR^{SP} \Rightarrow F_z < EHR^{SP} \cdot W_z + THR^{SP} \cdot Q_z + CHR^{SP} \cdot R_z \tag{5.41}$$

The *OTHR* represents the starting point for the generalization of the *FESR* to the trigeneration case outlined in Section 5.5.4.

It is interesting to consider the intrinsic performance characteristics of combined energy systems starting from the above defined indicators. In particular, considering the components of the array $\overline{ATHR_z}$, these entries alone would be suitable for evaluating energy systems with "parallel" generation model, as in *separate* production. However, energy production in *combined* trigeneration systems is ideally closer to a "*series*" or bottoming model (Figure 2.3). This can be pointed out by noticing that for *general* trigeneration cases the following relations hold

$$OTHR_z = \frac{1}{TEUF} \neq \frac{EHR_z \cdot W_z + THR_z \cdot Q_z + CHR_z \cdot R_z}{W_z + Q_z + R_z} \tag{5.42}$$

It can be noticed that in (5.42) *any* component of $\overline{ATHR_z}$ multiplied by the relative energy vector, yields the *whole* fuel $F_z = EHR_z \cdot W_z = THR_z \cdot Q_z = CHR_z \cdot R_z$ utilized also in the production of the other vectors. Thus, the amount at the right-hand side of the inequality (5.42) is three times higher than the inverse of *TEUF* (5.34), in which the fuel F_z appears only once. The production of the other energy vectors is, somehow, *hidden*. This "hiding" of the energy vectors is due to the concept itself of combined production and to the definition given of equivalent trigeneration efficiencies that all refer to the same *total* input. In fact, in the presence of parallel (separate) production, it is easy to show that

$$OTHR^{SP} = \frac{EHR^{SP} \cdot W_z + THR^{SP} \cdot Q_z + CHR^{SP} \cdot R_z}{W_z + Q_z + R_z} \tag{5.43}$$

This aspect is graphically illustrated in Figure 5.3.

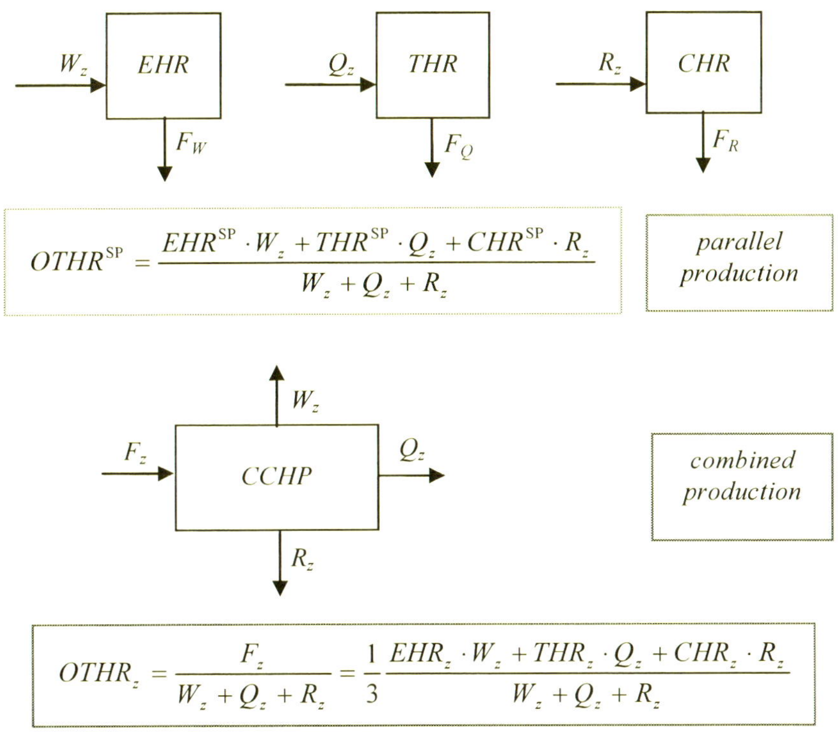

Figure 5.3. Comparison between *parallel* and *combined* production models for trigeneration.

5.5.4. Trigeneration Primary Energy Saving (*TPES*)

Performance assessment of trigeneration plants with absorption machines is presented in [Hav99] by introducing a *primary energy saving* (*PES*) index and in [HeB02] through a *Quality Index* (*QI*), defined as the *FESR* but applied to CCHP plant with a hot-water absorption group.

This approach can be extended to encompass, in general, all the possible trigeneration systems (with cogeneration, as well as trigeneration with absorption/adsorption machines, as relevant "sub-cases") [Man06]. In this outlook, the definition of primary energy saving is formulated in general terms, referring to any kind of trigeneration plants, by introducing the *TPES* (*Trigeneration Primary Energy Saving*) indicator [Man06][ChM07a]:

$$TPES = \frac{F^{SP} - F_z}{F^{SP}} \tag{5.44}$$

where F^{SP} is the total fuel energy input required in separate production (with equivalent reference efficiencies for the equipment) of the three energy vectors generated (electricity, heat and cooling), and F_z is the total trigeneration system fuel energy input.

One of the main upsides of this approach is that the index (5.44) takes *synthetically* into account all the vectors involved in the input/output flows, according to the black-box model approach, and as a consequence allows fast overall evaluation of the plant performance. As already mentioned for the *TEUF*, the net energy outputs and the overall energy input (including the fuel possibly needed to fire auxiliary boilers or parallel cooling equipment) must be considered in the analysis, so leading to a further and more specific definition:

$$TPES = 1 - \frac{F_z}{F^{SP}} = 1 - \frac{F_z}{\dfrac{W_z}{\eta_e^{SP}} + \dfrac{Q_z}{\eta_t^{SP}} + \dfrac{R_z}{\eta_e^{SP} \cdot COP^{SP}}} \tag{5.45}$$

As discussed above, W_z, Q_z, and R_z are the "net" *electricity-only*, *thermal-only* and *cooling-only* energy production seen *outside* the trigeneration plant black-box, and do not include the *internal* energy quotas. In the presence of energy inputs other than fuel energy (for instance, electricity from EDS, heat from DH network, or cooling from DCN), a similar generalized model as in (5.31) for the *FESR* can be applied, with the equivalent primary energy needed to produce the relevant energy vectors (according to the energy chain model) to be added to F_z in (5.45).

The entries η_e^{SP}, η_t^{SP} and COP^{SP} represent the references for the conventional SP models. In particular, an electric chiller is considered here as the benchmark technology, and the relevant energy chain model developed in Section 5.2 is applied to derive the equivalent primary energy used in SP for cooling generation. However, other reference technologies such as a GARG could be used as cooling reference. Similarly, a boiler is used as heat generation benchmark technology, but for instance an EHP could also be adopted. The formulation of the *TPES* with different SP benchmark technologies can be readily carried out on the basis of the energy chain models discussed.

The *TPES* (5.44) enables to synthetically account for all the input/output flows (including the shares sold or bought) in the DMG system, regardless of the actual equipment forming the plant, *i.e.*, regardless of how the final electricity, heat and cooling power are actually produced (which is the core of the black-box approach). This is a remarkably valid standpoint, based on a fuel energy saving reference, for carrying out unbiased comparative evaluation of alternative solutions.

5.5.5. Incremental Trigeneration Heat Rate (*ITHR*)

The *EIHR* has been defined in Section 5.4.4 for cogeneration systems with the aim of evaluating the net efficiency in producing electricity while cogenerating heat at the same time. However, in analysing an energy system that produces different energy vectors, the primary purpose may be different from the electricity production evaluation. For instance, the main interest could be on assessing the profitability of a district heating [CaB04][CuV02a] [CuV02b] or district cooling [LiW01][Dan06] system under different conditions. Thus, an analogous incremental approach has been followed in Section 5.4.4 to define the *TIHR* for thermal production. Following this rationale, in principle it is possible to introduce *three* incremental indicators for CCHP systems, which at the same time unify, reflect the symmetry and account for the production of the three energy vectors. These indicators can be grouped under the *Incremental Trigeneration Heat Rate* (*ITHR*) *array* [Man06][ChM06b] composed of *three* elements, each of which can assess, in an incremental fashion, the effectiveness of the production of an energy vector within *the whole* trigeneration system and *while* producing also the other energy vectors:

$$\overline{ITHR} = \begin{pmatrix} EITHR \\ TITHR \\ CITHR \end{pmatrix} \tag{5.46}$$

The entries in (5.46) can be defined through the same symbols used for the definition of the *TPES* (5.44), as

$$EITHR = \frac{F_z - \dfrac{Q_z}{\eta_t^{\mathrm{SP}}} - \dfrac{R_z}{COP^{\mathrm{SP}} \cdot \eta_e^{\mathrm{SP}}}}{W_z} \tag{5.47}$$

$$TITHR = \frac{F_z - \dfrac{W_z}{\eta_e^{\mathrm{SP}}} - \dfrac{R_z}{COP^{\mathrm{SP}} \cdot \eta_e^{\mathrm{SP}}}}{Q_z} \tag{5.48}$$

$$CITHR = \frac{F_z - \dfrac{W_z}{\eta_e^{\mathrm{SP}}} - \dfrac{Q_z}{\eta_t^{\mathrm{SP}}}}{R_z} \tag{5.49}$$

More specifically:

- the *EITHR* (*Electrical-side Incremental Trigeneration Heat Rate*) accounts for the "net" electricity-only production, conventionally putting aside the part of fuel that, in theory, would have been anyway needed to produce heat and cooling by means of separate production equipment, now displaced by trigeneration production;
- the *TITHR* (*Thermal-side Incremental Trigeneration Heat Rate*) accounts for the "net" heat-only production, conventionally putting aside the part of fuel that, in theory, would have been anyway needed to produce electricity and cooling by means of SP equipment;
- the *CITHR* (*Cooling-side Incremental Trigeneration Heat Rate*) accounts for the "net" cooling-only production, conventionally putting aside the part of fuel that, in theory, would have been anyway needed to produce electricity and heat by means of SP equipment.

The $\overline{ITHR}$ definition takes into account all the potential inputs and outputs to the plant (fuel and energy vectors from outside networks as inputs, the user's demand and energy vectors to outside networks as output), with the internal energy flows accounted for *within* the plant, according to the *black-box* approach. The SP benchmark technologies used, in particular for cooling production, are the same as for the *TPES*, but again different references could be adopted through the energy chain model. A common characteristic of these indices, likewise all the heat rate-based indices, is that lower values correspond to better solutions. *Negative* values are also possible and consistent with the "incremental" definition of the indices. Although this aspect may represent a drawback (whose origin is based on the same reasons discussed for the *TEUF*), at a planning stage comparison of a given CCHP configuration and SP benchmarks through incremental indices gives first hints on the energy performance characteristics for the *single* energy vector production. Alternative models more suitable for equipment design purposes could for instance use the approaches discussed in [Boe97][DiR07]. A number of energy assessment application examples of incremental indicators are provided in Chapter 7, whereas economic applications are outlined in Section and in Section 8.3.2.5 and in Section 8.3.4.

5.6. Performance Evaluation of Generic DMG Systems

5.6.1. Primary energy saving as the favourite assessment metric

As a result of the analyses carried out so far, it emerges how the complexity of the issues related to multi-generation systems calls for adoption of synthetic and effective indicators to characterize their energy performance.

According to the general approach discussed in this work, it is suitable to interpret the energy system as a *black-box* (Figure 5.4), in which the plant is characterized by equivalent performance indices on the basis of the only input-output energy flows, without entering into the detailed representation of the internal components. In turn, on the basis of these equivalent indices, further indicators can then be defined to assess the overall system performance. This approach is particularly effective within a DMG scenario, in which several plants might be connected together and might interact through energy networks. Synthetic models are needed for each plant, in order to be incorporated in the general approach.

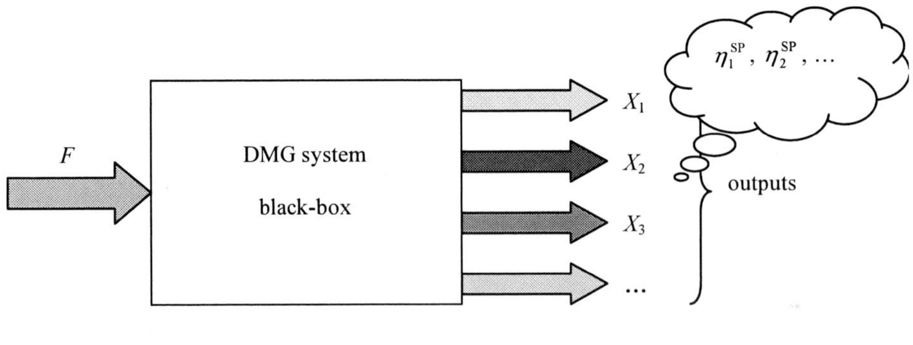

Figure 5.4. Black-box model and energy flows for multi-generation energy assessment.

The metric used in this section to assess the performance of a generic MG system is based on the primary energy saving related to relevant SP benchmark models. Indeed, as shown above, the energy saving is a meaningful indicator of the actual benefits from combined generation, and does not present possible critical drawbacks as other indicators. In addition, primary energy saving models are widely used for regulatory purposes in cogeneration [CaP05], so that the models illustrated here lend themselves well to extension to the assessment of multi-generation systems also for policy applications. Furthermore, this approach allows extension of the energy analysis to environmental issues [ChM08][ChM08b][MaC08][MaC08b], as also briefly discussed in Section 8.4, so as to derive a general and unified energy and environmental assessment framework for DMG systems.

5.6.2. The Poly-generation Primary Energy Saving (*PPES*) indicator for DMG energy systems and networks

For the multi-generation applications described below, the notation used for the various energy entries refers to calculations over given time spans (for instance, hourly, daily, annual ones) depending on the purpose of the study. The superscript p is used here to point out poly-generation (taken as synonym of the term multi-generation, as in Section 4.3.4), as it was used in [ChM08b].

In order to assess in a generalised way the MG system performance, the Poly-generation Primary Energy Saving (*PPES*) indicator has been defined as the relative fuel saving obtained in the MG system when compared to separate production. The *PPES* is expressed as [ChM08b]

$$PPES = \frac{F^{\mathrm{SP}} - F^{p}}{F^{\mathrm{SP}}} = 1 - \frac{F^{p}}{\sum_{(X,x)\in \boldsymbol{D}} \frac{X^{p}}{\eta_{x}^{\mathrm{SP}}}} \tag{5.50}$$

In (5.50), the set $\boldsymbol{D}$ contains the types of useful net energy outputs (included in the demand energy vector) from the poly-generation system. More specifically, the set $\boldsymbol{D}$ is composed of pairs (X, x), each of which contains an energy vector X and the corresponding type of energy x characterizing the useful output (demand).

Again, within the black-box approach only the energy flows visible from the *outside* are taken into account, regardless of what happens *inside* the individual boxes (Figure 5.4). For the purposes and the applications considered here, different forms of the same energy type, for instance hot water or steam in the case of heat generation, may also be deemed as *different* energy vectors (see for instance [ChM08b] for a quad-generation application).

The interpretation of the energy entries appearing in (5.50) is as follows:

- F^{SP} is the primary energy assessed through conventional SP benchmark models (opportunely defined in analogy to the models in Section 5.2);
- F^{p} is the fuel thermal input (primary energy) to the whole MG system; this term includes the fuel to feed CHP units, boilers, cooling generation equipment directly fed by fuel, and so forth;
- X^{p} represents the *net energy output* of a generic energy vector (*e.g.*, electricity W^{p}, heat Q^{p}, or cooling R^{p});

- the term η_x^{SP}, referred to the generic energy vector X produced, represents the *conventional* value of SP efficiency, evaluated according to a black-box model for conventional generation, as described in Section 5.2.

The *PPES* indicator can in turn be adopted for *networks* of DMG systems. Indeed, as mentioned above, it is possible to use the black-box approach for aggregating a number of DMG systems, with the purpose of evaluating the performance of the network-connected energy system as a whole (Figure 5.5) by considering the net input/output flows as entries in (5.50). Apparently, this approach is at the same time synthetic and effective, since, given the same overall final output (demand energy), it allows comparison of alternative DMG solutions regardless of the internal complexity of the energy system.

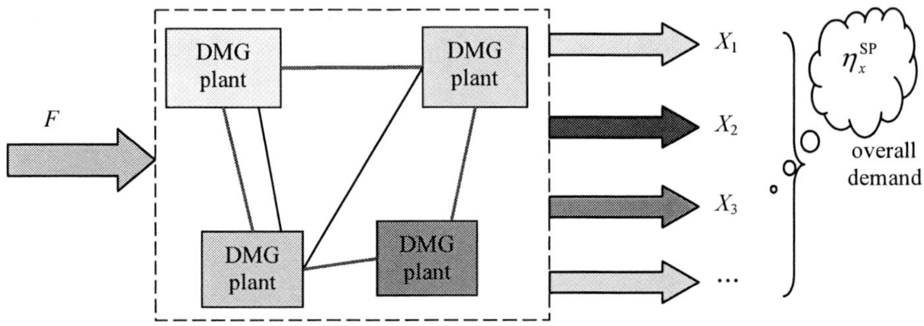

Figure 5.5. Equivalent black-box model for DMG network energy assessment.

5.6.3. Rationales for the selection of the separate production models

In the assessment methodologies that make reference to primary energy-based models, the equivalent characteristics of the SP benchmarks play a key role. In particular, conventional values must be assigned to the SP entries involved in the specific analyses, besides sorting out the issue of the most suitable reference model to adopt. Such numerical values may depend on the purpose of the analysis, and may represent matter of discussion, for instance within the development of a regulatory framework (as for the *PES* at the European level [CaP05][COG05]). In particular, the reference efficiency can consistently change the results of the performance assessment for the specific energy vector and for the whole system.

Generally speaking, two limit-case approaches may be typically envisaged for the SP equivalent models, namely, making reference to average performance and to state-of-the-art performance. However, further approaches can be followed, particularly regarding the SP electrical model, due to its key role in DMG evaluation and to actual generation mechanisms. Synthesizing the issue, it is possible to consider three basic cases to determine the equivalent plant model and η_e^{SP} value when addressing the evaluation of DMG compared to SP:

- η_e^{SP} is estimated on the basis of average values relevant to the operation of the *power system* as a whole. In this kind of assessment, average and practical indications are provided in terms of the avoided primary energy consumption (and relevant avoided environmental burden).

- η_e^{SP} is estimated on the basis of a "marginal power plant", according to the argument that DG sources are likely to displace the operation of the relatively more expensive and lower-efficiency power plants coming into operation, as dispatched according to a merit order.
- η_e^{SP} is estimated on the basis of new (centralized) plants that would be used if the DG systems were not installed [AlL02]. This approach is sound for comparison purposes and gives indications on the convenience of alternative *high-tech* generation paradigm, namely, centralized SP or decentralized DMG energy systems.

Of course, the approaches considered are all *approximated* to some extent, besides depending on the purpose of the study. In particular, if the aim is to address the actual performance with respect to the current centralized energy chain, an ideal analysis should consider the power system characteristics through the actual types of power plants (*marginal plants*) that are displaced time by time, possibly by simulations of the overall power system [VoD03][TsH07]. However, the presence of competitive energy markets might render an assessment based on marginal plant references extremely uncertain. Indeed, nuclear and some hydro sources are typically operated with flat profiles, to cover the base load demand, so that the load-following operation is left to other, less efficient types of plants. On the other hand, fast-changing energy market scenarios would boost the utilization of different primary energy sources and types of power plants, case by case. For instance, while low-efficiency and high-emission coal plants are claimed to be soon phased out in most countries, their utilization can be boosted by high gas prices that might make them more convenient even with respect to modern high-efficiency CCGT plants.

A further refinement of the analyses could also account for the performance variation trend for both DMG and SP reference systems, which would call for relevant information upon the trend forecast [MaC08].

5.7. Remarks on DMG Energy Performance Assessment Methodologies

A general and comprehensive methodology for energy assessment of DMG systems has been presented in this chapter. In particular, a number of indicators have been described to address the comparative evaluation of various multi-generation systems (including cogeneration and trigeneration as specific cases) with reference separate production technologies. The models developed are based on the modelling concepts of *energy chain* and *black-box*, which allow drawing significant and synthetic considerations on the performance characteristics of DMG systems for planning purposes. The methodologies described are based on first-law energy balances, and as such are to be intended only as indicative when used for detailed design studies. Indeed, above all for equipment design purposes, 2nd law-based models might be more suitable, as discussed. Nevertheless, the indicators presented can be important standpoints for analyses responding to today's energy framework issues, including security of supply, fuel saving, environmental issues, and interactions with the energy markets. In the last two cases, the primary energy saving models introduced lend themselves well to generalization to environmental issues, as mentioned in Section 8.4, as well as potential applications in emerging energy-related markets, as described in Section 8.3.

The indicators introduced can be evaluated over any time span, for instance on a yearly basis for integral performance assessment, rather than on an hourly basis for tracking the performance changes with respect to loading levels. Of course, the level of detail sought in the results also drives the level of detail for the input data. However, while evaluations based on rated or average values of the entries can be carried out for first and indicative planning analyses, more detailed evaluations should be based on the results of time-domain simulations [VoD03][Man06]. In particular, the time step used in the simulation should be short enough to catch the main phenomena, for instance the change in control strategies due to interaction with the market [ChM09b]. Furthermore, partial-load and off-design models of all the equipment should be implemented in the simulation codes. Detailed examples for a host of cogeneration and trigeneration plant configuration and evaluation cases are given in Chapter 6 and Chapter 7.

Chapter 6

COGENERATION ENERGY PERFORMANCE ASSESSMENT APPLICATIONS

Different energy assessment methodologies and indicators to be used for planning purposes have been introduced and discussed in Chapter 5. In this chapter, these indicators are applied to the analysis and evaluation of alternative solutions for cogeneration, taken as the simplest case of multi-generation. In particular, a number of given plant configurations are analyzed and evaluated, in order to exemplify with simple studies the energy chain and black-box models introduced. Parametric comparisons between different types of equipment that could typically compose a small-scale distributed energy system are illustrated. The tools adopted in the analysis are the incremental and the energy saving indicators. The use of the latter is particularly emphasized, given the importance of energy saving within a context aimed at maximizing environmental benefits and energy generation sustainability.

Besides providing indicative numerical results on the energy performance of cogeneration systems with current and perspective equipment, the aim of this chapter is to point out the *methodology* and the *approach* to energy evaluation, which should always take into account, besides the rated values, also actual operating conditions. In particular, a mathematical formulation of the considered indicators is provided for various plant configurations. Specific operational situations in terms of energy flows are modelled through *dispatch factors*. The results obtained depend on a number of factors, since specific analyses need to consider different load levels, different types of equipment, and finally different control strategies. *Parametric analyses* are thus run to highlight the sensitivity of the assessment results found to various operational configurations.

The exemplificative results, which point out a great potential for efficiency enhancement owing to cogeneration, represent the starting point for further practical studies for design of new composite DMG energy systems as well as comparison among existing ones.

6.1. ENERGY CHAIN MODEL APPLICATION TO HEATING GENERATION

6.1.1. Comparison between electric heat pumps and boilers

Electric heat pump-based HVAC applications have been increasing more and more in the past years, thanks to the high intrinsic efficiency of the EHP. Yet, in order to evaluate more deeply limits and potentiality of this technology, a comparative analysis between EHPs and traditional boilers can be carried out. The comparison is rendered more interesting by the fact that the energy input sources to the two devices are different and of different thermodynamic quality, which is handled by applying the "heat rate" approach discussed in Chapter 5.

Electric heat pump *COPs* are strongly related to the outdoor temperature (especially for air-source EHPs) and to the temperature of the ambient to heat up (Section 3.3.6). Therefore, a suitable approach for comparison purposes is to seek the *thermodynamic break-even* point, that is, the *energy break-even COP* for which, given the reference boiler efficiency (relatively more independent of external conditions), the EHP results more efficient than the boiler. Once this value of *COP* has been found, the planner can evaluate whether it is feasible (according to the temperature levels required by the user and available from the cold heat source) and proceed with further studies on the selection of the device.

For energy comparison purposes, the performance characterization for both devices is expressed in terms of primary energy input through the energy chain model discussed in Section 5.2.

The energy chain model for this analysis is shown in Figure 6.1, in which also transmission and distribution losses are implicitly included in the equivalent plant model of the power system.

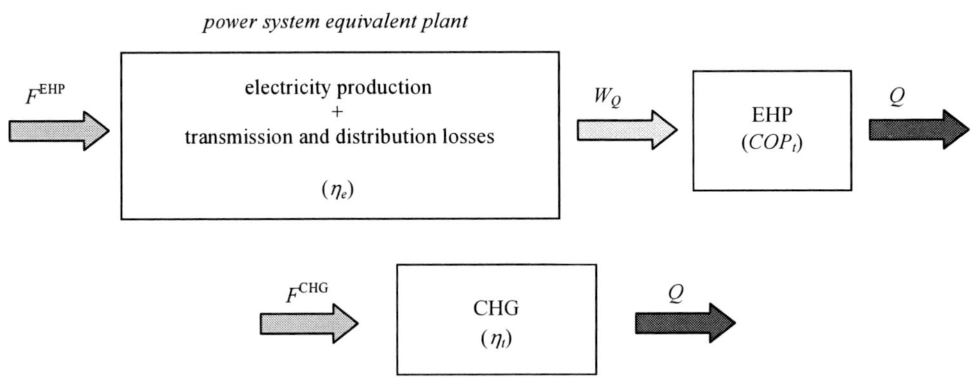

Figure 6.1. Energy chain models for EHP-CHG energy comparison.

The primary energy break-even point is expressed by the condition for which the two devices require the same fuel thermal input given the same thermal output. In other words, this corresponds to equal *THR* for the two devices:

$$THR^{CHG} = THR^{EHP} \tag{6.1}$$

Considering the expression (6.1) and the formulas in Figure 5.2, this expression yields the energy *break-even* COP_t*:

$$COP_t^* = \frac{\eta_t}{\eta_e} \tag{6.2}$$

As the equivalent power system efficiency η_e also includes electrical transmission and distribution losses, it is possible to consider, for instance, an average value of 0.40 for the Italian case; then, assuming a rated η_t of 0.9 for commercial boilers, the corresponding break-even COP_t* results equal to 2.25, a value that can be normally reached by an air-source or ground-source EHP even in relatively cold areas [Kre01][Dan06].

6.1.2. Primary energy saving analysis

It is also possible to calculate the potential energy saving a heat pump might bring with respect to a combustion heat generator, in the case it operated with *COP* values above the break-even point [CaR02]. The *FESR* (5.29) for the boiler-EHP system can then be defined as

$$FESR = \frac{F^{\text{CHG}} - F^{\text{EHP}}}{F^{\text{CHG}}} \tag{6.3}$$

where F^{CHG} and F^{EHP} are the primary energy input to produce the same given quantity of heat through the boiler and the heat pump, respectively (Figure 6.1). It is easy to show that

$$FESR = 1 - \frac{\eta_t}{\eta_e \cdot COP_t} \tag{6.4}$$

which can be plotted (Figure 6.2) as a function of the heat pump COP_t, with η_t as parameter and η_e assumed equal to 0.4.

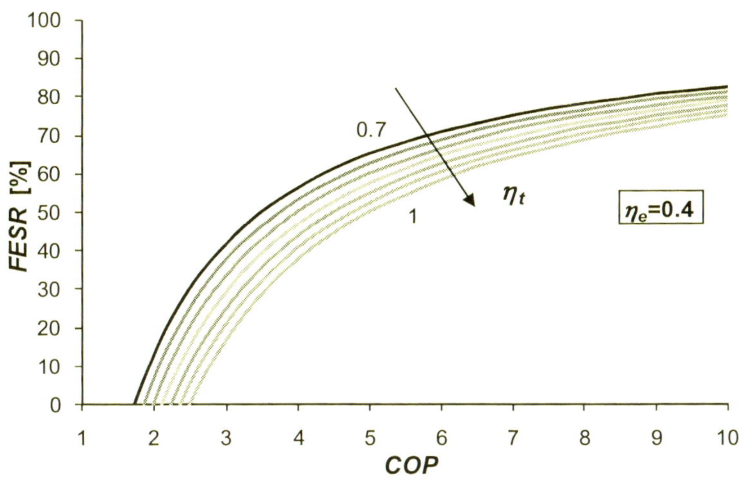

Figure 6.2. Primary energy saving in the comparison between electric heat pump and boiler.

Figure 6.2 shows the potential substantial effectiveness of an EHP over conventional combustion boilers in terms of energy saving for *COP* about 3 or 4. Pushing too far searching for *COPs* higher than 6÷7, instead, would not bring a significant incremental energy benefit. On the contrary, the EHP performs worse than the CHG for *COPs* lower than about 2 (break-even point). From similar charts it is also possible to assess the effect of a combined utilization of EHP-CHG heat generation system, in which either machine can operate in dependence on the *COP* reachable by the heat pump (which in turn depends on outdoor conditions and heating load requests).

It can also be interesting to evaluate the energy break-even *COP* as a function of the electrical efficiency η_e, again with η_t as the curve parameter (Figure 6.3). In this case, η_e might represent the power system equivalent efficiency in different countries or at different operating points (for instance, average or marginal power plants), or it could for instance model the electrical efficiency of a distributed power plant or DG system that feeds the electric heat pump. With even small-scale but high-efficiency prime movers such as ICEs, with electrical efficiency up to 0.45, the potential energy effectiveness of an EHP is substantial (although of course depending case by case on the operating point).

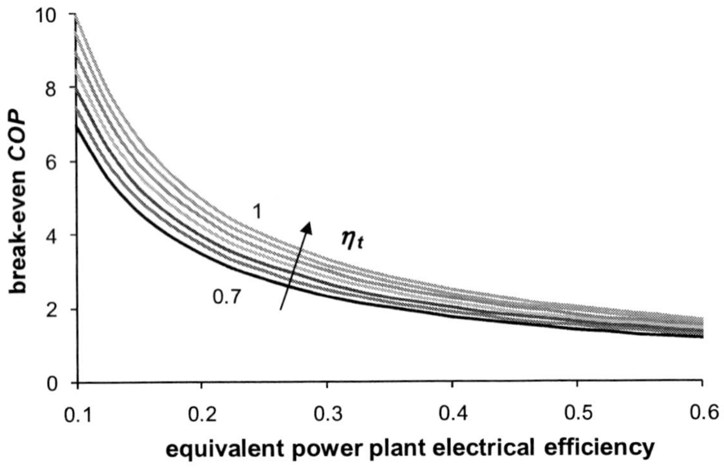

Figure 6.3. EHP-CHG energy break-even *COP* as a function of η_e.

6.1.3. Electric resistance heating

Electric resistance heating can be seen as a particular case of EHP with $COP = 1$ (Section 3.3.6.5). The definition of 1st law efficiency applied to this system would give a result of about 100% (in practice, except for parasitic losses, all the electrical input turns into desired thermal output). However, in spite of the apparent nearly "perfect" 1st law performance, on the above considerations and numerical analyses it stems that resistance heating can be efficient only if electricity is generated from renewable sources. Second law-based analyses also confirm the inefficiency of this kind of systems [Loz00]. In general, however, the possible convenience of electric resistance heating is mostly associated to specific economic conditions, namely, very low electricity prices, very low cost of investment, and/or unavailability of gas supply.

6.2. Heat-and-Electricity Cogeneration Assessment Examples

The classical cogeneration concept is related to the combined production of heat and electricity. As discussed in Chapter 5, among the several indices available to evaluate CHP plants, the *FESR* (or *PES*) plays a fundamental role. Indeed, the evaluation of fuel saving is also considered from a privileged standpoint by the European Regulation, because of the benefits brought to the society as a whole (environmental aspects and primary source saving) [EU04b][CaP05].

In this section, a detailed analysis of the *FESR* for different aspects is carried out, also focusing on some regulatory issues related to its evaluation.

6.2.1. General consideration on the *FESR*

The *FESR* expression (5.30) can also be written as

$$FESR = 1 - \frac{\lambda_y + 1}{EUF \cdot (\frac{1}{\eta_e^{SP}} + \frac{\lambda_y}{\eta_t^{SP}})} = 1 - \frac{\lambda_y + 1}{EUF \cdot (EHR^{SP} + \lambda_y \cdot THR^{SP})} \quad (6.5)$$

From (6.5) it is apparent that the primary energy saving depends on the cogeneration overall efficiency *EUF*, on the reference efficiencies for the SP, and on the useful heat to electricity cogeneration ratio:

$$\lambda_y = \frac{Q_y}{W_y} = \frac{\eta_Q}{\eta_W} \quad (6.6)$$

In terms of absolute heat rates, the *FESR* can also be expressed as

$$FESR = 1 - \frac{1}{\frac{THR^{SP}}{THR_y} + \frac{EHR^{SP}}{EHR_y}} \quad (6.7)$$

where EHR_y and THR_y are calculated with reference to the absolute (non-incremental) electrical and thermal efficiency of the cogeneration unit (Section 5.4.1).

In Figure 6.4 the *FESR* is plotted against the cogeneration ratio (in log-scale) for different values of *EUF* as parameter and for average efficiency values in the Italian system.

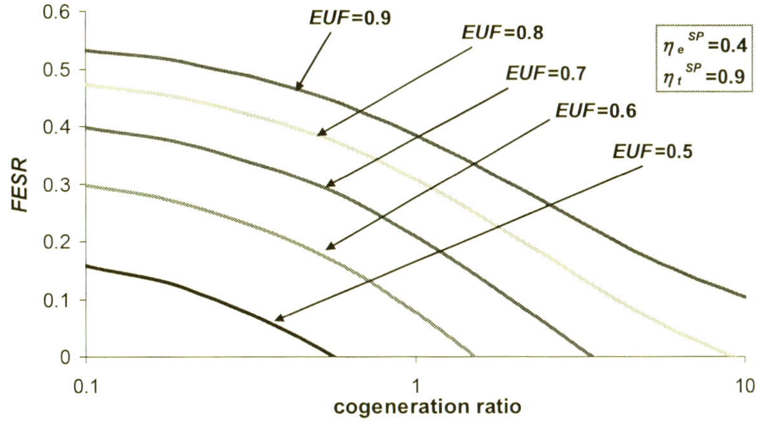

Figure 6.4. *FESR* in function of λ_y and *EUF*.

Whereas it is easy to work out the role of the *EUF* indicator in energy saving evaluation (increasing overall energy efficiencies corresponds to increasing energy saving), the dependence on the cogeneration ratio is not as clear. Indeed, the effectiveness to cogenerate has not been justified yet on a physical basis, which intrinsically lies behind the weakest point of separate production, namely, combustion heat generation. This may sound surprising, since the corresponding efficiency is in general very high (even close to unity, for condensing boilers). However, as already discussed, the definition of any efficiency is conventional, and the conventions utilised to define electrical and thermal efficiencies in separate production are substantially different. In fact, although both based on the 1st principle of thermodynamics, the quality of the energy produced (in terms of 2nd principle and exergy theory) is different [Bej97][DiR07].

Therefore, the reason why separate production of heat represents the weak point in the overall production system is that at the combustion stage heat is rendered available at a temperature far higher than the one at which it is used. There exists, as a consequence, a substantial waste of energy that could be used for instance to produce mechanical work (its exergy), which is pointed out by generation of irreversibility. When producing heat in separate production, *all* this high-valued (because of the high temperature) heat available at the combustion is "wasted" for thermal purposes at a far lower temperature. When producing in cogeneration, the heat available from combustion is effectively exploited to produce electricity when it is highly valuable (high temperature); afterwards, when at lower temperature (but still above the environment temperature), the remaining heat that has not been turned into work is used for thermal purposes [Ang98]. The "energy convenience" to cogenerate is implicitly hidden behind this "*series*" model of the combined production concept.

The *FESR* plotted in Figure 6.4 decreases with an increase of the cogeneration ratio. This can be addressed to the fact that, given a value of *EUF* and then of the energy transferred from the fuel to be transformed into heat and electricity, the more thermal energy is produced, the less the potential of exploiting the high temperature heat for electricity generation is used. In addition, from a merely numerical standpoint, η_t^{SP} is higher than η_e^{SP}, so that, in the comparison, heat production is somehow penalised with respect to electricity production.

The values assigned to η_t^{SP} and η_e^{SP} play an important role in the evaluation of the plant, as generally discussed in Section 5.6.3. In this respect, recent regulation developed in many European Countries requires application of the *FESR* to evaluate the actual performance of a CHP plant [EU04b][CaP05], and the selection of the specific SP benchmarks has been source of discussions [COG05].

6.2.1.1. What FESR?

In order to point out the sensitivity of the *FESR* evaluation to the reference efficiency parameters, three different cases with respect to the value assigned to η_e^{SP} are exemplified in the sequel, in line with some of the approaches that have been undertaken within different regulatory frameworks:

- η_e^{SP} is evaluated on the basis of the *best available technology* in the *same size range*: in this case, the energy saving is evaluated with respect to having otherwise used the best electrical technology for the separate production. This approach appears quite sound.
- η_e^{SP} is evaluated on the basis of the *best available technology, no matter the size*: therefore, the energy saving is evaluated from a "hypothetical" standpoint, simply requiring the highest possible performance in terms of energy saving. It is apparent that, compared with the highest possible electrical efficiency, the cogeneration thermal production must be considerable. This approach might be highly penalising for some kinds of technologies (see below).

- η_e^{SP} is evaluated on the basis of the *average* electrical efficiency of the power system: in this case, the *FESR* is a measure of the "avoided energy cost" to the society. This is also a sound approach.

Case 1 is taken into consideration by the Italian Authority [AEEG], while *Case 2* is suggested by the European Directive [EU04b]; *Case 3* is a general reference to energy saving evaluation, and is also considered by the Italian Authority for smaller-size plants [AEEG].

6.2.1.2. Primary energy saving break-even analyses

In order to readily compare these three approaches, a possible technique is to consider the *energy break-even condition* between cogeneration and separate production, namely,

$$FESR = 1 - \frac{1}{\dfrac{\eta_Q}{\eta_t^{SP}} + \dfrac{\eta_W}{\eta_e^{SP}}} = 0 \tag{6.8}$$

that yields

$$\eta_W^* = \eta_e^{SP} \cdot \left(1 - \frac{\eta_Q}{\eta_t^{SP}}\right) \tag{6.9}$$

The expression (6.9) can be interpreted as follows: if the prime mover electrical efficiency (at rated capacity, for instance) is higher than η_W^*, therefore there might be energy saving and further analysis can be gone through.

Following the lines drawn in [Mar98], relevant considerations, also taking into account the regulatory frameworks illustrated above, can be carried out by plotting the break-even condition (6.8) on a chart with the cogeneration thermal efficiency and the cogeneration electrical efficiency as the horizontal and the vertical axes, respectively (Figure 6.5). This kind of plot can analyse the potentiality of different prime movers with different cogeneration electrical and thermal efficiencies (in every operating point).

Figure 6.5 shows the break-even condition for three different η_e^{SP}, corresponding to *Case 1* (η_e^{SP} = 0.43, considering average natural gas ICEs as the reference technology for small-scale applications), *Case 2* (η_e^{SP} = 0.55, considering CCGT as state-of-the-art reference technology) and *Case 3* (η_e^{SP} = 0.40, about the average electrical production efficiency of the Italian power system, including losses). For all cases η_t^{SP} is assumed to be equal to 0.9 in order to better highlight the role of the SP electrical reference. To complete the chart, also parametric lines for different values of λ_y are given. These lines highlight the different behaviour of CHP systems with different generation characteristics (higher values of cogeneration ratio correspond to lower electrical efficiency and higher thermal efficiency).

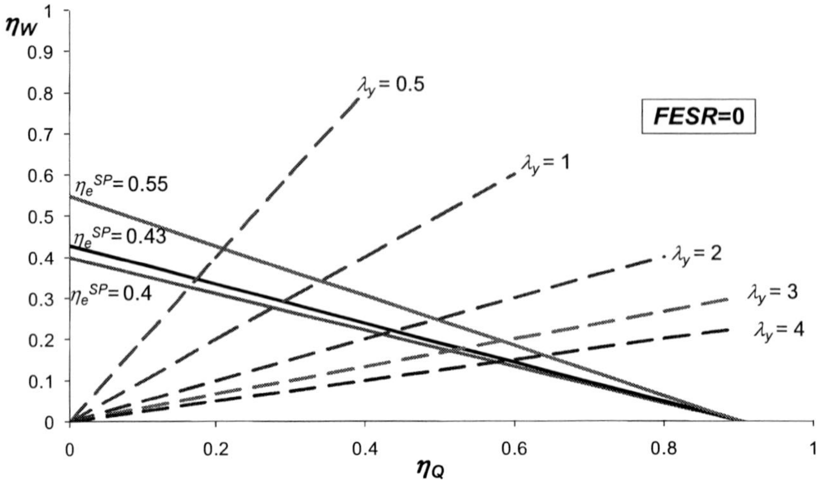

Figure 6.5. Energy saving break-even condition for cogeneration with respect to separate generation (η_t^{SP} = 0.9).

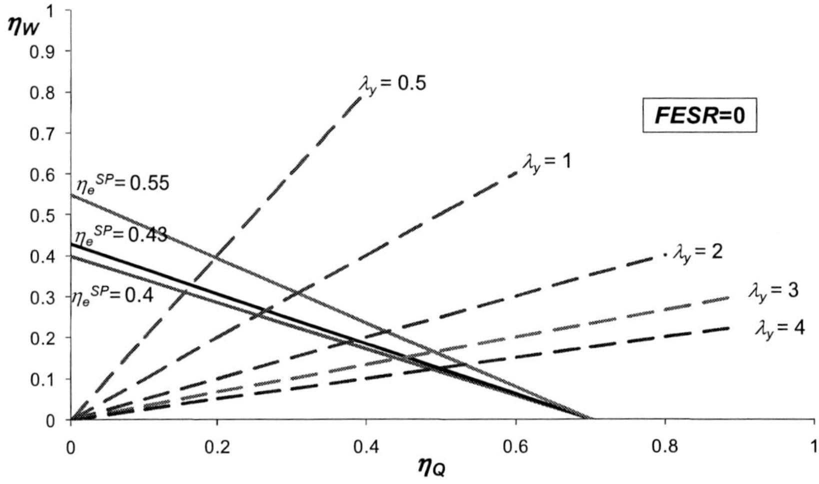

Figure 6.6. Energy saving break-even condition for cogeneration with respect to separate generation (η_t^{SP} = 0.7).

From Figure 6.5, *Case 2* is much more restrictive than the others. For instance, considering an MT with average electrical efficiency of 0.3, according to *Case 2* it should exhibit an average thermal efficiency higher than 0.4 in order to provide positive energy saving. Instead, in the other two approaches an average thermal efficiency of about 0.3 would be sufficient. For smaller electrical efficiency values, for all the different SP cases high values of thermal efficiency are needed to bring positive energy saving. For instance, considering a micro-CHP Stirling engine with electrical efficiency of about 0.1, its thermal efficiency should be higher than about 0.8 in order to get positive *FESR*. Practically, this corresponds to reaching the upper technical limits for the *EUF*. These results are due to the relatively high reference efficiency for thermal generation. Indeed, if η_t^{SP} is set to 0.7 (Figure 6.6), a value closer to actual average seasonal efficiency for household applications where micro-CHP are likely to be applied, energy saving can be reached for CHP thermal efficiency above around 0.6. Such efficiency figures can be easily achieved by current units.

While Figure 6.5 and Figure 6.6 plot the break-even condition (6.8), Figure 6.7 plots the *threshold conditions* to get certain values of *FESR*, namely, 0.1 and 0.2, in correspondence of given SP electrical efficiency, namely, η_e^{SP} = 0.4 and η_e^{SP} = 0.55. The thermal reference efficiency is set to η_t^{SP} = 0.9. As a major result, it can be noticed that with η_e^{SP} = 0.55 if the requested minimum *FESR* to achieve is higher than 10%, conditions very tough to reach technically (an extremely high level of heat recovery), or even thermodynamically prohibited (thermal efficiency higher than unity), would arise for the technologies characterized by lower electrical efficiency (for instance, MT without recuperator or micro-CHP SE). The performance of CHP systems improves if the thermal reference value is relaxed to 0.7 (Figure 6.8). In this case, even with electrical efficiencies below 0.2 it is possible to *feasibly* reach primary energy saving values of 0.2 or above if the thermal efficiency is high enough.

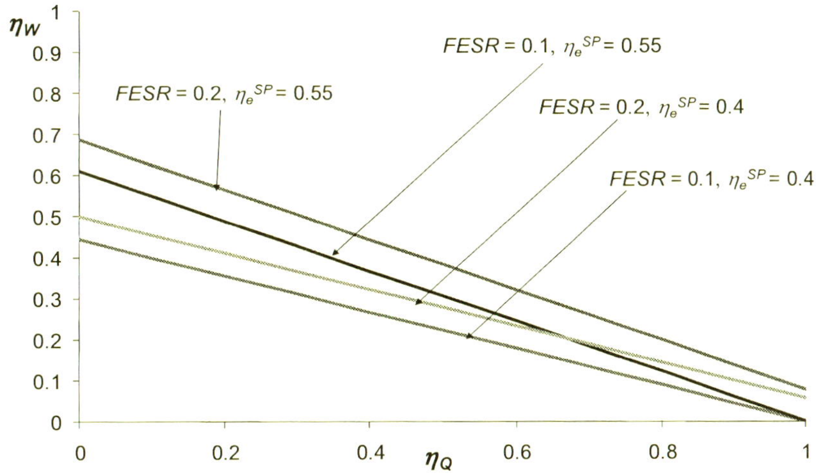

Figure 6.7. *FESR* thresholds for different electrical reference efficiency (η_t^{SP} = 0.9).

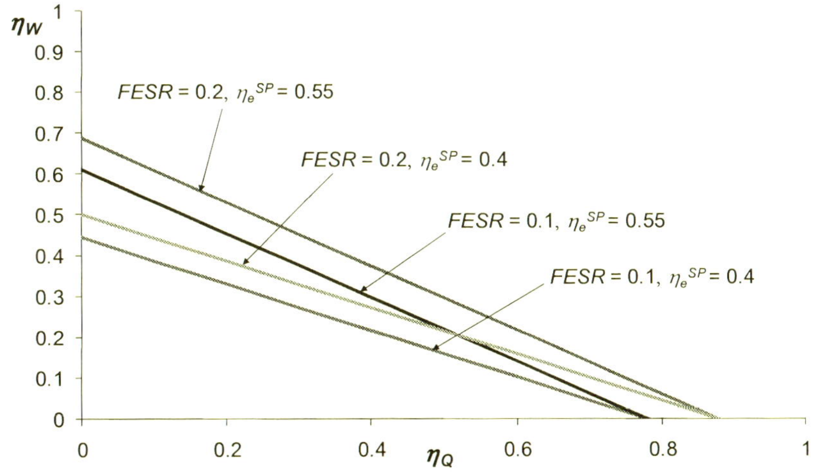

Figure 6.8. *FESR* thresholds for different electrical reference efficiency (η_t^{SP} = 0.7).

6.2.1.3. Some remarks on FESR applications

In general, from approaches and plots like the ones discussed above it is possible to assess the potential performance of different CHP systems with respect to different SP references. In particular, it is clear how it is crucial to suitably identify the benchmark technologies representing the baseline scenario for cogeneration assessment, which case by case depends on the purpose of the study and can be a matter of policy. In addition, it has to be highlighted that the analysis can be carried out over different time frames and for different operational configurations. For instance, the CHP efficiencies input to *FESR* analyses could be assessed as average values calculated on an annual basis as $\eta_Q = Q_y / F_y$ and $\eta_W = W_y / F_y$, where the relevant energy entries are now integral values from annual balances. On the other hand, CHP efficiencies could be evaluated hour by hour or even within shorter time frames. In this case, also off-design conditions could be appraised, by analysing the sensitivity of the energy saving performance to efficiency variations due to partial-load operation, changes in the outdoor temperature (above all for turbines and MTs), and so on.

6.2.2. CHP assessment through incremental indices

An evaluation of cogeneration plants on the basis of the incremental indices introduced in Section 5.4.4 is carried out in this section.

Figure 6.9 plots the *EIHR* (5.19) (that is, the classical *IHR*) against the prime mover electrical efficiency for different prime mover thermal efficiencies as curve parameters. An *EUF* upper limit of 0.9 has been considered. Comparison with the *EHR* (5.6), which would be the actual fuel consumption rate without considering the cogenerated heat, is also shown. It is apparent that, especially for low electrical efficiencies and higher thermal ones, the *EIHR* is quite lower than the corresponding *EHR*. This highlights the almost *obligation* to cogenerate, especially in the presence of low electrical efficiency, in order not to waste primary energy source. Furthermore, it can be seen how cogeneration of heat can bring an "actual" fuel consumption for electricity production lower (although only marginally) than for today's best technologies (*EHR* = 1.8 for η_e^{SP} = 0.55) even with the lowest electrical efficiency considered, as long as a good thermal efficiency is achieved.

In general, for small-scale applications, both ICEs (considering average characteristic efficiency values close to η_W = 0.4 and η_Q = 0.4) and MTs (with characteristic efficiency values close to, in the best cases, η_W = 0.3 and η_Q = 0.5) can achieve quite good performance in terms of "actual" fuel consumption for producing electricity. Also micro-CHP systems, for instance based on SEs, with electrical efficiencies lower than 0.2 could be energy-effective, but only provided that a high thermal recovery is possible. The results found are consistent with the outcomes of the analyses carried out above through the *FESR* indicator.

While the evaluation of cogeneration plants through the *IHR* somehow represents a classical approach [Hor97][EDC01][CaP05][EPAww], the dual incremental evaluation of heat production is not widespread. Potential applications might for instance address the environmental impact of distributed CHP systems with respect to the separate production, considering that the energy sources to be potentially displaced are represented by the boilers spread over the territory [ChM06c]. Apart from the possible applications, for the sake of illustration Figure 6.10 and Figure 6.11 show the *TIHR* (5.22) as a function of the prime mover electrical and thermal efficiency and for two different values of η_e^{SP} reflecting the issues discussed for *FESR* assessment. In this case, the results are compared with the absolute heat rate values from heat separate production through boilers ($THR^{SP} = 1.1$ and $\eta_t^{SP} = 0.9$).

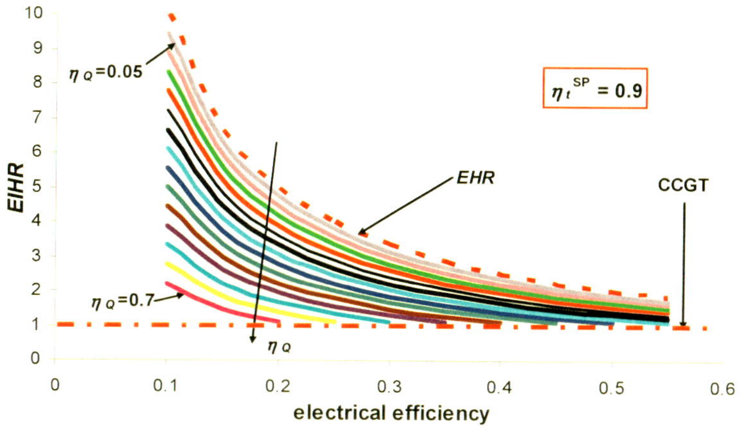

Figure 6.9. CHP plant *EIHR* for different values of electrical and thermal efficiency ($\eta_t^{SP} = 0.9$).

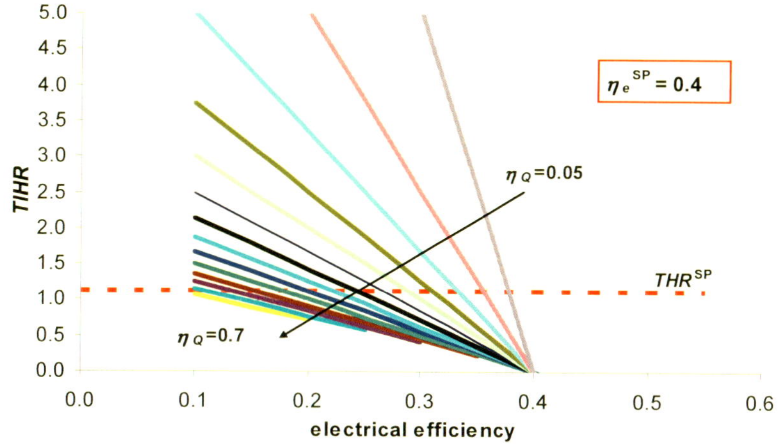

Figure 6.10. CHP plant *TIHR* for different values of electrical and thermal efficiency ($\eta_e^{SP} = 0.4$).

As general considerations on the results with $\eta_e^{SP} = 0.4$ (Figure 6.10), with a minimum amount of heat cogenerated ($\eta_Q = 0.1$, for instance) the *TIHR* is lower than the THR^{SP} if the prime mover electrical efficiency is higher than about 0.36 (easily feasible for small-scale ICEs under rated conditions). In order to obtain $TIHR < THR^{SP}$ with electrical efficiencies of about 0.3 (MTs, for instance), the thermal recovery should be at least about 25% of the fuel primary input. Considering $\eta_e^{SP} = 0.55$ (Figure 6.11), the same objective $TIHR < THR^{SP}$ can be reached, for electrical efficiencies about 0.3, with thermal recovery at least about 45% (close to the limit, with today's MT technology). Again, the ideal condition $TIHR < THR^{SP}$ would be easier to satisfy for ICEs (considering $\eta_W = 0.4$) with thermal efficiencies higher than about 0.25. However, as also pointed out with reference to the regulatory framework [COG05][AEEG], the value 0.55 appears to be too high when evaluating the performance of small-scale CHP prime movers (that, with respect to the "bigger" relatives, are penalised because of the well-known scale factors, but bring other benefits, for instance in terms of flexibility).

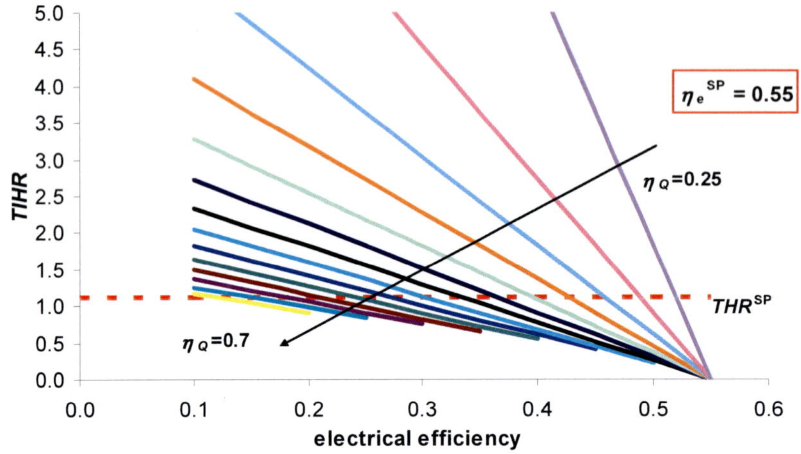

Figure 6.11. CHP plant *TIHR* for different values of electrical and thermal efficiency ($\eta_e^{SP} = 0.55$).

6.3. Heat and Electricity Cogeneration: CHP Coupled to EHP

6.3.1. Primary energy saving model for a composite CHP-EHP scheme

The conditions for which EHPs can be more energy-effective than classical combustion heat generators have been discussed in Section 6.1. In general, the characteristics of a heat pump can be profitably exploited when there is need of thermal power and excess of electricity produced. Thus, it is interesting to evaluate the energy performance of a DMG scheme where thermal power is supplied by both a cogeneration group and an EHP, which in turns is driven by cogenerated electricity. As discussed in Section 4.5.5, such a scheme can provide high flexibility to the plant, since the equivalent cogeneration ratio in Figure 4.14 can vary to a large extent. Hence, this plant configuration can be economically profitable in dependence of variable energy prices, besides being potentially energy-effective, as shown below.

The general formula representing the primary energy saving for an EHP coupled to the power grid or a generic prime mover is expressed by (6.4). If we suppose that the EHP electrical source is composed of a CHP system, and that not all the electricity produced is used to drive the EHP, the composite energy system so obtained operates in *electricity-and-heat cogeneration* mode (Figure 6.12).

The energy flows in the composite energy system in Figure 6.12 are modelled through the electricity-to-heat dispatch factor α_Q [Man06], already introduced in Section 4.6.4.3, indicating the relative shares of electricity going to supply the EHP as opposed to the electrical demand (see also Figure 4.17). Hence, the relevant energy balances are such that the quota $W_d = (1-\alpha_Q)W_y$ of the cogenerated electricity W_y goes to supply the electrical needs, whereas the share $W_Q = \alpha_Q W_y$ feeds the EHP. Therefore, the overall heat produced to satisfy the thermal demand Q_d (in case at different thermal levels, such as those possibly obtainable from CHP heat recovery and EHP production) and the electrical demand W_d can be written as

$$Q_d = Q_y + Q^{\mathrm{EHP}} = \eta_Q \cdot F_y + COP_t \cdot W_Q = \eta_Q \cdot F_y + \alpha_Q \cdot \eta_W \cdot COP_t \cdot F_y \qquad (6.10)$$

$$W_d = (1-\alpha_Q) \cdot \eta_W \cdot F_y \qquad (6.11)$$

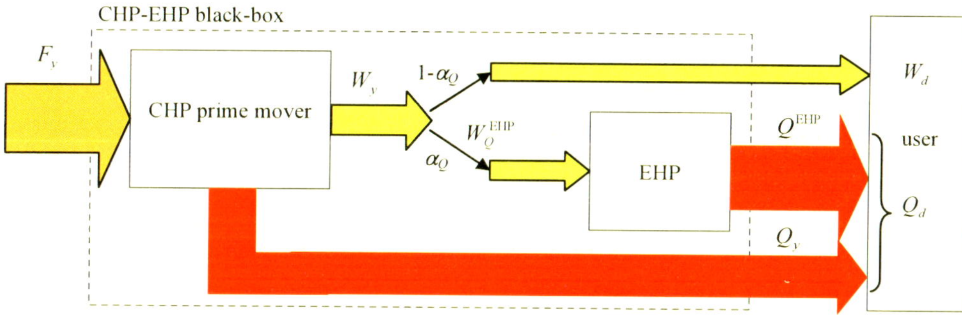

Figure 6.12. Black-box model of a composite CHP-EHP system.

By comparing the thermal fuel input F_y with the F^{SP} needed if the same W_d and Q_d were generated in separate production, after few calculations it is possible to obtain the *FESR* as [Man09]

$$FESR = \frac{\Delta F}{F^{\mathrm{SP}}} = 1 - \frac{F_y}{F^{\mathrm{SP}}} = 1 - \frac{\eta_t^{\mathrm{SP}} \cdot \eta_e^{\mathrm{SP}}}{(\eta_t^{\mathrm{SP}} \cdot \eta_W + \eta_e^{\mathrm{SP}} \cdot \eta_Q) + \alpha_Q \cdot \eta_W (\eta_e^{\mathrm{SP}} \cdot COP_t - \eta_t^{\mathrm{SP}})}$$

$$= 1 - \frac{1}{\left(\frac{\eta_W}{\eta_e^{\mathrm{SP}}} + \frac{\eta_Q}{\eta_t^{\mathrm{SP}}}\right) + \alpha_Q \frac{\eta_W}{\eta_e^{\mathrm{SP}}}\left(\frac{\eta_e^{\mathrm{SP}}}{\eta_t^{\mathrm{SP}}} COP_t - 1\right)}$$

$$= 1 - \frac{1}{\left(\frac{EHR^{\mathrm{SP}}}{EHR_y} + \frac{THR^{\mathrm{SP}}}{THR_y}\right) + \alpha_Q \frac{EHR^{\mathrm{SP}}}{EHR_y}\left(\frac{THR^{\mathrm{SP}}}{EHR^{\mathrm{SP}}} COP_t - 1\right)} \qquad (6.12)$$

The expression (6.12) is in line with the black-box representations for energy assessment discussed in Chapter 5, in that the external energy flows are used as input entries to the *FESR* calculation, and the internal flows are opportunely modelled (through dispatch factors).

6.3.1.1. Numerical examples

The *FESR* (6.12) enables to get a picture of the energy saving potential of a CHP-EHP composite system with different equipment characteristics and various operational conditions. In this respect, Figure 6.13 to Figure 6.16 plot the CHP-EHP *FESR* as a function of α_Q for different reference efficiency values and CHP performance characteristics. In particular, instead of referring to specific case study applications, analyses like the ones run provide general indications on the performance that can be expected from the composite system when it operates under various configurations. Indeed, the values of α_Q depend on the demand load levels and on the control strategy adopted. More specifically, the dispatch factor can also be seen as an *energy-shifting* operator [Man09], as it represents the variable that drives the equivalent cogeneration ratio changes seen from the user's side (Section 4.5.5).

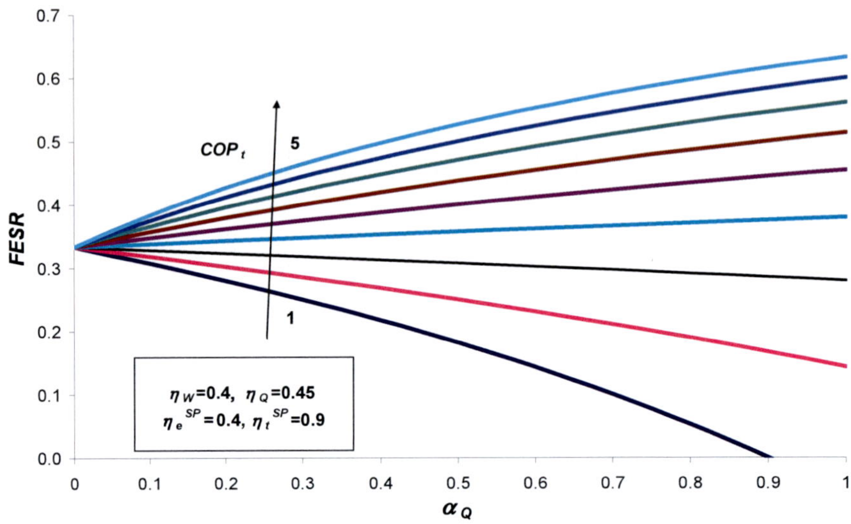

Figure 6.13. *FESR* for CHP-EHP cogeneration (η_W=0.4, η_Q=0.45, η_e^{SP}=0.4, η_t^{SP}=0.9).

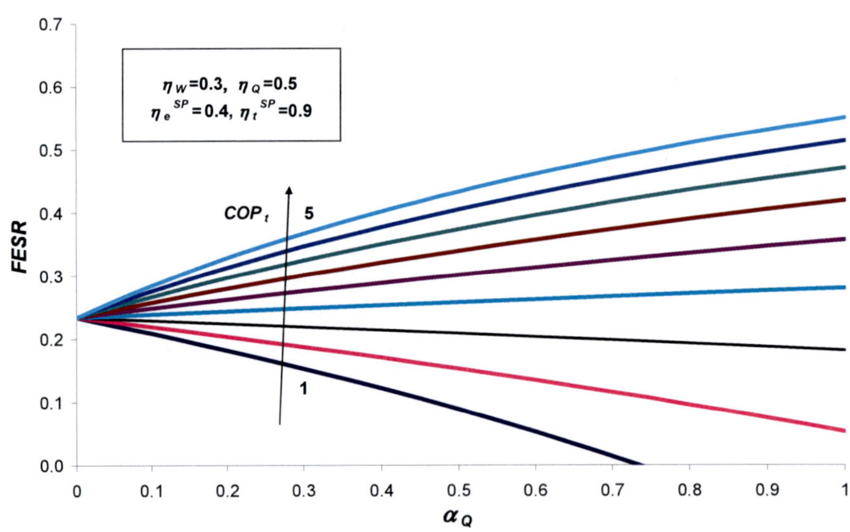

Figure 6.14. *FESR* for CHP-EHP cogeneration (η_W=0.3, η_Q=0.5, η_e^{SP}=0.4, η_t^{SP}=0.9).

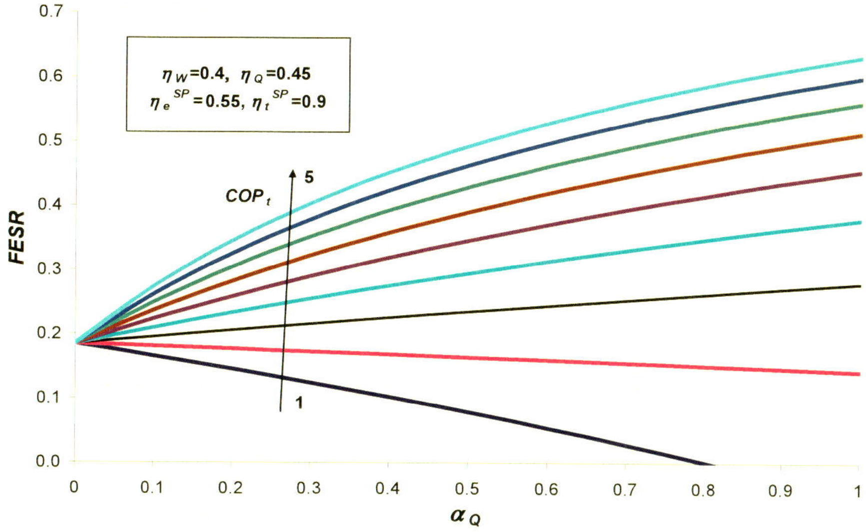

Figure 6.15. *FESR* for CHP-EHP cogeneration (η_W=0.4, η_Q=0.45, η_e^{SP}=0.55, η_t^{SP}=0.9).

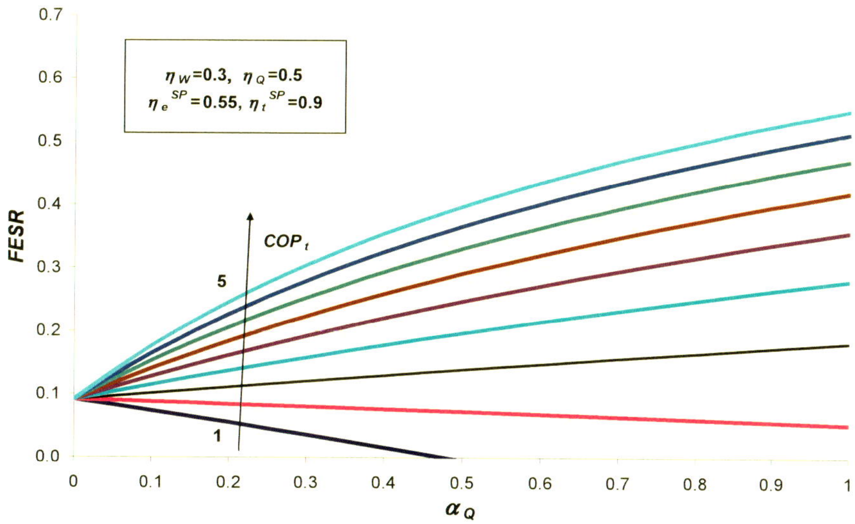

Figure 6.16. *FESR* for CHP-EHP cogeneration (η_W=0.3, η_Q=0.5, η_e^{SP}=0.55, η_t^{SP}=0.9).

The pair (η_W = 0.4, η_Q = 0.45) could be considered as representative of typical values for ICEs, whereas the pair (η_W = 0.3, η_Q = 0.5) can refer to MTs. The energy saving obtaining with ICEs is about 10% higher for both reference configurations (η_e^{SP} = 0.55 and η_e^{SP} = 0.4).

In general, the *FESR* decreases when the reference electrical efficiency increases. However, the worsening becomes lower for higher values of α_Q, since less and less electricity is produced, in favour of thermal production.

Depending on the *COP* value, the *FESR* increases or decreases with α_Q, as it can be easily found by deriving it with respect to α_Q:

$$\frac{\partial(FESR)}{\partial\alpha_Q} \geq 0 \Rightarrow COP_t \geq COP_t^* \tag{6.13}$$

with

$$COP_t^* = \frac{\eta_t^{SP}}{\eta_e^{SP}} \tag{6.14}$$

As COP_t* depends only on the reference efficiency values, it can be evaluated *a priori*. Hence, for *COPs* above this threshold, increasing the quota of heat produced in the EHP would bring higher energy saving. Taking $\eta_t^{SP} = 0.9$, with η_e^{SP} equal to 0.55 and 0.4, COP_t* becomes 1.6 and 2.25, respectively. In fact, if the electrical reference becomes higher, it becomes easier to obtain energy saving by producing more thermal power. Both values (1.6 and 2.25) are commonly reachable by using commercial air-source or water/ground-source heat pumps even in cold climates. However, in general the EHP performance is a function of the operating temperatures, as discussed in Section 3.3.6. Hence, if the plant is required to supply heat at different thermal levels, the heat pump should be used to supply the lower quality one in order to maximise its performance.

The expression (6.12) is the most general formula for a CHP-EHP cogeneration system [Man09], which includes as particular sub-cases the limit case of *no heat production* from the EHP ($\alpha_Q = 0$) and *heating-only production mode* for the whole energy system ($\alpha_Q = 1$, $W_d = 0$):

$$\alpha_Q = 0 \Rightarrow \frac{\Delta F}{F^{SP}} = 1 - \frac{\eta_t^{SP} \cdot \eta_e^{SP}}{\eta_t^{SP} \cdot \eta_W + \eta_e^{SP} \cdot \eta_Q} = 1 - \frac{1}{\dfrac{\eta_W}{\eta_e^{SP}} + \dfrac{\eta_Q}{\eta_t^{SP}}} \tag{6.15}$$

$$\alpha_Q = 1 \Rightarrow \frac{\Delta F}{F^{SP}} = 1 - \frac{\eta_t^{SP}}{\eta_Q + \eta_W \cdot COP_t} \tag{6.16}$$

The expression (6.15) is the classical formula to evaluate a CHP prime mover, whereas the expression (6.16) represents the *FESR* for a CHP-EHP system under heating-only scheme, shown in Figure 6.17.

For the heating-only configuration, the energy break-even COP_t * with respect to the separate production is

$$COP_t^* = \frac{\eta_t^{SP} - \eta_Q}{\eta_W} \tag{6.17}$$

The numerical value of this break-even *COP* is now much lower than from (6.2) when there is no heat cogenerated (indeed, although illustrated with reference to the power system, the case in Section 6.1.1 could be more generally referred to an EHP coupled to an electricity-only DG thermal prime mover).

Figure 6.18 shows the *FESR* obtained with respect to $\eta_t^{SP} = 0.9$ in heating-only mode for different values of thermal and electrical efficiencies of the CHP unit at constant $EUF = 0.8$. In this way, an indicative comparison is carried out on a "fair" basis for prime movers with different rated cogeneration characteristics but equal exploitation of the fuel primary energy in input. Given the overall efficiency, the electrical efficiency has a weight much higher than the electrical one to get fuel saving. In particular, with *COP* in the range 3 to 4, *FESR* values of about 50% can be reached with efficiencies common for MTs or ICEs, whereas the break-even COP_t* is below 1.5.

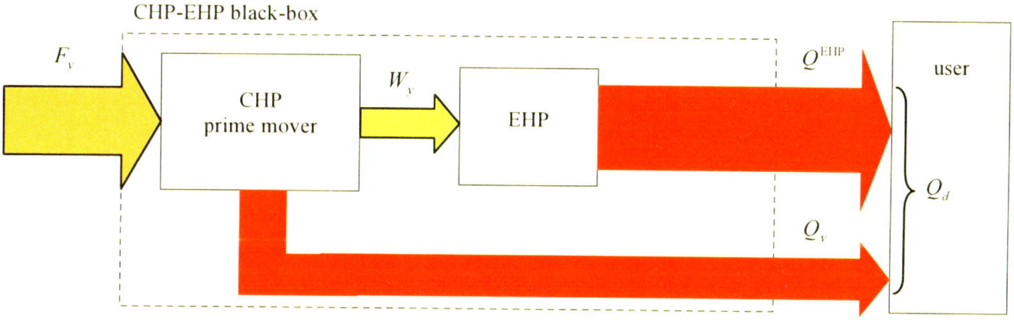

Figure 6.17. Composite CHP-EHP system under heating-only mode.

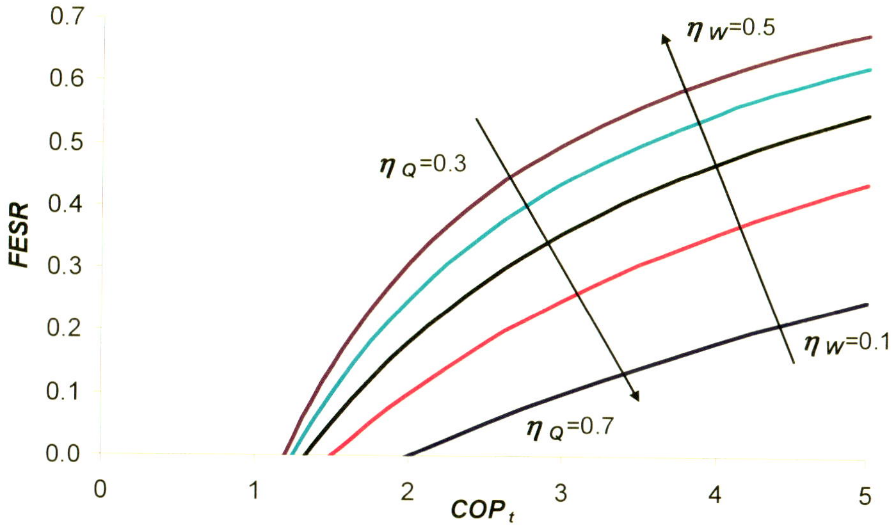

Figure 6.18. *FESR* for CHP-EHP heating-only scheme (η_t^{SP}=0.9).

6.3.2. Incremental indicators for CHP-EHP assessment

The black-box approach lends itself well also for incremental analyses. Indeed, if the focus is on the *net* external energy flows, it is possible to express equivalent incremental indicators for the overall combined CHP-EHP scheme as a whole as a function of the performance characteristics of the equipment. In the general electricity-and-heat cogeneration mode, and with reference to Figure 6.12, it is possible to write [Man06][ChM06c]

$$EIHR = \frac{F_y - \dfrac{(Q_y + Q^{\text{EHP}})}{\eta_t^{\text{SP}}}}{(1-\alpha_Q)\cdot W_y} = \frac{1}{\eta_W(1-\alpha_Q)}\left(1 - \frac{\eta_Q + \alpha_Q \eta_W COP_t}{\eta_t^{\text{SP}}}\right) \quad (6.18)$$

$$TIHR = \frac{F_y - \dfrac{(1-\alpha_Q)\cdot W_y}{\eta_e^{\text{SP}}}}{Q_y + Q^{\text{EHP}}} = \frac{1-(1-\alpha_Q)\dfrac{\eta_W}{\eta_e^{\text{SP}}}}{\eta_Q + \alpha_Q \eta_W COP_t} \quad (6.19)$$

The expressions (6.18) and (6.19) allow fast evaluation of the CHP-EHP combined system in its *actual* operating points, namely, CHP thermal and electrical efficiency, as well as EHP COP_t, evaluated at partial load, under the actual outdoor and indoor conditions, and finally considering the specific load configuration described by the dispatch factor.

The evaluation of the incremental indicators (6.18) and (6.19) can provide useful information on the comparative assessment of the production efficiency from different prime movers and combined energy systems. The general reference is again given by the SP (non-incremental) indicators (5.10) and (5.11). In particular, the conditions to obtain cogeneration production effectiveness are given by

$$EIHR < EHR^{\text{SP}} \quad (6.20)$$

$$TIHR < THR^{\text{SP}} \quad (6.21)$$

6.3.2.1. Numerical applications: performance evaluation of different CHP prime movers coupled with an EHP

Figure 6.19 plots the *EITHR* (6.18) of the CHP-EHP combined system against the dispatch factor α_Q, while the prime mover electric and thermal efficiencies are given as curve parameters, with a constant *EUF* equal to 0.8 for all cases. The effectiveness for the heat pump is set to $COP_t = 3$, representing a reasonable average value consistent with today's technologies, if weather conditions are not too severe.

In general, the effectiveness in net production of electricity increases with increasing α_Q and can even become negative when the electricity share to produce heat overcomes a certain value (thanks to the multiplicative effect in the heat production given by the heat pump). However, the higher is α_Q (and thus the higher the electricity production effectiveness), the lower is the electricity produced. In addition, there is a quite apparent *EIHR* drop (performance improvement) when passing from the curve with η_W = 0.1 to the one with η_W = 0.2, whereas the marginal benefits decreases while further increasing the electrical efficiency. As a general reference, again it is interesting to consider an EHR^{SP} = 1.8 (for $\eta_e^{\text{SP}} = 0.55$).

Figure 6.20 shows analogous results for the *TIHR* (6.19). In this case, the performance for the incremental index is always better than for the corresponding "non-incremental" one (THR^{SP} = 1.1 if η_t^{SP} = 0.9). In addition, for the higher values of electrical efficiency (and the lower values of thermal one), the index increases with α_Q, somehow due to the loss in electricity production efficiency while fostering heat production, whereas it is quite flat for lower electrical efficiencies.

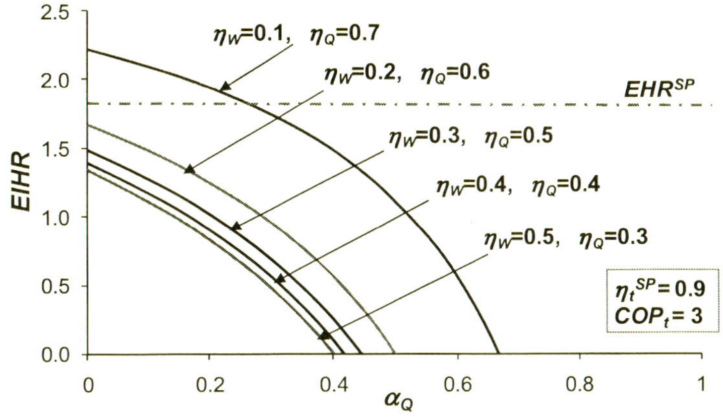

Figure 6.19. *EIHR* for CHP-EHP systems as a function of α_Q ($COP_t = 3$, $\eta_t^{SP} = 0.9$).

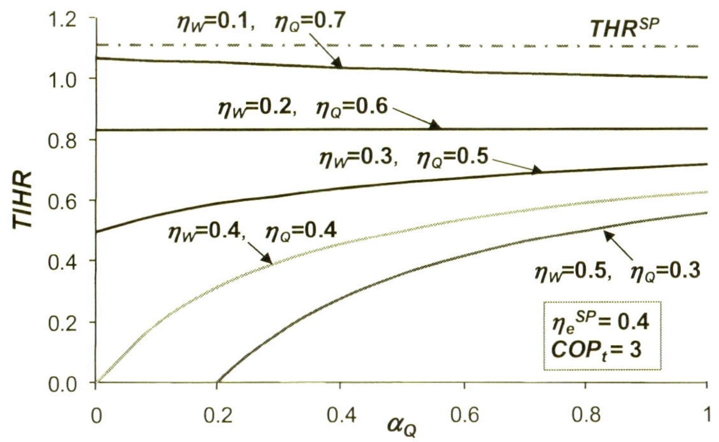

Figure 6.20. *TIHR* for CHP-EHP systems as a function of α_Q ($COP_t = 3$, $\eta_e^{SP} = 0.4$).

As a general conclusion on the parametric analyses for the CHP-EHP system, with the equipment considered ($COP_t = 3$) both *EIHR* and *TIHR* decrease (leading to performance improvement) when adopting a combined CHP-EHP scheme with respect to the results obtained for CHP-only production illustrated earlier. This outcome is due to the implicitly high-efficiency characteristics of the EHP in producing thermal power by exploiting "free" heat from outdoor sources (air, water or ground), and is consistent with the above results based on primary energy saving assessment.

Chapter 7

ENERGY PERFORMANCE ASSESSMENT OF TRIGENERATION ALTERNATIVES

In this chapter, some of the assessment methodologies introduced above are applied to trigeneration systems. More specifically, different trigeneration plant configurations are analyzed, with various pieces of equipment that typically might compose a small-scale distributed multi-generation energy system. The tools used in the analyses are the *TPES* and the $\overline{ITHR}$ indicators. In particular, the *TPES* plays a key role within the current framework in which energy saving issues are of primary interest. In addition, considering that power systems and distributed resources are evolving fast, it is crucial to run *direct* comparisons between combined and conventional generation for possible scenarios and energy system solutions, as enabled by the *TPES*.

As in the previous chapter, the approach to DMG evaluation is structured in such a way as to encompass the characteristics of the plant components in actual operating conditions. Interactions among plant components are again described by using *dispatch factors*. The higher complexity of the plant structures calls for dealing with more sophisticated control strategies, in which the dispatch factors may play the role of control variables. Parametric analyses are extensively used to represent various solutions for multi-generation system operation, and to establish sound comparisons between the alternative plant configurations analysed and classical centralized generation means.

The results presented in this chapter provide indicative information on multi-generation energy efficiency at a planning stage. As such, they are a good starting point before moving further in practical design studies, as well as for high-level energy comparisons among energy system alternatives.

7.1. CONSIDERATIONS ON REFERENCE MODELS

As a general comment about the definition of the separate production reference plant models and relevant efficiencies, the analyses carried out in this chapter mostly refer to η_e^{SP} values equal to 0.4 (Italian average of the bulk power system) and 0.55 (CCGT, in case with allowance for transportation losses), in line with the methodologies and rationales discussed above (and used, in particular, in Chapter 6 for the *FESR* calculations). Values up to η_e^{SP} = 0.6 might be considered for "state-of-the-art" CCGT. The same rationales can also be applied for extended assessment of energy and environmental performance of MG systems [ChM08b]. Insights on "marginal plant operation" can also be found in [MaC08b].

For heat generation, the typical thermal efficiency reference values considered are of the order of 0.8÷0.9 (rated values for residential and small industrial applications).

Regarding cooling generation, in general it might be arguable which reference model to adopt (see Section 5.3.3), and, if adopting a reference electric chiller model, which value to assign to the COP^{SP}. In this respect, a value equal to 3 is mostly considered in the analyses, as an average reference figure for typical residential applications in several countries, while values up to 5 might be considered closer to upper boundaries (modern larger plants for centralized air conditioning applications). Nonetheless, *COP* evaluation should always be considered with respect to the *actual* conditions in which the plant operates.

Apart from the numerical figures used here to provide meaningful indications on actual energy system characteristics and performance, the generality of the methodologies and tools introduced is such that it is easy to extend the analyses to specific studies, in case on the basis of time-domain plant simulations.

7.2. Cogeneration of Cooling and Electricity (Seasonal Trigeneration)

There are several applications in which the final uses refer only to electricity and cooling loads. Hence, in a number of practical cases cogeneration of heat and power finally results in a *cogeneration of cooling and power*, as all the produced heat is used to feed absorption (or also adsorption) chillers, for instance for summertime air conditioning (*e.g.*, see [MaP00] [BaK02][CaP03][CoG03][Hwa04]). These cases are often addressed to as *seasonal trigeneration*, and may be in practice evaluated in terms of primary energy saving relative to separate production of electricity and cooling by exploiting the general *TPES* model.

The model for "cogeneration" of cooling and electricity is shown in Figure 7.1.

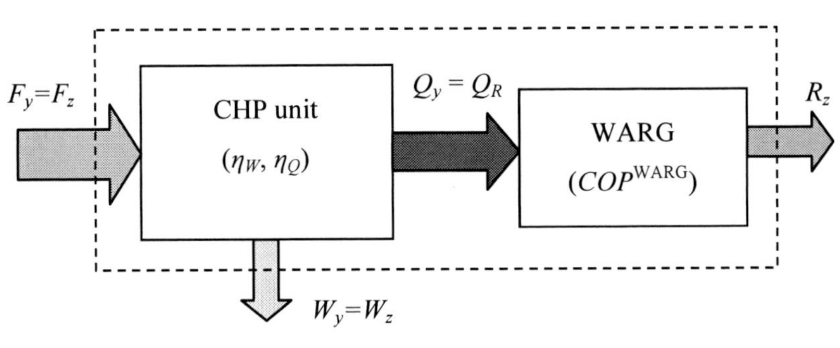

Figure 7.1. Scheme for electricity-and-cooling cogeneration assessment.

7.2.1. Primary energy saving break-even conditions

From Figure 7.1, in the overall CHP-WARG system black-box the fuel input $F_y = F_z$ is converted into $Q_y = Q_R$ (thermal input that fires the absorption chiller to produce the overall cooling output R_z) and $W_y = W_z$. Hence, in the input-output plant model the entries to be considered are F_z, R_z and W_z, where the subscript *z* points out trigeneration.

The relevant energy balances yield

$$R_z = COP^{\mathrm{WARG}} \cdot Q_R = COP^{\mathrm{WARG}} \cdot \eta_Q \cdot F_z \tag{7.1}$$

$$W_z = \eta_W \cdot F_z = \eta_W \cdot \frac{R_z}{COP^{\mathrm{WARG}} \cdot \eta_Q} \tag{7.2}$$

The same quantity of R_z and W_z generated in separate production would require a fuel input given by

$$F^{\mathrm{SP}} = F_W^{\mathrm{SP}} + F_R^{\mathrm{SP}} = \frac{W_z}{\eta_e^{\mathrm{SP}}} + \frac{R_z}{\eta_e^{\mathrm{SP}} \cdot COP^{\mathrm{SP}}} \tag{7.3}$$

In order to obtain primary energy saving, after some elaborations it can be found that the *break-even condition*, expressed for instance in terms of electrical efficiency of the CHP group, is [HeB02]

$$\eta_W{}^* = \eta_e^{\mathrm{SP}} - \eta_Q \frac{COP^{\mathrm{WARG}}}{COP^{\mathrm{SP}}} \tag{7.4}$$

The break-even condition (7.4) for η_W is plotted in Figure 7.2 against η_Q. In this case, reference values of η_e^{SP} equal to 0.4 and 0.6 (last generation CCGT) are considered. In addition, three values are set for the absorption chiller, namely, *COP* equal to 0.65 (single-effect machine), 1.1 (double-effect) and 1.5 (triple-effect). In order to assess the role played by the electric chiller reference COP^{SP}, two values are considered, equal to 3 and 5, respectively on the left-hand side and on the right-hand side of Figure 7.2.

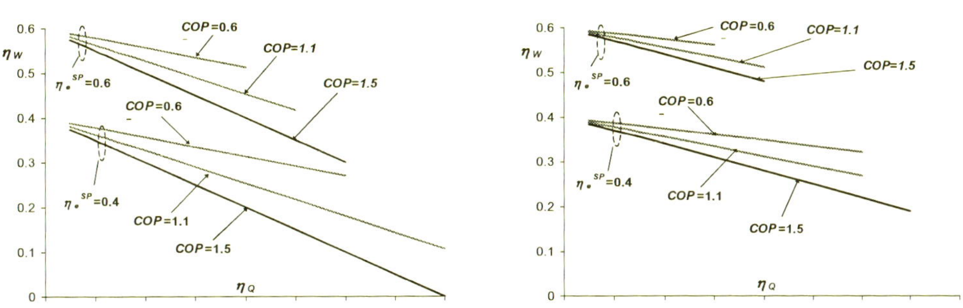

a) $COP^{\mathrm{SP}} = 3$ b) $COP^{\mathrm{SP}} = 5$

Figure 7.2. Energy break-even conditions for electricity-and-cooling cogeneration systems.

Once fixed the SP reference efficiencies, from plots like the ones in Figure 7.2 it is possible to evaluate, while planning the electricity-and-cooling DMG system or while operating it, the possibility of obtaining or not energy saving given the actual characteristics of absorption group and cogenerator, as well as of the cooling and electrical loads. However, also the link between the CHP group and the absorption group should be checked. Indeed, depending on the type of linking system used (Section 3.3.4) not all the cogenerated heat might be of quality high enough to feed the absorption chiller; therefore, only a share of the recuperated heat could be used to fire the chiller, and thus not all the points of the map could actually be feasible.

On the basis of the plots in Figure 7.2, further analyses can be run to assess the potential energy saving of available CHP prime movers. For instance, with reference to the general range of characteristics of typical MTs (η_W = 0.2 ÷ 0.3, η_Q = 0.4 ÷ 0.5) and ICEs (η_W = 0.3÷0.4, η_Q = 0.4÷0.45), MTs can never operate with energy saving with respect to separate production if the electrical efficiency reference value is set to 0.6, for both COP^{SP} values. Instead, the MT can yield energy saving with COP^{SP} = 3 and η_e^{SP} = 0.4, but only for high thermal efficiencies and adopting double-effect or triple-effect chillers. However, in this case the relevant thermal value corresponding to characteristics able to fire the chiller should be considered. The general hint is that the MT, even in comparison with relatively low SP efficiencies, should run at high loads (so as not to incur in partial-load efficiency penalties) and then operate with high efficiencies in order to prompt some energy saving.

ICEs can perform in general better than MTs, owing to the higher electrical efficiency. However, also for an ICE if η_e^{SP} = 0.6 no energy saving can be obtained, also considering that not all the thermal efficiency range is actually exploitable in terms of heat generation to fire double-effect and triple-effect chillers (Section 3.3.4). Instead, considering η_e^{SP} = 0.4, for both values of COP^{SP} energy saving can be obtained with performance commonly boasted by current units.

7.2.2. Primary energy saving assessment

Once established that energy saving can be obtained, the actual energy saving values in every operating point, and thus with the relevant thermal and electrical efficiency value, can be calculated by applying the general *TPES* definition (5.45) with no heat generation, so yielding

$$TPES = \frac{F^{SP} - F_z}{F^{SP}} = 1 - \frac{1}{\dfrac{\eta_W}{\eta_e^{SP}} + \dfrac{COP^{WARG}}{COP^{SP}}\dfrac{\eta_Q}{\eta_e^{SP}}} = 1 - \frac{1}{\dfrac{EHR^{SP}}{EHR_z} + \dfrac{CHR^{SP}}{CHR_z}} \tag{7.5}$$

where EHR_z is calculated on an absolute (non-incremental) basis and CHR_z is an equivalent cooling heat rate resulting from the energy chain model of Section 5.2.6 applied to the CHP-WARG system:

$$EHR_z = \frac{1}{\eta_W} \tag{7.6}$$

$$CHR_z = \frac{1}{COP^{WARG} \cdot \eta_Q} \tag{7.7}$$

Some graphical results are shown in Figure 7.3, where the *TPES* is calculated with single-effect and triple-effect absorption chillers, for given separate production reference efficiencies of η_e^{SP} = 0.4 and COP^{SP} = 4, an intermediate value between average and upper boundary.

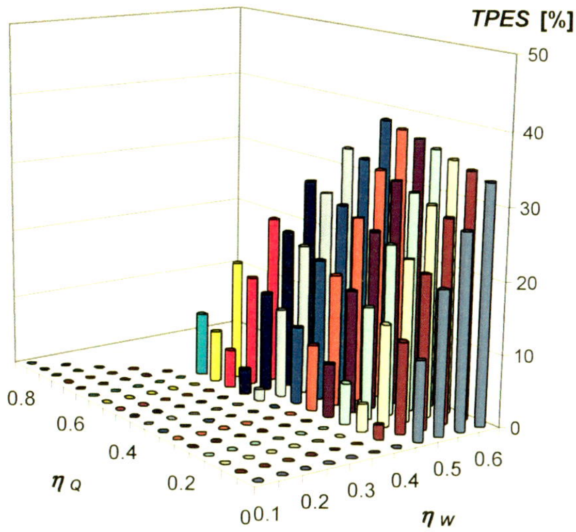

a) $COP^{\mathrm{WARG}} = 0.65$

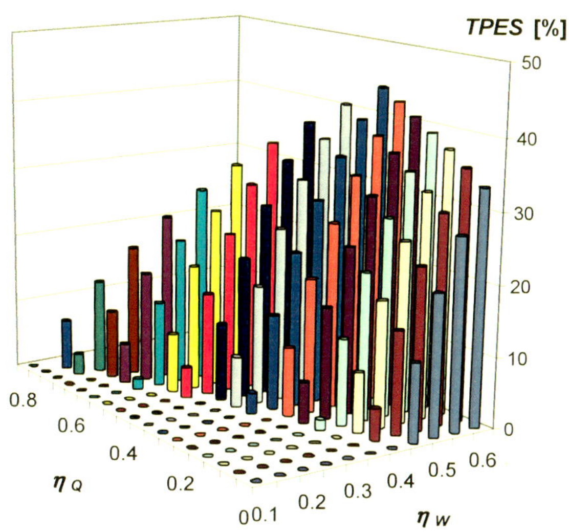

b) $COP^{\mathrm{WARG}} = 1.5$

Figure 7.3. *TPES* plot (with $\eta_e^{\mathrm{SP}} = 0.4$, $COP^{\mathrm{SP}} = 4$) for electricity-and-cooling cogeneration, with: a) single-effect absorption chiller ($COP^{\mathrm{WARG}} = 0.65$); b) triple-effect absorption chiller ($COP^{\mathrm{WARG}} = 1.5$).

Considering single-effect chillers (Figure 7.3a), energy saving up to about 40% can be obtained, but above all for high electrical efficiency levels. However, an MT with electrical efficiency of 0.3 could yield no energy saving at all, while an improved one with $\eta_W = 0.35$ could yield *TPES* up to 9%, but only operating with $\eta_Q = 0.35 \div 0.55$. An ICE could perform better, reaching up to 25% of energy saving, but operating with very high levels of both electrical and thermal efficiencies (equal to 0.45). Also, for these two technologies all the efficiency points considered are consistent with the firing characteristics of the single-effect chiller, for which hot water at a temperature around 90 °C might be sufficient (Section 3.3.4).

On the other hand, a triple-effect chiller might bring quite higher energy savings. However, not all the points shown in Figure 7.3b are consistent with the chiller firing characteristics if steam at high temperature were needed (Section 3.3.4). In particular, considering typical steam generation characteristics of MTs and ICEs, *TPES* values up to 20% and 25% could be reasonably reached. A possible alternative for MTs would be to fire directly the chiller with the exhaust gases [MoA98][MaC01][MoC01][ViB07].

For higher reference values η_e^{SP}, for instance equal to 0.5, basically the DMG systems considered could never feature a positive *TPES* with this plant configuration, with any absorption chiller technology. The situation would not change substantially even if lowering the COP^{SP} value from 4 to 3. Further analyses on energy saving for general trigeneration cases entailing seasonal trigeneration as a sub-case are presented below.

7.2.3. Incremental assessment

Alternative evaluations of the CHP-WARG seasonal trigeneration operation could be run through the $\overline{ITHR}$ (with no allowance for *thermal* production), so yielding

$$EITHR = \frac{\eta_e^{SP} \cdot COP^{SP} - \eta_Q \cdot COP^{WARG}}{\eta_e^{SP} \cdot COP^{SP} \cdot \eta_W} \tag{7.8}$$

$$CITHR = \frac{\eta_e^{SP} - \eta_W}{\eta_e^{SP} \cdot COP^{WARG} \cdot \eta_Q} \tag{7.9}$$

Figure 7.4 and Figure 7.5 plot the *EITHR* for different efficiency values of the cogenerator and different *COPs* for the WARG, namely 0.65 (single-effect) and 1.1 (double-effect). With the lower *COP* value, the electricity production "effectiveness" break-even point relative to separate production ($\eta_e^{SP} = 0.4$) is obtained with prime mover electrical efficiency of about 0.3 combined with thermal efficiency of about 0.4. With $COP^{WARG} = 1.1$, the results are much better, and a prime mover with $\eta_W = 0.3$ and $\eta_Q = 0.25$ is already more effective than in SP, in terms of equivalent electricity generation. However, again limitations in the chiller firing characteristics might apply (Section 3.3.4).

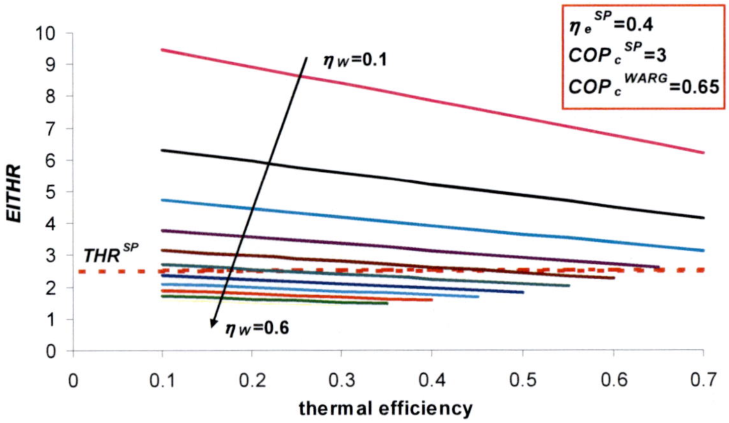

Figure 7.4. *EITHR* for seasonal trigeneration as a function of η_W and η_Q ($\eta_e^{SP} = 0.4$, $COP^{SP} = 3$, $COP^{WARG} = 0.65$).

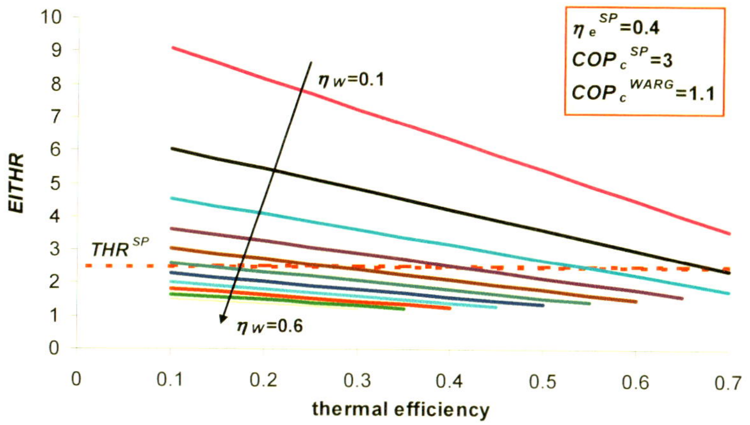

Figure 7.5. *EITHR* for seasonal trigeneration as a function of η_W and η_Q (η_e^{SP} = 0.4, COP^{SP} = 3, COP^{WARG} = 1.1).

Figure 7.6 and Figure 7.7 plot the *CITHR* again for different efficiency values of the cogenerator and different *COPs* for the WARG (0.65 and 1.1). In both cases, cooling production is very effective, so that for electrical efficiencies above 0.4 the indicator becomes even negative. By comparing the *CITHR* with the CHR^{WARG} obtained if the WARG were fired by a separate production thermal source (energy chain model in Figure 5.2), it can be appreciated from Figure 7.6 and Figure 7.7 how, when operating in combination with CHP units and thus exploiting cogenerated heat, the WARG is far more effective than when direct fired. In addition, for single-effect WARG, with electrical efficiencies about 0.3÷0.35 and thermal ones in the range 0.2÷0.4, it is possible to perform even better than the separate production with an electric chiller (CHR^{SP} = 0.83, if considering η_e^{SP} = 0.4 and COP^{SP} = 3). These results are consistent with the energy saving break-even analyses discussed in Section 7.2.1.

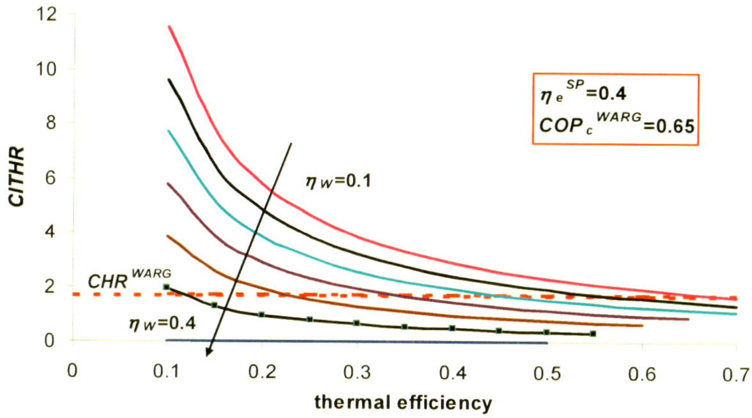

Figure 7.6. *CITHR* for seasonal trigeneration as a function of η_W and η_Q (η_e^{SP} = 0.4, COP^{WARG} = 0.65).

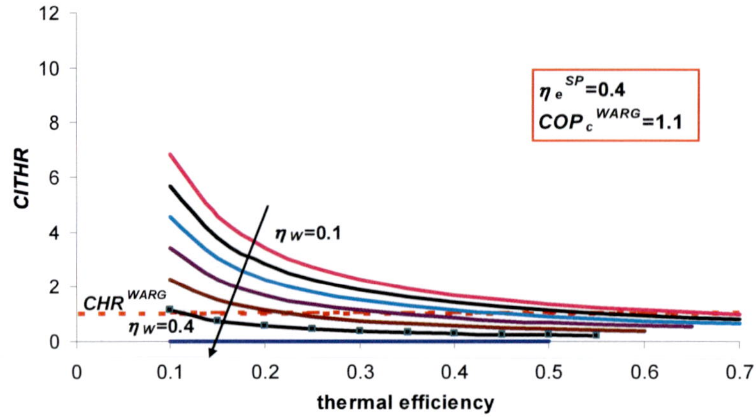

Figure 7.7. *CITHR* for seasonal trigeneration as a function of η_W and η_Q ($\eta_e^{SP} = 0.4$, $COP^{WARG} = 1.1$).

7.3. Trigeneration of Electricity, Heat and Cooling Power in a CHP-WARG Scheme

With respect to the cogeneration case, difficulty in trigeneration assessment arises due to the increased number of variables. In this section, parametric models are then developed for assessments based on the *TPES* and the $\overline{ITHR}$ indicators, which can be the starting point for further analyses.

As a matter of nomenclature, we point out again that in general the subscript y is used for the entries related to the cogenerative prime mover, while z refers to the entries related to the *overall* trigeneration plant black-box. However, provided that no loss of clarity occurs, the notation is kept as "light" as possible, avoiding excessive use of subscripts and superscripts when the context is clear enough.

7.3.1. Trigeneration plant model and energy flows

In Section 7.2, the case of a CHP plant where all the produced heat goes to feed an absorption chiller, so making up electricity-and-cooling cogeneration, has been discussed. This can be seen as a particular case of a trigeneration system where electricity, heat and cooling power are produced at the same time starting from a unique fuel thermal source. Indeed, the full model of a CHP plant coupled with an absorption chiller (the same model also applies to an adsorption chiller) driven by cogenerated heat is shown in Figure 4.16, and is recalled here in Figure 7.8 with the relevant notation.

More specifically, in Figure 7.8, Q_z is the actual plant thermal output, Q_R is the cogenerated thermal input that fires the absorption chiller to produce the cooling output R_z, and $W_y = W_z$ is the cogenerated/trigenerated electricity.

Also in this case, we recall the relevant *dispatch factor* α_R [Man06] for heat-to-cooling production (see Section 4.6.4.1) in order to describe the internal energy flows. More specifically, α_R represents the share of the overall cogenerated Q_y used to produce Q_R, that is, $Q_R = \alpha_R Q_y$. Hence, the total cogenerated heat Q_y splits into two contributions, for which the following expressions apply

$$Q_y = \alpha_R \cdot Q_y + (1-\alpha_R) \cdot Q_y = Q_R + Q_z = \frac{R_z}{COP^{\text{WARG}}} + Q_z = \left(\frac{\alpha_R}{1-\alpha_R} + 1 \right) \cdot Q_z \qquad (7.10)$$

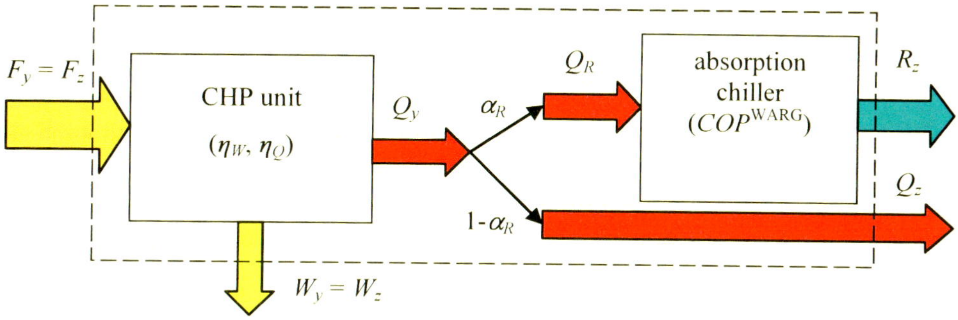

Figure 7.8. Scheme for simultaneous trigeneration of heat, cooling and electrical power.

Therefore, the fuel thermal power input can be written as

$$F_z = F_y = \left(\frac{R_z}{COP} + Q_z \right) \frac{1}{\eta_Q} = \frac{1}{1-\alpha_R} \frac{Q_z}{\eta_Q} = \frac{W_z}{\eta_W} \qquad (7.11)$$

The fuel thermal power needed for generating the same outputs in separate production is then

$$F^{\text{SP}} = \frac{W_z}{\eta_e^{\text{SP}}} + \frac{R_z}{COP^{\text{SP}} \cdot \eta_e^{\text{SP}}} + \frac{Q_z}{\eta_t^{\text{SP}}} = \frac{\eta_W}{\eta_Q} \frac{Q_z}{1-\alpha_R} \frac{1}{\eta_e^{\text{SP}}} + \frac{COP^{\text{WARG}}}{COP^{\text{SP}}} \frac{\alpha_R}{1-\alpha_R} \frac{Q_z}{\eta_e^{\text{SP}}} + \frac{Q_z}{\eta_t^{\text{SP}}} \qquad (7.12)$$

7.3.2. Energy saving break-even conditions

It is easy to show that the break-even condition for fuel saving with respect to separate production ($F_z = F^{\text{SP}}$) can be expressed, for instance, in terms of the CHP electrical efficiency as [HeB02]

$$\eta_W^* = \eta_e^{\text{SP}} - \left(\alpha_R \frac{COP^{\text{WARG}}}{COP^{\text{SP}}} + \frac{\eta_e^{\text{SP}}}{\eta_t^{\text{SP}}} (1-\alpha_R) \right) \cdot \eta_Q \qquad (7.13)$$

so that there is energy saving if the prime mover features an electrical efficiency higher than η_W^*.

The expression (7.13) is the most general formula for CHP-WARG systems. It is interesting to notice as the *limit cases* with α_R equal to 0 and 1 correspond respectively to the discussed "sub-cases" of classical *electricity-and-heat cogeneration* and *electricity-and-cooling cogeneration*:

$$\alpha_R = 0 \Rightarrow \eta_W^* = \eta_e^{SP} \cdot \left(1 - \frac{\eta_Q}{\eta_t^{SP}}\right) \quad (7.14)$$

$$\alpha_R = 1 \Rightarrow \eta_W^* = \eta_e^{SP} - \eta_Q \frac{COP^{WARG}}{COP^{SP}} \quad (7.15)$$

which coincide respectively with the relations (6.9) and (7.4) found earlier.

Figure 7.9 shows the plot of η_W^* in function of η_Q, with α_R as the curve parameter. The reference values for the separate production are $\eta_e^{SP} = 0.4$, $\eta_t^{SP} = 0.9$, and $COP^{SP} = 3$. The COP^{WARG} is set to 0.65 (single-effect). The plot takes into account an upper limit of about 90% for the $EUF = \eta_W + \eta_Q$. Feasibility areas for typical families and characteristics of MTs and ICEs available in the market are shown as well.

As a result of the efficiencies considered, the higher is the share of heat used to produce cooling power, the higher is the break-even electrical efficiency to reach energy saving. On the other hand, the break-even point curves decrease if the cogeneration thermal efficiency increases.

A similar plot is shown in Figure 7.10 for $COP^{WARG} = 1.1$ (double-effect chiller). In this case, the break-even conditions are easy to reach with respect to a single-effect WARG. However, again not all the thermal power recoverable might actually be of sufficient thermodynamic quality to fire the chiller. As a further point to recall, the thermodynamic requirements for the hot water/steam/exhausts feeding the chiller might also impact on the thermal efficiency of the engine (the higher the required temperature, the less the heat recovered, the less the thermal efficiency). This aspect is further illustrated in Section 7.3.4. In addition, a more accurate model, suitable for off-design simulations, might also consider different values for the COP^{WARG}, variable with the system operating point (and therefore with thermal and electrical efficiencies of the prime mover). In general, with such plots and the actual feasibility areas of the cogenerative prime mover, it is possible to evaluate in every condition whether (and to what extent) or not the energy system allows energy saving.

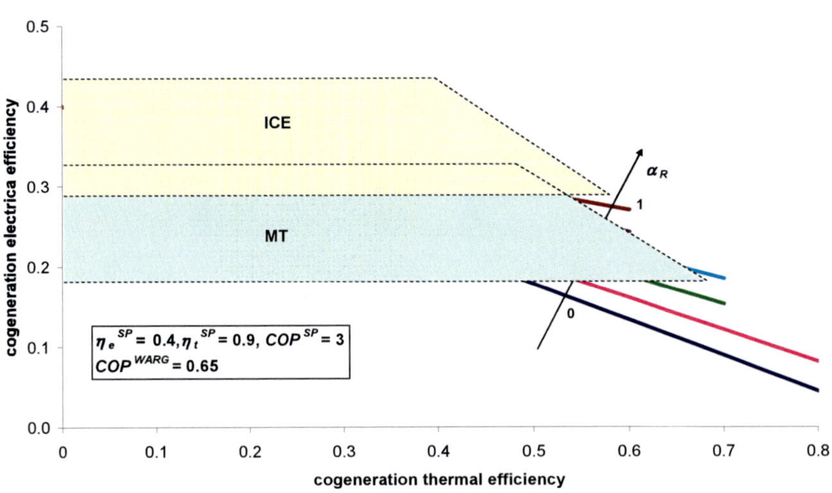

Figure 7.9. Energy saving break-even conditions for trigeneration with single-effect chiller.

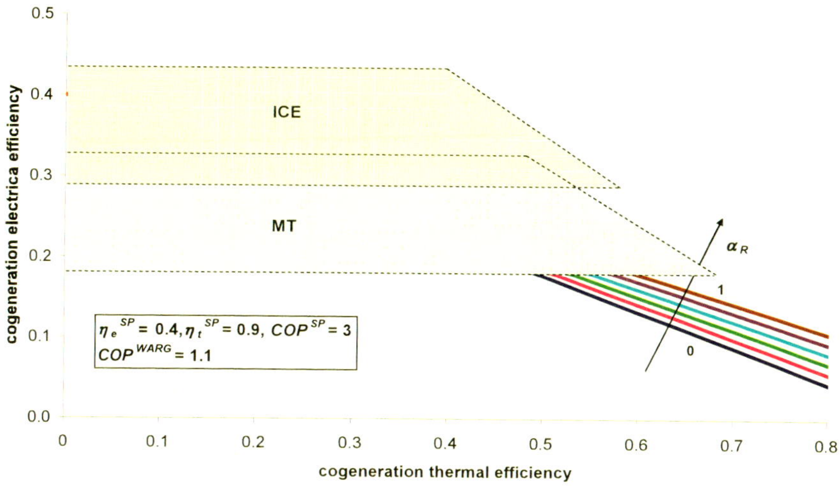

Figure 7.10. Energy saving break-even conditions for trigeneration with double-effect chiller.

Similar plots might be drawn for other reference figures for separate production (*e.g.*, with $\eta_e^{SP} = 0.55$ and/or higher COP^{SP}) and help the designer evaluate the energy saving potentiality of the trigeneration system at a planning level.

7.3.3. Trigeneration primary energy saving assessment

From (7.11) and (7.12), it is easy to calculate the *TPES* in the case of trigeneration with a CHP unit and an absorption/adsorption chiller as

$$TPES = \frac{F^{SP} - F_z}{F^{SP}} = 1 - \frac{1}{\dfrac{\eta_W}{\eta_e^{SP}} + \dfrac{\eta_Q}{\eta_t^{SP}} - \alpha_R \cdot \left(\dfrac{\eta_Q}{\eta_t^{SP}} - \dfrac{COP^{WARG}}{COP^{SP}} \dfrac{\eta_Q}{\eta_e^{SP}} \right)} \qquad (7.16)$$

Again, α_R values equal to 0 and 1 lead to the relations found for heat-and-electricity and cooling-and-electricity, respectively.

In Figure 7.11, the *TPES* is plotted as a function of α_R for given separate production reference figures ($\eta_e^{SP} = 0.4$, $\eta_t^{SP} = 0.9$, and $COP^{SP} = 3$) and different values of η_W and η_Q of the CHP unit, and of the COP^{WARG} (0.65 and 1.1). In particular, the pair ($\eta_W = 0.40$, $\eta_Q = 0.45$) could be representative of rated conditions for an ICE, whereas the pair ($\eta_W = 0.30$, $\eta_Q = 0.50$) could refer to the rated conditions of an excellent microturbine. Potentially, an ICE is more profitable than an MT in terms of energy saving, thanks to the higher weight of the electrical efficiency over the thermal one. In addition, the curves corresponding to a higher COP^{WARG} perform much better than the relative ones at lower efficiency, although not all the points of the curve corresponding to $COP^{WARG} = 1.1$ might be feasible.

The same plots are evaluated in Figure 7.12, but for a different value of COP^{SP}, namely, equal to 4. The "negative" impact, obvious consequence of an increase in a reference value for an SP parameter, is more relevant for the MT characteristics. Indeed, for MTs the thermal production (and consequently the cooling production from absorption groups) has a higher weight than for ICEs, which are more "unbalanced" towards electrical production. In general, the selection of the cooling production reference should be evaluated, as already said, on the basis of the application (condenser and evaporator temperatures). Of course, also the *COP* of the WARG should be evaluated under the actual working conditions corresponding to these temperatures.

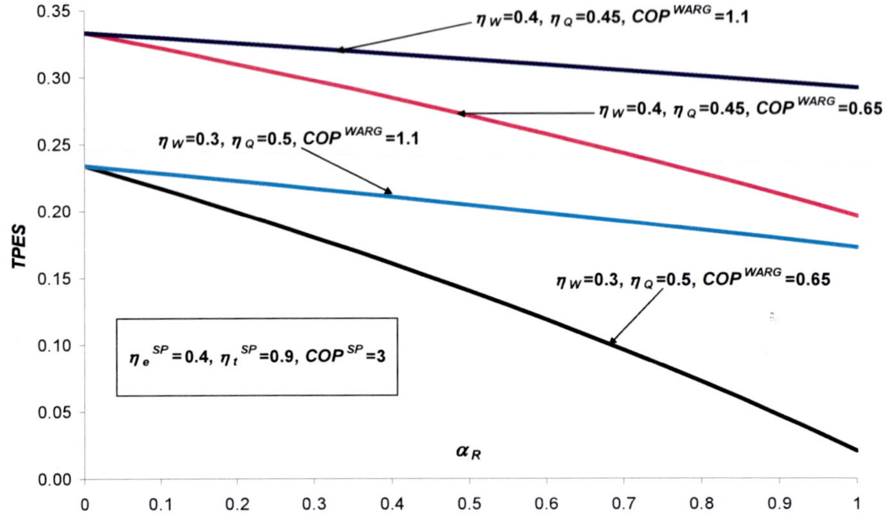

Figure 7.11. *TPES* as a function of α_R (η_e^{SP}=0.4, η_t^{SP}=0.9, and COP^{SP}=3).

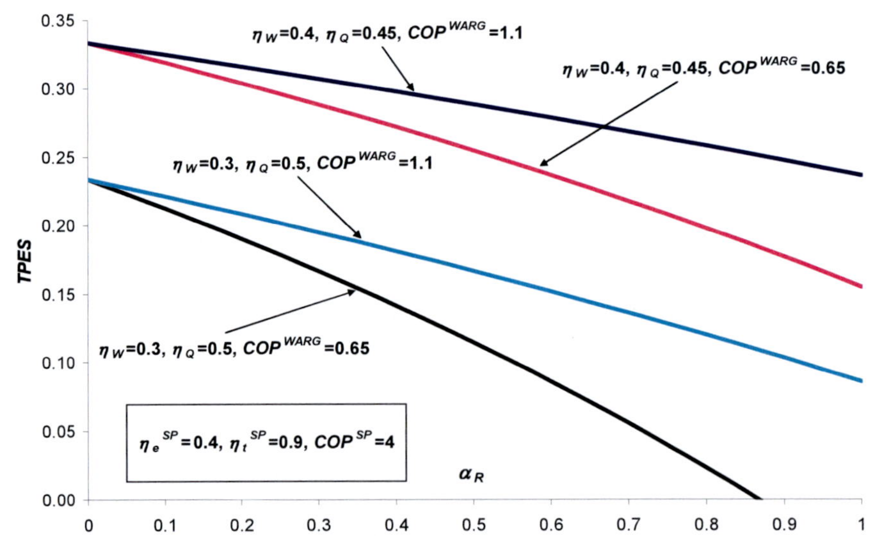

Figure 7.12. *TPES* as a function of α_R (η_e^{SP}=0.4, η_t^{SP}=0.9, and COP^{SP}=4).

Figure 7.13 and Figure 7.14 show the same type of plots, with 3 and 4 as COP^{SP} reference values, respectively, and with electrical efficiency reference value set to 0.55. Now, all the energy saving values in any condition are much worse than in the previous cases analysed, and the performance gap between ICEs and MTs is consistently shrunk. Indeed, when the electrical reference is made "tougher", the technology more "pushed" towards electrical production (the ICE) is more penalised than the technology (relatively) more inclined towards thermal production (the MT). In this sense, the MT would further benefit from a reduction in the thermal production reference value.

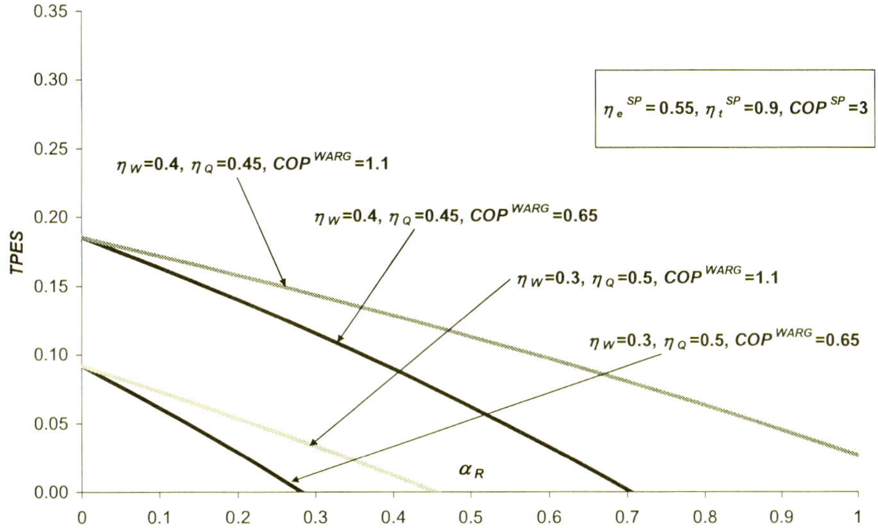

Figure 7.13. *TPES* as a function of α_R (η_e^{SP}=0.55, η_t^{SP}=0.9, and COP^{SP}=3).

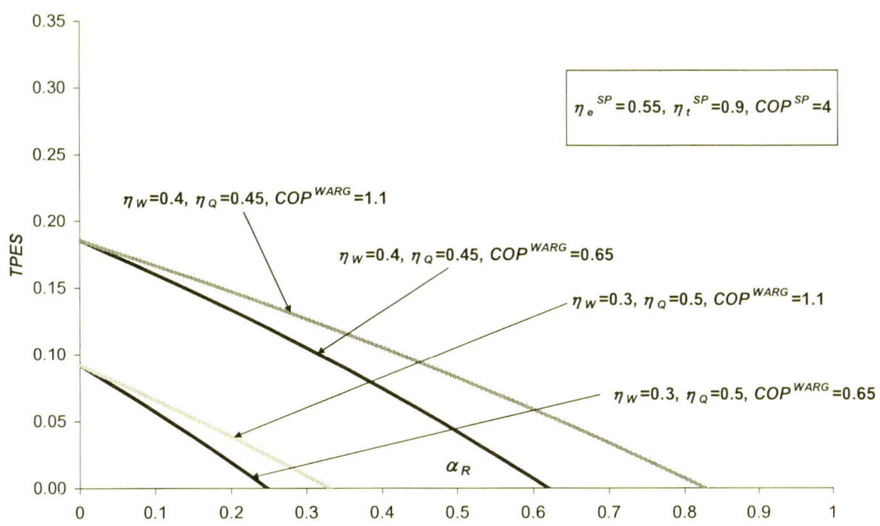

Figure 7.14. *TPES* as a function of α_R (η_e^{SP}=0.55, η_t^{SP}=0.9, and COP^{SP}=4).

7.3.4. Further issues related to CHP-chiller coupling

As discussed in Section 3.3.4, the firing source characteristics play a significant role in a WARG, with minimum firing temperatures necessary to obtain the rated chiller performance. From this point of view, summarizing the concepts of Section 3.3.4.4, typically water-lithium bromide single-effect *absorption* chillers ($COP \cong 0.7$) need hot water sources at about 80÷90 °C, while double-effect absorption chillers ($COP \cong 1.2$) need steam sources at about 130÷140 °C; triple-effect chillers ($COP \cong 1.5$) under commercialization need firing temperatures above 200 °C, and as such are likely to be mostly direct-fired. *Adsorption* chillers ($COP \cong 0.4$) may operate with lower-temperature sources, of about 50 °C. These firing temperature constraints also impact on the thermal efficiency of the CHP unit to be coupled to the chiller, since the higher is the temperature of the heat to provide, the lower is the thermal efficiency. As a consequence, the overall CCHP system performance is affected by the chiller thermal constraints.

In order to address this issue in more details, let's refine our model and consider three different chillers, namely, an *adsorption* chiller, a *single-effect absorption* chiller and a *double-effect absorption* chiller, coupled to natural gas-fuelled ICEs or MTs. Typical efficiency values for the equipment, referred to electrical sizes in the range 100÷500 kW_e, are reported in Table 7.1. Different values of η_W and η_Q of the CHP unit correspond to typical figures for the technologies under study. In addition, different thermal efficiencies are considered when the given CHP unit is used to fire the different chillers, estimated from manufacturers' data and accounting for the actual efficiency drop due to higher-temperature heat generation. In particular, for an ICE firing a double-effect WARG, only the heat from the exhaust gases can generate steam, so the relevant thermal efficiency corresponds only to this share, while the overall thermal efficiency for hot water production also benefits of the heat recovery shares from cooling and lubricant circuits (Section 3.1.2). The case with triple-effect chillers would correspond to a higher decrease in thermal efficiency if steam is the input. However, in case also direct-firing through exhaust gases might be considered, above all for MTs and relatively larger GTs, with relevant thermal efficiency penalty.

With reference to the considered equipment, the *TPES* is plotted in Figure 7.15 (MT) and Figure 7.16 (ICE). The separate production references refer to average values, namely, equal to $\eta_e^{SP} = 0.4$, $\eta_t^{SP} = 0.8$, and $COP^{SP} = 3$.

Table 7.1. Typical efficiency values for small-scale trigeneration system components

CHP unit	*electrical efficiency*	*thermal efficiency*		
		adsorption chiller ($COP = 0.4$)	single-effect WARG ($COP = 0.7$)	double-effect WARG ($COP = 1.2$)
MT	0.3	0.55	0.5	0.4
ICE	0.35	0.55	0.5	0.2

7.3.5. General comments on the *TPES* results

From the results shown in Figure 7.15 and Figure 7.16, it emerges that exploiting adsorption chillers is most effective when a little share of cooling is produced, corresponding to relatively low α_R values, owing to the higher thermal efficiency exhibited by the CHP units with respect to firing the other types of chillers. However, when α_R increases, higher energy benefits are brought by single-effect and double-effect absorption chillers, owing to their higher *COP*. A double-effect WARG represents indeed the best option for electricity-and-cooling generation for the MT case, while the ICE-related performance is affected by the relatively smaller thermal efficiency for steam generation. However, in the model used to assess the ICE-double-effect WARG configuration, only the share for firing the chiller has been considered in the *TPES* calculation; accounting for the production of hot water for thermal purposes from jacket coolant and lubricant circuits would increase the overall performance.

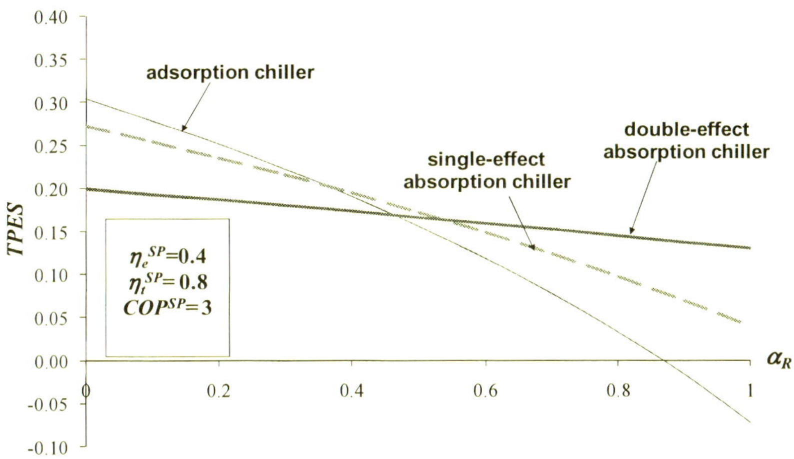

Figure 7.15. *TPES* with an MT with different types of chillers and off-design characteristics.

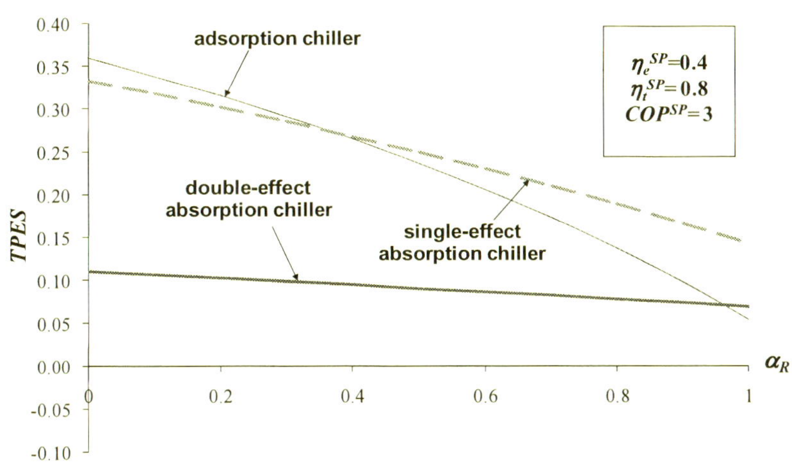

Figure 7.16. *TPES* with an ICE with different types of chillers and off-design characteristics.

From the figures analysed, it can be seen how in general the energy saving decreases by increasing the share of cooling generation. This is due to the efficiency values considered, in particular for electric reference chillers and adsorption/absorption chillers. In general, higher efficiency values for absorption chillers, for instance from triple-effect ones, could improve the energy saving brought by the overall trigeneration scheme (provided that there are no excessive thermal efficiency penalties in the CHP). However, although the *TPES* value decreases with higher cooling production (and then, by passing from sheer cogeneration of electricity and heat to sheer cogeneration of electricity and cooling), it must be acknowledged that waste heat is being effectively used, which otherwise would have been disposed. In this respect, CHP-WARG schemes are more efficient than alternative schemes such as CHP-CERG ones, as for instance pointed out in [RoL05] through 2nd law-based approaches. On the other hand, analyses based on energy saving assessment through the *TPES* indicator highlight the energy performance of the DMG system with respect to separate production. Hence, negative or marginally positive values of energy saving for the DMG system might hint to opt for separate production high-efficiency solutions rather than combined production. Of course, this all depends on the reference SP model adopted and, accordingly, on the purpose of the study.

It is also important to reckon that these general considerations are mostly applicable for a first rough evaluation. Actual evaluation of the plant needs to be carried out by resorting to time-domain simulations, taking into account the time-related load variations and the possible control strategies, so that a map of performance, for different operation point of the energy system, may be drawn. A more suitable evaluation of the *TPES*, therefore, should be based on annual values of the entries in the formulas, of course calculated after running simulations with a suitable (possibly hourly) time step.

Likewise, the comparison between MTs and ICEs cannot be exhaustive on the basis of *a priori* evaluation of the different rated characteristics of the two technologies. Indeed, the main differences arise when coming to talk about operational characteristics, especially for the MT. Other considerations, such as the potential higher modularity of MTs over ICEs, should be better considered case by case. For instance, if a cluster of MTs were at disposal, some of them might be operated to drive absorption chillers, whereas the rest might be used only for thermal cogeneration. In this way, α_R would be equal to 1 for the "chiller-drivers" and to 0 for the others, so that the overall characteristics for the single machine as seen in the figures considered could not be taken as granted for the whole cluster.

7.3.6. Incremental trigeneration assessment

From the point of view of incremental indices, it is easy to prove the following relations:

$$EITHR = \frac{F_z - \dfrac{(1-\alpha_R)\eta_Q F_z}{\eta_t^{\mathrm{SP}}} - \dfrac{\alpha_R COP^{\mathrm{WARG}}\eta_Q F_z}{\eta_e^{\mathrm{SP}} COP^{\mathrm{SP}}}}{\eta_W F_z}$$

$$= \eta_W^{-1} - (1-\alpha_R)\frac{\lambda_y}{\eta_t^{\mathrm{SP}}} - \alpha_R \frac{COP^{\mathrm{WARG}}}{COP^{\mathrm{SP}}}\frac{\lambda_y}{\eta_e^{\mathrm{SP}}} \tag{7.17}$$

$$TITHR = \frac{F_z - \dfrac{\eta_W F_z}{\eta_e^{\mathrm{SP}}} - \dfrac{\alpha_R COP^{\mathrm{WARG}}\eta_Q F_z}{\eta_e^{\mathrm{SP}} COP^{\mathrm{SP}}}}{(1-\alpha_R)\eta_Q F_z} = \frac{\eta_Q^{-1} - \dfrac{\lambda_y^{-1}}{\eta_e^{\mathrm{SP}}} - \alpha_R \dfrac{COP^{\mathrm{WARG}}}{COP^{\mathrm{SP}}}\dfrac{1}{\eta_e^{\mathrm{SP}}}}{(1-\alpha_R)} \tag{7.18}$$

$$CITHR = \frac{F_z - \dfrac{\eta_W F_z}{\eta_e^{\mathrm{SP}}} - \dfrac{(1-\alpha_R)\eta_Q F_z}{\eta_t^{\mathrm{SP}}}}{\alpha_R COP^{\mathrm{WARG}}\eta_Q F_z} = \frac{\eta_Q^{-1} - \dfrac{\lambda_y^{-1}}{\eta_e^{\mathrm{SP}}} - \dfrac{(1-\alpha_R)}{\eta_t^{\mathrm{SP}}}}{\alpha_R COP^{\mathrm{WARG}}} \tag{7.19}$$

Plots of the *EITHR* (7.17) against α_R, with the cogeneration ratio λ_y as the curve parameter and different values of η_W, are shown in Figure 7.17 and Figure 7.18 for a single-effect chiller. Figure 7.19 shows the same kind of plots, with η_W equal to 0.45 for a double-effect chiller.

As general considerations on the pictures shown, the plant, from the electricity production point of view, performs better with lower values of α_R and higher values of λ_y. In particular, the dependence on α_R is more accentuated for higher values of the cogeneration ratio and becomes lower when adopting a double-effect absorption chiller rather than a single-effect one. The explanations of these characteristics can be inferred from the general concepts discussed and are left to the interested reader.

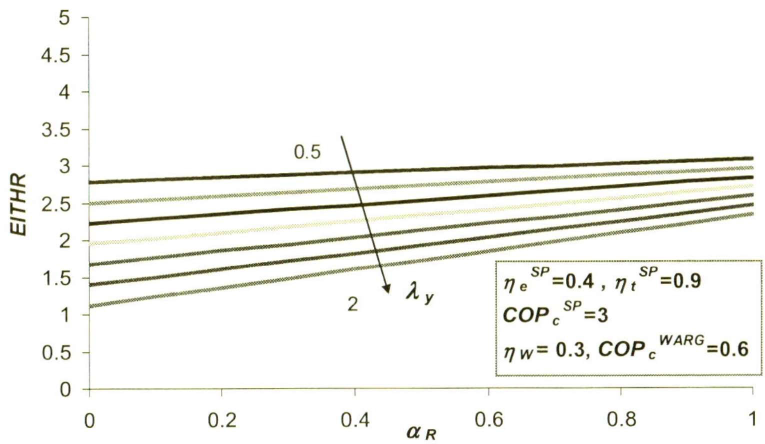

Figure 7.17. CHP-WARG trigeneration system *EITHR* as a function of α_R and λ_y (η_e^{SP} =0.4, η_e^{SP} =0.9, COP_c^{SP}=3, COP_c^{WARG}=0.6, η_W =0.3).

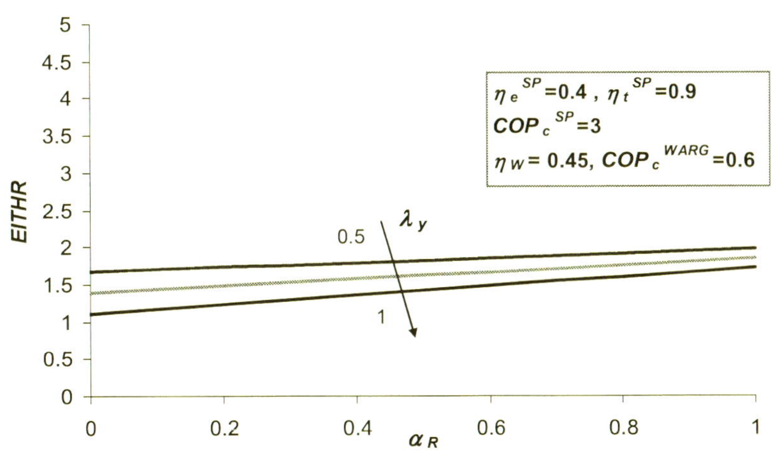

Figure 7.18. CHP-WARG trigeneration system *EITHR* as a function of α_R and λ_y (η_e^{SP} =0.4, η_e^{SP} =0.9, COP_c^{SP}=3, COP_c^{WARG}=0.6, η_W =0.45).

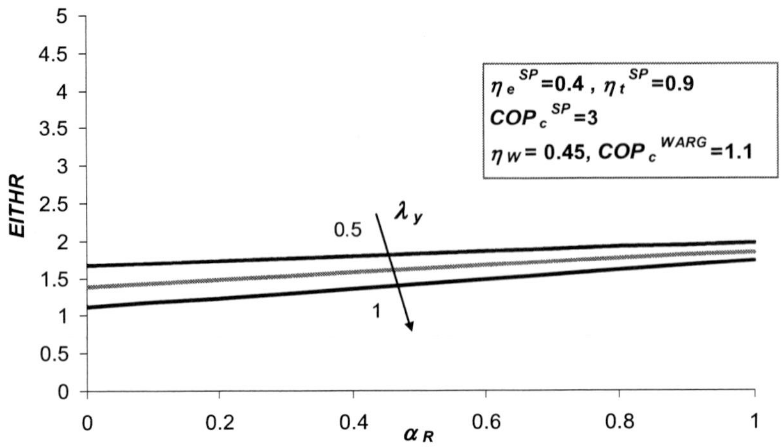

Figure 7.19. CHP-WARG trigeneration system *EITHR* as a function of α_R and λ_y (η_e^{SP} =0.4, η_e^{SP} =0.9, COP_c^{SP}=3, COP_c^{WARG}=1.1, η_W =0.45).

7.4. Trigeneration and Parallel Cooling Production: CHP-DFC Systems

7.4.1. Plant model and energy balances

Parallel (separate) cooling generation in direct-fired chillers can be a sound alternative in the presence of specific energy market conditions and depending on specific loading configurations, according to what discussed in the lambda analysis framework (Section 4.5.1). Focusing on engine-driven chillers, the trigeneration plant would be composed of two or more prime movers, of which one is an engine completely "dedicated" to cooling production. The same black-box model could be used for both a CHP-GARG and a CHP-EDC plant (Section 5.2.6). However, EDCs exhibit the practical possibility of easily cogenerating heat from the engine driving the compression chiller, which allows enhanced energy performance with respect to conventional heating and cooling systems [SuW04][LaN06b][Sun07]. This might be an interesting solution in those cases where a certain overall power is required, so that the share of the capacity among more units is a viable option also for reliability reasons. For instance, suitable applications refer to cases in which consistent cooling power and relatively high thermal power is required simultaneously (*e.g.*, in supermarket applications [MaT02]). Therefore, instead of connecting a group of prime-movers (namely ICEs) to the plant electrical hub and use a part of the electricity to drive a CERG, one of them is directly connected to the chiller, so optimising the off-design performance and avoiding the conversion losses from mechanical to electrical energy.

A DMG system based on CHP and EDC units is shown in Figure 7.20. The CHP unit produces heat and electricity, whereas the EDC unit produces cooling and cogenerated heat from the engine exhausts and jacket coolant circuits [ASH00]. The EDC is characterized by its cooling *COP* and its recovery thermal efficiency η_Q^{EDC} (Section 3.3.7.2). Considering the overall system as a black box, an effective model for DMG analysis consists of representing the relevant energy flows as a function of the shares of fuel (natural gas) that supply the CHP unit and the EDC. In particular, it is possible to introduce a *fuel-to-cooling* dispatch factor [Man06], again dubbed α_R, representing the relative share of natural gas going to supply the EDC. Hence, $\alpha_R = 1$ means that all the fuel supplies the EDC, while $\alpha_R = 0$ means that all the fuel supplies the CHP unit.

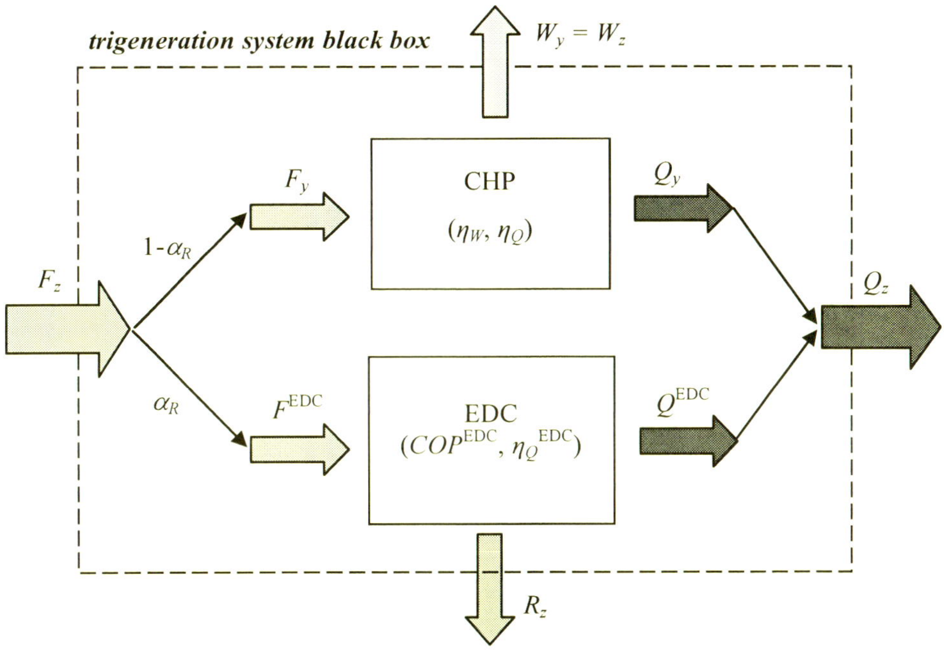

Figure 7.20. Black-box model for parallel cooling trigeneration scheme with CHP and EDC.

From the relevant energy balances, and considering that $W_y = W_z$ and $Q_y + Q^{EDC} = Q_z$, the overall fuel thermal input is expressed as

$$F_z = F_y + F^{EDC} = (1-\alpha_R)\cdot F_z + \alpha_R \cdot F_z = \frac{W_z}{\eta_W} + \frac{R_z}{COP^{EDC}} = \frac{Q_y}{\eta_Q} + \frac{Q^{EDC}}{\eta_Q^{EDC}} \tag{7.20}$$

whereas the input in the case of separate production (considering the aggregation of heat from EDC and CHP units) would be

$$F^{SP} = \frac{W_z}{\eta_e^{SP}} + \frac{Q_y + Q^{EDC}}{\eta_t^{SP}} + \frac{R_z}{\eta_e^{SP}\cdot COP^{SP}} \tag{7.21}$$

7.4.2. Primary energy saving assessment

Given the separate production references, also in this case the *TPES* can be written in function of only the relevant plant efficiencies and the dispatch factor, as

$$TPES = 1 - \frac{1}{\frac{\eta_W}{\eta_e^{SP}} + \frac{\eta_Q}{\eta_t^{SP}} - \alpha_R \cdot \left(\frac{\eta_Q - \eta_Q^{EDC}}{\eta_t^{SP}} + \frac{\eta_W - \frac{COP^{EDC}}{COP^{SP}}}{\eta_e^{SP}} \right)} \tag{7.22}$$

A preliminary evaluation of the energy saving brought by adopting a CHP-EDC system could be carried out for instance by plotting the *TPES* in function of the dispatch factor α_R. In particular, the dispatch factor can be related to different load configurations, and more specifically to the cooling/electricity output ratio, according to:

$$\frac{R_z}{W_z} = \frac{COP^{EDC}}{\eta_W} \frac{\alpha_R}{1-\alpha_R} \tag{7.23}$$

Additional parametric analyses on the entries of (7.22) could give indications on the energy performance of adopting different equipment with different energy characteristics. For instance, it is possible to draw *TPES* curves with different values for the CHP thermal and electrical efficiencies as the curve parameters. For a comparative evaluation, the *EUF* can be kept constant, as done previously, for instance equal to 0.8 (a reasonable value for hot water production for most technologies at a small-scale level). In this way, again, a rough comparison among available DG technologies (such as MTs and ICEs) can be run. The values COP^{EDC} = 1.5 and η_Q^{EDC} = 0.4 (for hot water generation) can be used as average figures for commercial equipment.

The outcomes of the analysis are shown in Figure 7.21, plotted considering the same reference efficiencies for separate production as in Section 7.3.4.

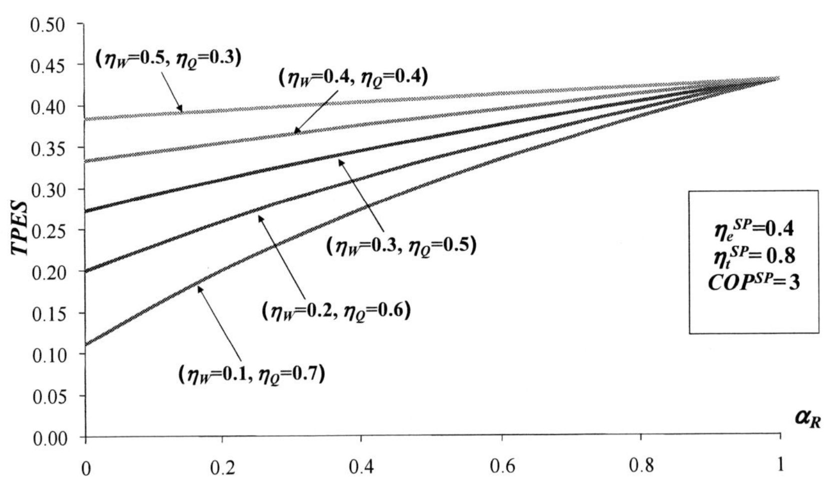

Figure 7.21. Primary energy saving characteristics of DMG systems with CHP and EDC units.

The performance of the combined DMG system is in general significantly good, owing to the excellent cooling generation performance of the EDC and to the additional benefit brought by the thermal recovery from the chiller, besides the intrinsic cogeneration benefits. More specifically, the *TPES* increases when α_R increases, so that this kind of plant might be suitable for installations with relatively high cooling production also from an energy saving standpoint. The maximum value $TPES \cong 0.43$ is reached at $\alpha_R = 1$ (no electricity generation and cogeneration of heat and cooling from the EDC only). Availability of more efficient EDC (with *COP* as high as about 2 [ASH00]) could even improve the *TPES* and thus the overall energy saving from the DMG system.

7.4.3. DMG applications with CHP-GARG combination

CHP-GARG systems could be analysed with the same black-box model as for EDC, but with different efficiency values. In particular, normally GARGs are used with no heat recovery, although it is possible to have heat recovery condensers. In this respect, strictly speaking it would be more adequate to talk about *actual* "trigeneration" only when heat is recovered from the separate cooling plant.

While there might be thermal constraints for triple-effect absorption chillers coupled to CHP units, for direct-fired chillers there will be no such hurdles to their diffusion, so that most probably they will occupy a more important role in the market when triple-effect technologies will be commercially available [Meu02][McE06].

In a CCHP plant made up of a CHP system and a direct-fired cooling plant (Figure 7.22), it is possible to analyse independently the cogeneration side and the cooling side. In particular, the CHP side can be analysed following the lines drawn in Section 6.2. Details on the comparison between GARG (and more in general DFC) and electrical (reference) cooling plants can be found for instance in [MaC08b].

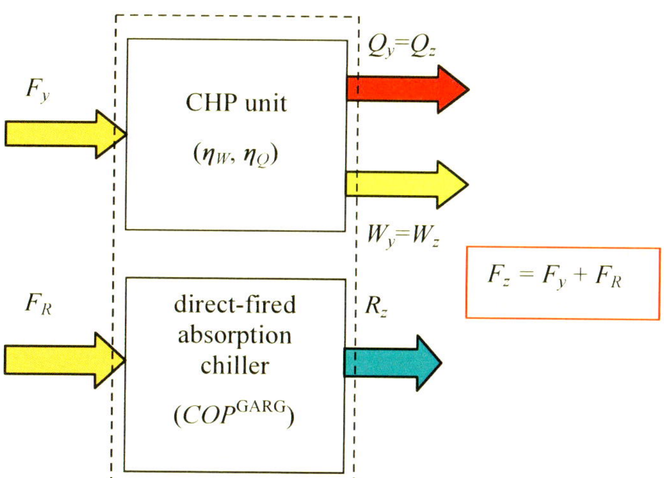

Figure 7.22. Black-box model for trigeneration with direct-fired absorption chiller.

On the other hand, by adopting the black-box model formulation of the *TPES* and with reference to the energy balances in Figure 7.22, it is straightforward to write:

$$TPES = 1 - \frac{F_z}{F^{SP}} = 1 - \frac{F_y + F_R}{F^{SP}} = 1 - \frac{\dfrac{W_z}{\eta_W} + \dfrac{R_z}{COP^{GARG}}}{\dfrac{W_z}{\eta_e^{SP}} + \dfrac{Q_z}{\eta_t^{SP}} + \dfrac{R_z}{\eta_e^{SP} \cdot COP^{SP}}} =$$

$$= 1 - \frac{1 + \dfrac{R_z}{W_z}\dfrac{\eta_W}{COP^{\mathrm{GARG}}}}{\dfrac{\eta_W}{\eta_e^{\mathrm{SP}}} + \dfrac{\eta_Q}{\eta_t^{\mathrm{SP}}} + \dfrac{R_z}{W_z}\dfrac{\eta_W}{\eta_e^{\mathrm{SP}} \cdot COP^{\mathrm{SP}}}} \quad (7.24)$$

A parametric analysis on the entries involved in the formula (7.24) could be carried out, for instance by plotting the *TPES* as a function of the group R_z/W_z, which basically indicates the relative size of the CHP plant and cooling plant. Figure 7.23, Figure 7.24 and Figure 7.25 show this kind of plots for difference separate production references and with different electrical and thermal efficiencies (with overall efficiency kept constant at EUF = 0.85) as curve parameters.

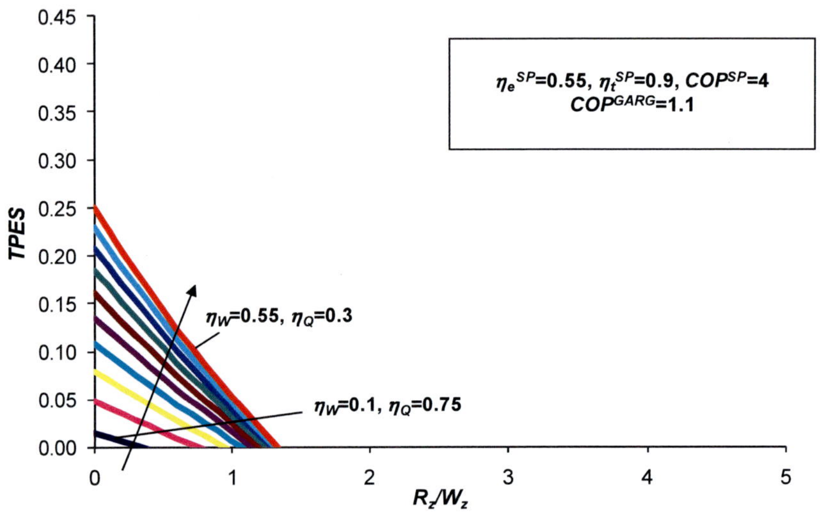

Figure 7.23. CHP-GARG plant *TPES* (η_e^{SP}=0.55, η_t^{SP}=0.9, and COP^{SP}=4).

As a general result, the energy saving strongly depends on the group R_z/W_z, especially for higher COP^{SP}. In particular, whereas in Figure 7.23 and Figure 7.24 the *TPES* decreases when R_z/W_z increases, in the case with the lowest COP^{SP} considered (equal to 2) and η_e^{SP} = 0.4 (Figure 7.25), for the curves with lower electrical efficiency and higher thermal one the *TPES* (7.24) increases along with R_z/W_z. Such low COP^{SP} for the reference CERG might occur in severe outdoor conditions for air-cooled units, whereas the performance of a direct-fired machine is relatively less dependent on the outdoor conditions. For the higher levels of COP^{SP} and η_e^{SP} (Figure 7.25), there are break-even values of R_z/W_z beyond which the energy saving become negative, also depending on the parametric CHP efficiencies.

It is important to point out that the load levels play a key role in the evaluation of the *TPES* (7.24). In particular, the role of the relative values of electrical and cooling load is apparent through the value of the group R_z/W_z, whereas the thermal load is implicitly considered, in that it is related to the electrical one by the cogeneration ratio. In practice, the group R_z/W_z plays here the same role played by the dispatch factor α_R in the CHP-WARG and CHP-EDC trigeneration schemes studied above.

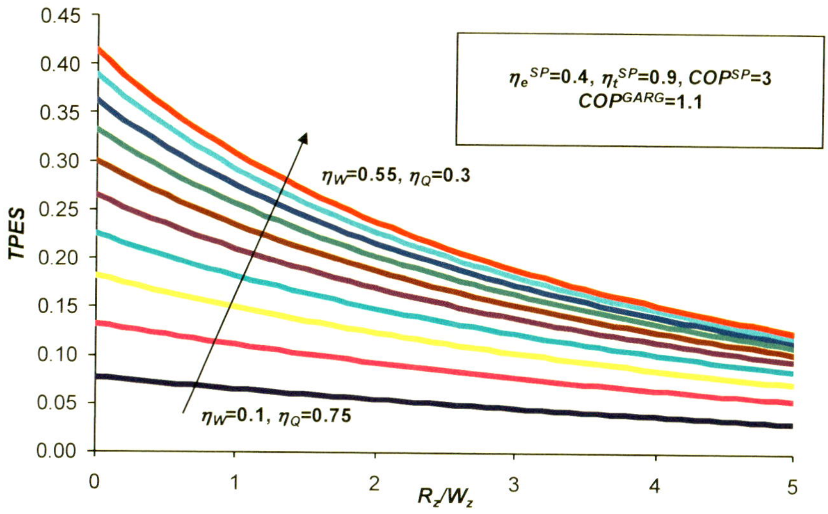

Figure 7.24. CHP-GARG plant *TPES* (η_e^{SP}=0.4, η_t^{SP}=0.9, and COP^{SP}=3).

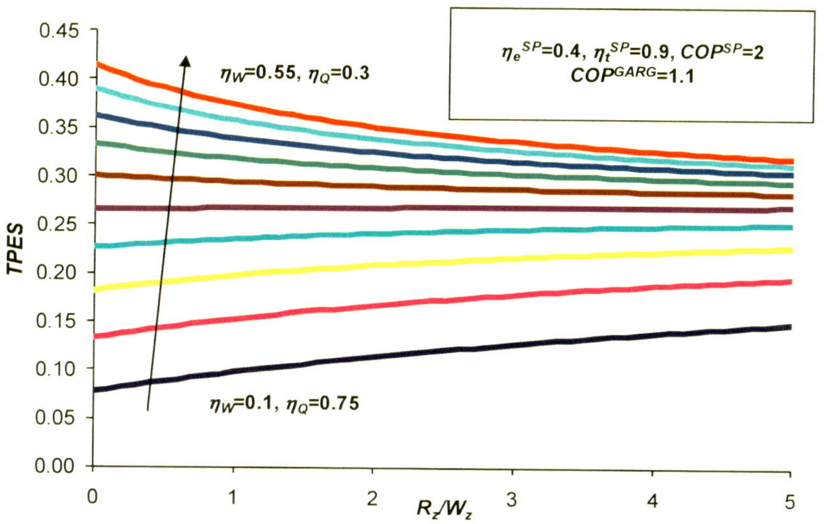

Figure 7.25. CHP-GARG plant *TPES* (η_e^{SP}=0.4, η_t^{SP}=0.9, and COP^{SP}=2).

7.5. Composite Trigeneration Systems

As discussed in Section 3.4.1, heat pumps, and in particular EHPs, can be effectively used to raise the temperature of low-grade heat recovered from a HRC in a cooling generation plant to a level suitable for applications such as air conditioning or domestic hot water generation. Hence, in the presence of simultaneous trigeneration loads, interesting combined schemes for enhanced energy saving could be for instance adopted by coupling a CHP unit to both an EHP, fed by cogenerated electricity, and an absorption chiller, fed by cogenerated heat, as described below.

7.5.1. Plant scheme description and energy models

Figure 7.26 [ChM07c] shows a schematic model in which the EHP operation is set to the heating mode, and upgrades the thermal level of the heat that is recovered from the WARG. For practical configurations, for instance [Hav99] illustrates schemes where the heat feeding the WARG is recovered from the jacket coolant or from the exhausts of an ICE. The various pieces of equipment are described as black boxes characterized through the relevant performance indicators. The interaction among the different machines is described through dispatch factors. Thus, in this case, α_Q and α_R represent the electricity-to-heat (CHP to EHP) and the heat-to-cooling (CHP to WARG) dispatch factors, respectively.

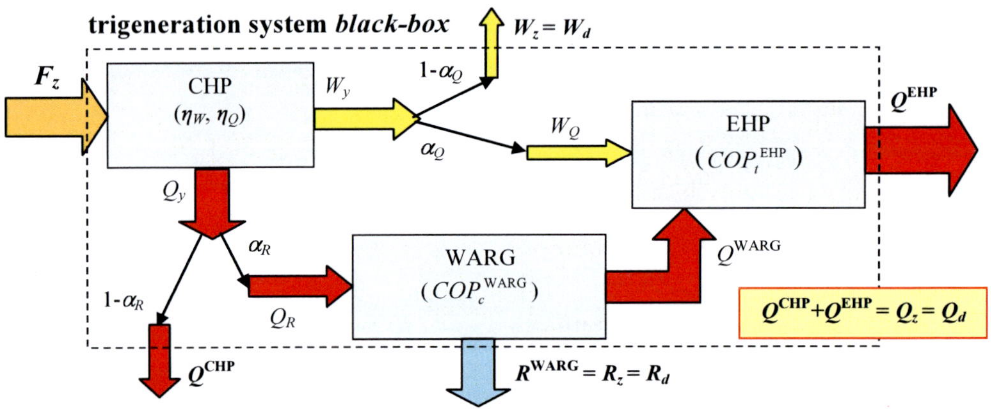

Figure 7.26. Composite trigeneration system black-box model.

With reference to Figure 7.26, and with clear meaning of the notation used, the relevant energy balances yield:

$$W_y = W_Q + W_z = \alpha_Q \cdot W_y + (1-\alpha_Q) \cdot W_y = \eta_W \cdot F_y \tag{7.25}$$

$$Q_y = Q_R + Q^{CHP} = \alpha_R \cdot Q_y + (1-\alpha_R) \cdot Q_y = \eta_Q \cdot F_y \tag{7.26}$$

$$Q_z = Q^{CHP} + Q^{EHP} \tag{7.27}$$

and in addition

$$COP_t^{EHP} = \frac{Q^{EHP}}{W_Q}, \; COP_c^{WARG} = \frac{R^{WARG}}{Q_R} \tag{7.28}$$

Considering the overall system as an equivalent black-box, with net trigenerated energy vectors W_z, Q_z, and R_z (Figure 7.26) it is easy to work out the expression for the *TPES* [ChM07c]:

$$TPES = 1 - \frac{1}{\frac{\eta_W}{\eta_e^{\mathrm{SP}}}(1-\alpha_Q) + \frac{\eta_Q(1-\alpha_R) + \alpha_Q \eta_W COP_t^{\mathrm{EHP}}}{\eta_t^{\mathrm{SP}}} + \frac{\alpha_R \eta_Q COP_c^{\mathrm{WARG}}}{\eta_e^{\mathrm{SP}} COP_c^{\mathrm{SP}}}} \tag{7.29}$$

In analogy to the other mathematical formulations for CCHP scheme assessments illustrated earlier in this chapter, the expression (7.29), function of the plant equipment efficiencies, of the SP references, and of the pair of variables (α_R, α_Q), allows the evaluation of the *TPES* under every operational situation. In particular, α_R and α_Q are subject to change under actual operating conditions, as they depend on both loading levels and control strategies adopted. It is thus possible to draw *performance maps* that give the *TPES* values in every possible operating point of the plant, taking into account, in case, partial-load and in general off-design characteristics of the equipment.

It is also interesting to point out that, by setting $\alpha_R = 0$ or $\alpha_Q = 0$, the energy saving expressions (6.12) and (7.16) worked out for CHP-EHP and CHP-WARG schemes are respectively found.

7.5.2. Energy saving performance for different equipment characteristics and operational points

Figure 7.27 shows a plot of the *TPES* in function of the dispatch factor α_R, with α_Q as the curve parameter. In this specific case, the characteristics of the CHP unit correspond to the ones of a typical ICE ($\eta_W = 0.4$, $\eta_Q = 0.45$). The COP^{WARG} is assumed equal to 0.65 (single-effect unit). The COP_t of the EHP is assumed equal to 4. As discussed in Section 3.3.6, the EHP performance depends upon the operational temperatures for condenser and evaporator and increases by increasing the heat source temperature. In this composite scheme, heat is typically recovered from the absorption chiller HRC at temperatures of the order of 30 °C. Therefore, the possibility of providing to the EHP heat discarded from the WARG may lead to EHP performance better than in usual applications, so that in principle even higher values could be envisaged. Regarding the SP reference efficiencies, average Italian values are again assumed, namely, $\eta_e^{\mathrm{SP}} = 0.4$, $\eta_t^{\mathrm{SP}} = 0.9$, and $COP^{\mathrm{SP}} = 3$. As a general result, the *TPES* decreases when α_R increases, whereas it increases along with α_Q.

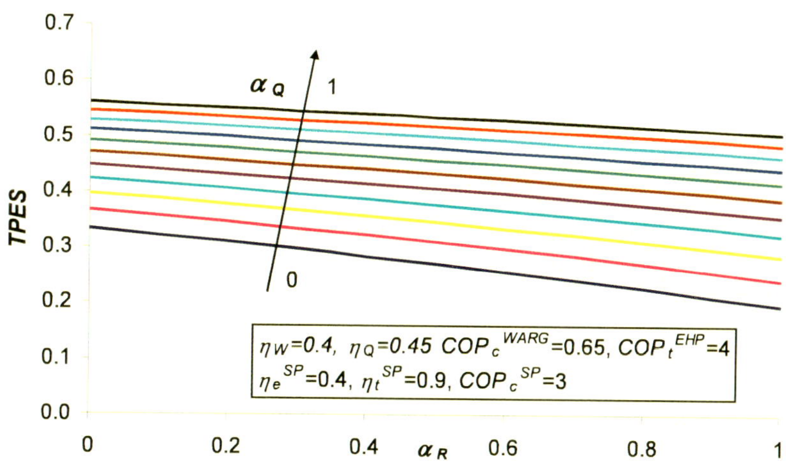

Figure 7.27. CHP-EHP-WARG plant *TPES* (η_W=0.4, η_Q=0.45, COP_c^{WARG}=0.65, COP_t^{EHP}=4).

Figure 7.28 shows the same kind of plot, but for a unit with CHP characteristics more similar to an MT (η_W = 0.3, η_Q = 0.5). The overall energy saving is 10% less than in the previous case, which is due above all to the lower electrical efficiency, besides the lower *EUF*, and the drop when α_R increases is more accentuated than before. However, an MT could be profitably used for firing double-effect (and even triple-effect) absorption chillers, as discussed in Section 3.3.4. In this respect, Figure 7.29 shows the situation occurring with *COP* = 1.1. Now, the energy saving performance with respect to α_R is much better, as the negative slope in dependence on α_R is very low. In practice, using the cogenerated heat for heating or cooling purposes affects only marginally the overall *TPES*, while the maximum reachable values do not change with respect to the previous case, and depend only upon the quota of cogenerated electricity used to supply the EHP. If a triple-effect machine were to be used, the *TPES* curve slopes could become even positive with respect to α_R, as it can be easily calculated.

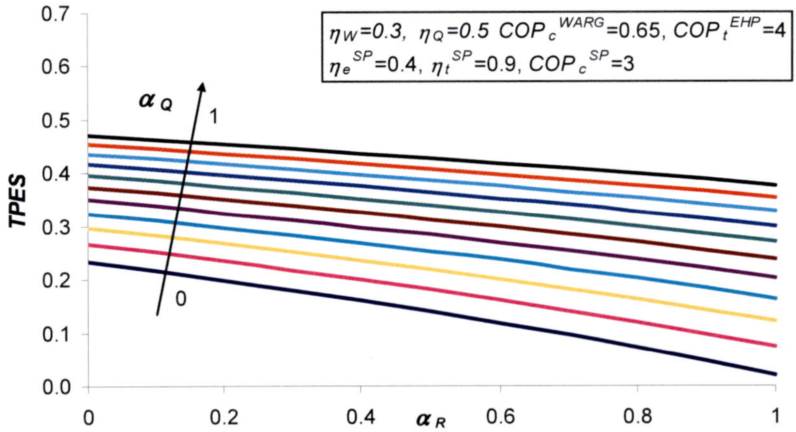

Figure 7.28. CHP-EHP-WARG plant *TPES* (η_W=0.3, η_Q=0.5, COP_c^{WARG}=0.65, COP_t^{EHP}=4).

Sensitivity analyses of the *TPES* values with respect to different values of SP reference efficiencies can also be carried out. These analyses are not addressed here. The related developments and discussions are left to the reader.

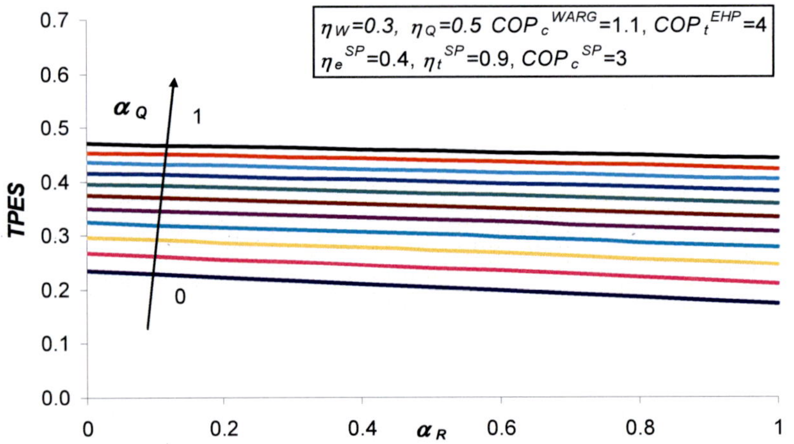

Figure 7.29. CHP-EHP-WARG plant *TPES* (η_W=0.3, η_Q=0.5, COP_c^{WARG}=1.1, COP_t^{EHP}=4).

7.6. DMG Case Study Analysis Based on Time-Domain Simulations

In order to apply the above concepts to a practical case study, let's first consider (*Case I*) the same base load pattern models and equipment used in the lambda analysis case study in Section 4.7.

Hourly-based time-domain simulations have been run for the three different trigeneration plants with the different types of cooling equipment (*Case 1* – GARG; *Case 2* – CERG; *Case 3* – WARG), also considering three different control strategies for the CHP prime mover (*A* – electrical load-following; *B* – thermal load-following; *C* – engine always on), described in Section 2.2.2. The relevant performance indicators (*TPES* and $\overline{ITHR}$ components), computed on the basis of the aggregated hourly energy over one year, are shown in Table 7.2. The results are also compared with the separate production figures. For electricity-related assessment, the best option (minimum *EITHR*) would be a CERG plant run under thermal load-following mode. Likewise, the best option for an effective cooling and/or heat production (minimum *TITHR* and *CITHR*) is a WARG plant run under always-on mode. The latter case represents also the best alternative in terms of energy savings (highest *TPES*). To some extent, this indicates that in this case the benefits from efficient cooling and heat production overcome the benefits from efficient electricity production (in terms of primary energy performance). This is consistent with the general concepts pointing out the critical importance of effective production of heat as well as cooling, possibly through recovery schemes

Table 7.2. Performance indicators calculated from the aggregated hourly energy over one year – *Case I*

indicator	*separate production*	*CERG* A	B	C	*WARG* A	B	C	*GARG* A	B	C
EITHR	2.5	2.165	**1.312**	2.252	2.349	2.095	2.238	2.328	2.154	2.298
TITHR	1.11	0.848	0.958	0.675	0.984	0.772	**0.601**	0.966	1.001	0.718
CITHR	0.62	0.138	0.341	-0.182	0.390	-0.001	**-0.318**	0.357	0.421	-0.101
TPES	---	0.061	0.036	0.073	0.029	0.071	**0.081**	0.034	0.025	0.062

Let us now suppose (*Case II*) that the assigned loads are unchanged, except for the thermal one which is punctually doubled. The generation plants are unchanged (except for adding two additional boilers to supply the augmented heat load). The results of the new time-domain simulations and index calculation are shown in Table 7.3. Again, the CERG plant in thermal load-following mode is very efficient for electricity production (lowest *EIHR*), whereas the best option for cooling and thermal production and overall energy saving is the CERG plant under always-on control. Also the GARG plant performs quite well with these load conditions.

From the analyses, it emerges how the actual energy performance of the DMG system depends on several factors, such as equipment, load profiles and control strategies, as pointed out in [ChM07a]. In particular, in the example shown the solution based upon an absorption chiller (that in principle allows the highest heat recovery and should thus lead to high energy saving, as in *Case I*) passes from being optimal to sub-optimal when the thermal loading level increases. Indeed, the overall thermal level (base heat plus heat to fire the chiller) now overtakes the CHP thermal capacity, making it necessary to produce part of the needed heat through boilers. This leads to lose part of the benefits from cogeneration, and other more efficient plant solutions take over. At the same time, apart from the best solution, it is also interesting to notice that for *Case II* the energy saving for all alternatives is higher than in *Case I*, indicating that all the equipment capacity is better deployed.

Table 7.3. Performance indicators calculated from the aggregated hourly energy over one year – *Case II*

indicator	*separate production*	*CERG*			*WARG*			*GARG*		
		A	B	C	A	B	C	A	B	C
EITHR	2.5	1.774	**1.373**	1.879	2.233	1.861	2.019	2.025	1.633	1.962
TITHR	1.11	0.823	0.663	**0.567**	0.999	0.713	0.644	0.912	0.685	0.588
CITHR	0.62	-0.431	-1.031	**-1.389**	0.211	-0.848	-1.102	-0.112	-0.950	-1.308
TPES	---	0.106	0.149	**0.154**	0.041	0.126	0.126	0.074	0.138	0.141

Of course, also an economic assessment should be run to provide meaningful indications on the plant feasibility [ChM06]. The choice of the best economic solution, in the two cases analysed, would depend on the tariff and market conditions, besides the cost of the equipment. For instance, if selling electricity were highly profitable, the CERG in thermal load-following configuration could be adopted (higher electrical production efficiency, in spite of lower energy savings). On the other hand, expensive fuel or non-profitable electricity sale would call for adopting the best energy saving alternatives. Therefore, having different cooling generation components in parallel (as in the case study in Section 4.7) and using one rather than another according to the fuel/electricity spot prices might also be a sound economic alternative [CaP06]. In general, whether energy and economic objectives are conflicting or not depends on the specific energy prices, load configurations, and control strategy [ChM06][ChM06b][ChM09b].

7.7. Remarks on the Energy Performance Assessment Examples

As a general comment on the analyses carried out so far, it can be seen how combined multi-generation schemes can exhibit good energy performance and bring potentially extremely high energy saving with respect to conventional generation means in various operational points and for different types of equipment. In addition, the possibility of feeding different machines for producing different energy vectors gives great freedom and opportunities to the energy planner, as pointed out by the lambda analysis framework (Chapter 4). Although no generalization can be carried out and every single plant should be specifically evaluated case by case, the numerical results obtained in this chapter and in Chapter 6 exemplify how DMG systems could be a key driver in the evolution of the energy scenario towards solutions allowing higher fuel conversion efficiency (and thus potentially reduced environmental impact). At the same time, the thorough evaluation of complex MG systems can be a challenging task, but the approaches, the models and the indicators presented in this work can provide powerful and synthetic indications for the energy planner.

Chapter 8

EXTENDED DISTRIBUTED MULTI-GENERATION APPLICATIONS

The manifold models, analyses and results shown in the previous chapters have focused on fuel-based and electricity-based distributed technologies and on their interactions in composite multi-generation systems. This chapter contains further illustrations on how the tools and methodologies developed could be used to address more extensive cases, with various openings in different directions.

A first direction concerns the possible coupling of multi-generation technologies and renewable energy sources. This kind of coupling can be addressed by using the energy efficiency indicators introduced, with benefits referred to the reduction in the amount of fuel needed to satisfy the user's demand. An illustrative application is provided in this chapter on a CCHP-PV system.

Further extensions refer to the assessment of the possible impact on the energy networks from production of different types of energy vectors through different multi-generation plant solutions. This appraisal is carried out on the basis of the multi-generation *lambda analysis* concepts in order to point out the effects of the *load transformation* from one form of energy into another. The effects of load transformations are evaluated on the basis of specific criteria such as energy saving, energy network loading, and economic performance. The results of this kind of assessment can be used to assist the selection of the most suitable equipment depending on the nature of the demand and on the interconnection to the energy networks.

From another viewpoint, multi-generation systems exhibit a potential for economic profitability enhancement if they were allowed participation to energy-related markets other than the electricity market, such as aimed at trading for instance *white certificates* (energy efficiency market) and *emission allowances* (emission trading schemes). In order to set the conceptual bases for exploiting such potential, a dedicated economic model is introduced to assess the effects of exploiting cogeneration and trigeneration systems within a framework entailing various energy-related markets. A specific application is considered to show that the system efficiency enhancements due to exploiting multi-generation, associated to the trading of energy-related commodities, can lead to a significantly reduced electricity production cost. The economic models developed illustrate interesting applications of the cogeneration and trigeneration incremental indicators introduced in Chapter 5.

Environmental aspects represent a natural follow-up of the energy models and analyses illustrated hitherto, and complete the comprehensive "*triple E*" picture mentioned in Section 1.5. While the focus of this book has been on energy aspects, few hints on environmental aspects are provided in this chapter to show how it is possible to extend the energy models and analyses developed to CO_2 emission assessment. In addition, the main issues relevant to the distinction between local and global environmental impacts are discussed.

A number of additional openings and extensions to the multi-generation issues specifically addressed in this book are summarized in the final section, together with the concluding remarks on the concepts, models, and analyses presented.

8.1. Integrated Assessment of Multi-Generation Systems with Renewable Sources

8.1.1. Integrated CCHP-PV systems

The black-box approach adopted in the analyses illustrated in the previous chapters for cogeneration and trigeneration applications can be easily adopted to represent more extended cases such as DMG systems containing renewable sources. An example referred to the integration of CCHP and PV systems is described in this section.

The potential of producing clean energy from PV systems is often dramatically limited by high investment costs, which could be consistently reduced in the case of larger installations and flatter exploitation of the energy produced over the year [FtA06]. Therefore, a relatively large PV system combined with cogeneration or trigeneration installations for *district energy system* applications could exhibit economic potential higher than classical small-scale PV installations. This kind of CCHP-PV solutions could represent a viable economic alternative for providing reduction of energy produced from fossil fuels, increase of energy produced from clean sources, and overall reduction of greenhouse gases [AsM07].

The energy saving performance of a combined MT-PV district energy system for trigeneration applications is analysed in this section from the energy saving point of view. In this case, the *FESR* indicator (5.30) is selected to assess the energy performance of the combined systems with respect to separate production, with the aim of highlighting the impact of PV and absorption machines (in case also adsorption machines might be used) on the CHP prime mover performance. More specifically, the "classical" *FESR* indicator is modified in order to take into account the benefits brought by electricity production from the PV system. The PV system model is reduced to a black box and incorporated in the model of the integrated energy system. A case study application is illustrated to highlight some numerical aspects.

8.1.1.1. PV system components and models

PV systems connected to the electrical network are essentially composed of arrays of PV modules and a Power Conditioning Unit (PCU) [SiE95][MaC03]. The role of the PCU is manifold. On the one hand, the PCU provides Maximum Power Point Tracking (MPPT) [MaD02] on the current-voltage (*I-U*) characteristic of the PV array. For this purpose, the MPPT system shuts down in case of high solar irradiance or high temperature of the power electronics devices. Furthermore, the PCU is aimed at providing optimal conversion efficiency from the DC side to the AC side, also under partial-load conditions, as well as power factor adjustment. The objective is to transfer only active power (at the fundamental harmonic) to the user/grid and to allow minimum harmonic distortion on the AC current waveforms, minimum ripple on the DC waveforms, and reliable protection against islanding operation with respect to the power grid.

Different models of the PV array/PCU system can be drawn in order to work out the *I-U* curves under various conditions of solar irradiance and cell temperature [SiE95][MaC03].

However, if the focus is concentrated on the power and energy conversion [PeL07], it is possible to consider simple models entailing all the PV system components. More specifically, it is possible to determine overall efficiency models in function of the solar irradiance G_i and of the cell temperature T_i (the subscript *i* points out that it is a relevant *input* to the model). In particular, given a certain temperature T_i, the conversion efficiency of the PV array (with crystalline silicon material) remains almost constant above a certain level of solar irradiance. The unit efficiency decreases with increasing cell temperature. This is illustrated in Figure 8.1, drawn with reference to typical manufacturers' data, in which the efficiency curves are plotted in function of the solar irradiance G_i, considering the cell temperature T_i as a parameter (with $T_3>T_2>T_1$). Figure 8.2 shows typical partial-load characteristics of the PCU efficiency η^{PCU} against the PV AC power output at unity power factor.

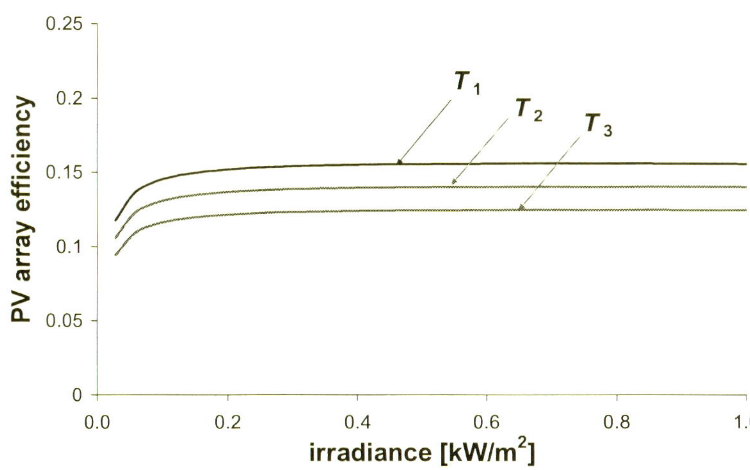

Figure 8.1. Typical PV array efficiency characteristics ($T_3>T_2>T_1$).

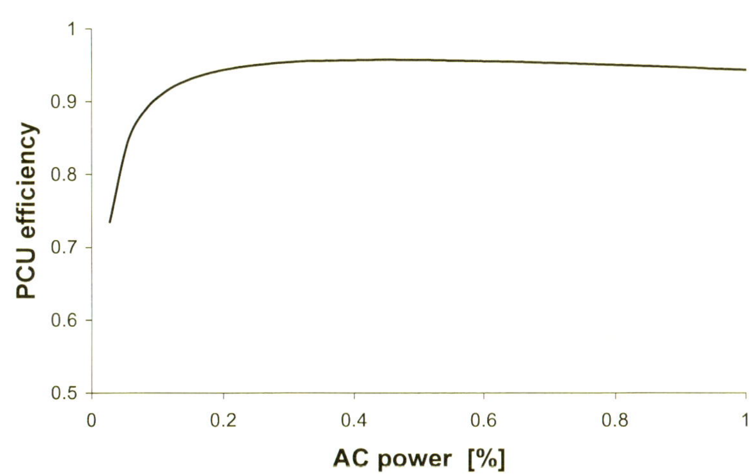

Figure 8.2. Typical PCU partial-load efficiency characteristic.

Under the black-box model approach, Figure 8.3 shows a PV system scheme consistent with the models presented in Figure 8.1 and Figure 8.2. The energy input is the solar irradiance, and the energy output is the electrical power at the AC side. In addition, also the cell temperature is shown as input to the model, to point out the efficiency dependence on such variable. This kind of modelling approach allows fast evaluation of the overall performance of the PV system within a composite energy generation system, and exemplifies how it is generally possible to entail RES technologies within the models illustrated in this book and mostly developed with reference to fuel-based energy systems.

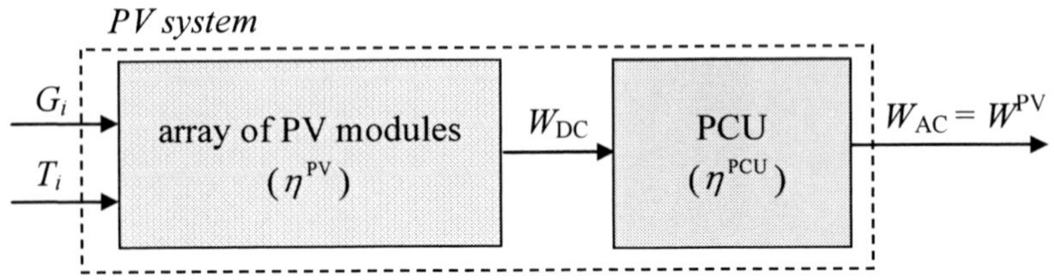

Figure 8.3. Black-box model for PV systems.

8.1.2. Energy saving in composite trigeneration systems with PV generation

A specific model can be built to assess the energy saving brought by adopting trigeneration systems with electricity production from PV arrays. Figure 8.4 shows the black-box representation of the combined system. The demands of electricity W_d, heat Q_d and cooling R_d are provided by the CHP-CGP-PV combined energy system. In the specific example presented here, the CGP is composed of a CERG *or* a WARG, as detailed in the sequel.

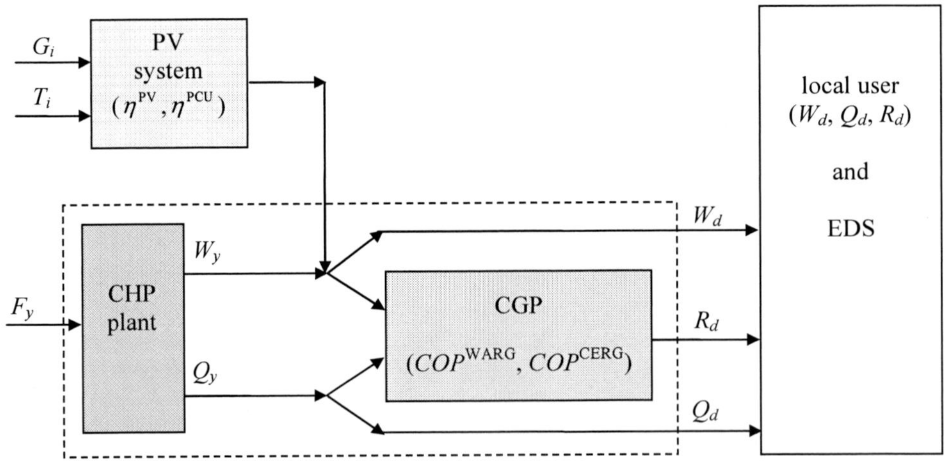

Figure 8.4. Black-box representation of the combined CCHP-PV system.

With specific reference to the CHP plant, it is possible to write down the formulation of the *PES* indicator, corresponding to the *FESR* (5.30), as

$$PES = \frac{\Delta F}{F^{SP}} = \frac{F^{SP} - F_y}{F^{SP}} = 1 - \frac{F_y}{\dfrac{Q_y}{\eta_t^{SP}} + \dfrac{W_y}{\eta_e^{SP}}} \tag{8.1}$$

In the trigeneration case, the primary energy saving through the *PES* indicator can be estimated by considering that the term Q_y in (8.1) also accounts for the heat used to feed the absorption chillers, besides the one used for thermal purposes (for instance, hot water production and space heating). Indeed, a CHP/WARG trigeneration system enables one to get higher energy saving than a CHP-only system, being the overall cogenerated heat exploited to a larger extent in this case (seasonal trigeneration), as widely discussed above.

Utilisation of PV energy to produce electricity can further lead to save important amounts of primary energy from non-renewable sources. In fact, the presence of PV systems corresponds to decreasing the entry W_y in (8.1) while obtaining the same desired energy output, owing to the fact that part of this output is supplied (either directly at the user's load or indirectly as electricity input to the CERG) through the electricity W^{PV} produced by the PV unit. Depending on the control strategy of the CHP unit, this reduction in the CHP electrical output can for instance lead to reduce the thermal output (electrical load-following case) and thus the input fuel F_y. This could help reduce the wasted heat in periods of relatively low equivalent thermal load. Alternatively, PV production could allow coverage of additional electrical load or sale of electricity to the grid (thermal load-following case), and thus exploitation of the same CHP fuel input F_y in a more profitable way.

As discussed in Section 6.2.1, the *PES* indicator can be evaluated over any time span, for instance hourly, daily, weekly, and so on, depending on the purpose of the study. In this specific application, as in most integral assessment cases, the energy saving is assessed on an annual basis, taking into account the various aspects that can occur over this time frame, and in particular the seasonal effects due to PV and cooling generation.

8.1.3. Case study example

8.1.3.1. Case study description

Let us consider the real case of a district energy system composed of a block of houses and offices in a North Italian city [ChM07d]. A simplified analysis is carried out by representing two typical months (an average *summer* month and an average *off-summer* month). These two months are representative of typical load variations, depending in particular on the effects of air conditioning (summer) and space heating (off-summer). The threefold energy demand considered (electricity, heat for domestic hot water production and off-summer space heating, and cooling for summer air conditioning) is shown in Table 8.1.

Table 8.1. Trigeneration user's seasonal loads for the case study application

month	electrical load [kWh_e]	thermal load [kWh_t]		cooling load [kWh_c]
		hot water	space heating	
off-summer	99,500	50,000	290,000	---
summer	94,500	40,000	---	75,000

The PV modules used in the district energy system under analysis (peak power 420 kW, surface 3000 m^2) supply a monthly electrical energy of 60,000 kWh_e and 25,000 kWh_e in the summer month and in the off-summer month, respectively [ChM07d]. A rooftop system is adopted, consisting of 30°-tilted modules [BaC06], with average PV system efficiency (encompassing the efficiencies of both PV modules and PCU, Figure 8.3) of 11%.

The following alternatives for supplying the loads are considered:

- a cluster of ten CHP MTs of rated capacity 100 kW_e each;
- a cluster of ten CHP MTs of rated capacity 100 kW_e each, with the additional rooftop PV system indicated above.

Each MT exhibits a rated electrical efficiency of 0.3 and a rated thermal efficiency of 0.45. The efficiencies under off-design conditions are assumed constant and equal to their rated values.

The cooling load of each plant configuration can be supplied by one of the following alternatives:

- a hot water-fed absorption chiller of rated cooling capacity 450 kW_c, with rated $COP^{WARG} = 0.6$ (assumed constant under off-design conditions);
- an electric chiller of rated cooling capacity 450 kW_c, with rated $COP^{CERG} = 3$ (assumed constant under off-design conditions).

The overall thermal load (domestic hot water, space heating, and in case hot water for feeding the absorption chiller) not covered by the CHP unit is covered by auxiliary boilers with 1 MW_t of rated thermal capacity each and rated efficiency of 0.9, assumed constant under off-design operation.

The CHP units operate at full load, and are switched off in the weekend days (8 days per month) and also during the night (7 hours per day off, and 17 hours per day on) in order to maximize the plant economic profitability (although this aspect is not tackled here). In this way, when the request is lower the wasted heat is also minimised. Electricity exceeding the demand is injected into the electrical grid.

The energy performance of the composite system is assessed by working out the *PES* values for the above plant configurations in the two typical summer and off-summer months.

The following cases are considered:

- CHP: cluster of MTs with electric chiller for cooling production in the summertime;
- CHP+PV: cluster of MTs with electric chiller and PV system;
- CCHP: cluster of MTs with absorption chiller for cooling production in the summertime;
- CCHP+PV: cluster of MTs with absorption chiller and PV system.

For all the above cases, the operating number of MTs is variable from 1 to 10. The separate production reference efficiencies are set to $\eta_e^{SP} = 0.4$ and $\eta_t^{SP} = 0.9$.

8.1.3.2. Case study results and discussion

Figure 8.5 shows the *PES* outcomes of the various analyses. From Figure 8.5, it is possible to evaluate the impact of adopting high-efficiency generation techniques such as trigeneration, the impact of adopting PV generation system on the overall plant energy economy, and finally the best plant configurations in terms of primary energy saving.

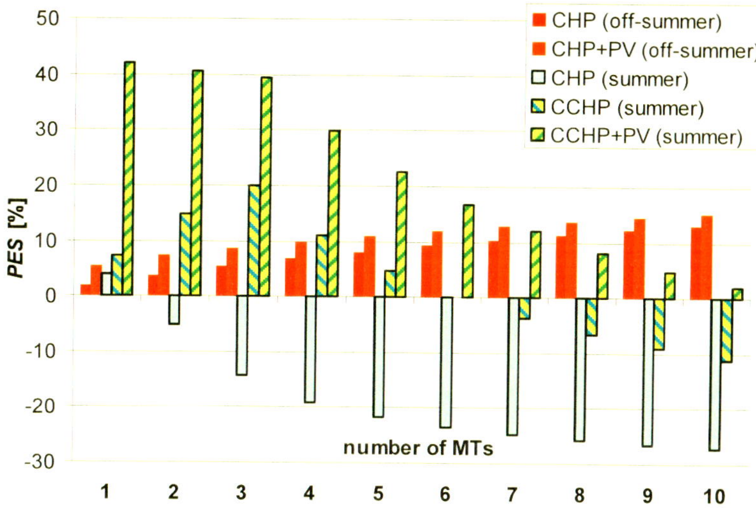

Figure 8.5. Primary energy saving characteristics for different plant configurations in different periods of the year.

For the case under analysis, a CHP-only system based on MTs would never be energy-efficient in the summer month (negative *PES*), apart from the case with only one MT in operation. Instead, adopting a trigeneration system with a cluster of three MTs could bring as much as 20% of primary energy saving with respect to the separate production of the same energy vectors, owing to the longer useful exploitation of cogenerated heat (otherwise wasted) to fire the chiller. In addition, adopting also a PV system along with trigeneration could consistently increase the *PES* in the summer month owing to the high time correlation between solar radiation and cooling load request, with basically the same values of as much as about 40% for one-, two-, or three-unit configurations. Thus, in practice in this case the best option would be to install and run only one machine (because of economic reasons). Instead, as far as the operating characteristics in the off-summer month are concerned, it is mostly convenient to run as many units as possible, since the *PES* increases essentially in a linear fashion with the number of operating MTs. In addition, the impact of the PV production in terms of energy saving is much smaller than in the summer month, as it would be expected considering the smaller amount of electricity produced and the higher importance of the cogenerated heat in the off-summer period with respect to the summer period.

As far as general results over a whole year are concerned, at first approximation it is possible to construct the summary of the load configurations by properly weighting the two available months. The overall energy saving evaluation is then carried out by weighting the results according to the relevant number of months for which each configuration is applied. For instance, by weighting the results of Figure 8.5 considering eight times the equivalent off-summer month (without cooling demand), and four times the equivalent summer month (with cooling demand), the corresponding results are reported in Figure 8.6. With these assumptions, the most convenient configuration in terms of *PES* is the one with three MTs, with an absorption chiller for cooling generation (trigeneration), and with the PV system. In particular, in this case the adoption of the PV system almost doubles the overall energy saving benefits. However, *PES* values only slightly smaller than the best case can be reached in the configurations with one or two MTs. In these cases, the energy benefits brought by adoption of the PV system are even more apparent, with a *PES* increasing of about four times (one

MT) or of about three times (two MTs) passing from a CCHP configuration to a CCHP-PV configuration.

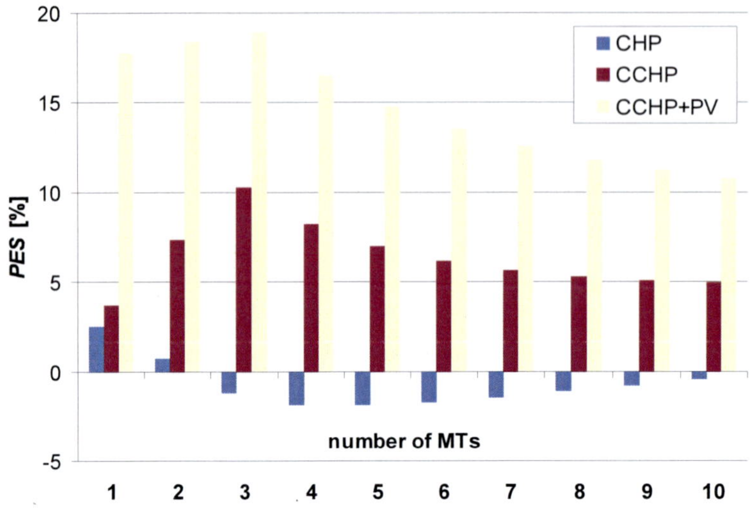

Figure 8.6. Overall primary energy saving characteristics for different plant configurations over one year.

Some indicative considerations can be carried out from the results of the simple example shown here, clearly without presumption of generality. The incremental energy benefits obtained in the CCHP configuration passing from one MT to three MTs are consistent, and so are also the incremental benefits passing from the CCHP to the CCHP-PV configuration with only one MT. Instead, the incremental benefits passing from one to three MTs in CCHP-PV configuration are slight. Thus, the additional energy benefits from installing a PV system besides the CCHP system with one MT may be sound also from an economic standpoint, while the additional installation of other MT units would not be economically justifiable. In any case, in this outlook the analyses run, focused on the energy aspects, should be backed up by the economic convenience of setting up a high-efficiency combined system like the one discussed. In this respect, district applications also allow relatively better economy-of-scale with respect to small PV systems. In addition, high market cost of the equipment (above all PV modules) could be in part counterbalanced by profitably valuing possible financial incentives [MuO97] that could be obtained within emerging *energy-related markets* [Ack07][ChM07b] owing to the enhanced energy benefits brought by combined energy systems. In particular, proved energy saving could correspond to *white certificates* tradable in the *energy efficiency market* [BeH06], while the electricity quota produced by the PV system could be traded as *green certificates* in the *renewable energy market* [BeH06]. In addition, primary energy saving may correspond to avoided CO_2 emissions, also depending on the specific power system of a given country [Meu02]. Thus, an extension of the *emission allowance* trading scheme [Bod06][TsH07] to smaller applications than the ones currently allowed to participate in the relevant market could further improve the economic profitability of a high-efficiency low-emission composite DMG energy system like the one presented here. Further details on the possible profitability of DMG solutions in energy-related markets are provided in Section 8.3.

8.2. DMG Systems and Interactions with External Networks

8.2.1. Equivalent demands on the energy network side

This section illustrates through a simple case study the transformation of different energy vectors forming the user demand into equivalent electricity and heat loads seen from the cogeneration side and, in turn, into electricity and fuel demands seen from the energy network side [ChM08c]. The methodology shown is a practical application of the overall approach to DMG modelling, planning and assessment with respect to energy network interaction and impact based on the concepts of the *lambda analysis* introduced in Section 4.3.

The effects of load transformations in a multi-generation plant interacting with external networks can be interpreted according to different evaluation criteria for carrying out, in particular, energy saving, environmental, and economic analyses. The results obtained highlight the main differences emerging from the adoption of various DMG alternatives, and are particularly relevant for trigeneration applications, with the aim of coping with the skyrocketing cooling demand throughout the world. In this respect, the application of the load transformation principles highlights additional flexibility for diversifying the supply of the various energy sources to satisfy a given demand. Increased flexibility can indeed be one of the keys for reducing energy system criticality and vulnerability, of growing concern in the latest years.

With specific regard to cooling generation, the increasing demand of HVAC in several countries has brought about higher and higher power flows in the electrical grid, making the power system heavily loaded in the Summer period, up to causing congestions and blackouts [MaR05][PoK07]. This trend could justify a potentially growing interest towards DFCs, in order to reduce the power grid vulnerability. In this respect, cooling generation from DFCs occurs *locally*, on the user's site, thus avoiding a certain amount of electricity flowing in the grid, reducing the risk of congestions and blackouts, as well as reducing transmission and distribution losses. Under this perspective, shifting from electricity-driven chillers to gas-driven chillers could have a similar impact as that of DG on distribution networks. In fact, DFCs contribute to reduce the electrical network loading by decreasing the quota of electricity that would be needed for cooling generation in conventional CERGs. This aspect can be framed within the lambda analysis approach for different DMG solutions.

Further economic benefits could take place if cooling power is generated through DFCs, since air conditioning is mostly needed in the electricity peak hours (central hours of summertime days). Of course, increasing diffusion of gas-fed chillers could consistently increase the flows in the gas transmission and distribution systems, as a consequence of the load shifting phenomenon mentioned above. However, in many countries the gas network is normally sized on the winter loading, when the request for natural gas for heating is the highest, and in these countries it is unlikely that the gas request for HVAC in the summertime could overcome the gas demand occurring in winter.

Within this framework, the load transformation concept could add significant value in addressing the impact of DMG systems (and CCHP in particular) on energy networks. In this respect, adoption of a CERG for cooling production in a CCHP plant corresponds to *electrically loading* the cogenerator (Section 4.5.2) and, if the CHP electrical production is not sufficient to supply the CERG, also the EDS. Hence, in terms of network impact a CERG may impact on both the GDS (by additionally loading the gas-fed cogenerator) and the EDS. With the same reasoning, an Indirect-Fired Absorption/Adsorption Chiller (IFAC) would "transfer" all the cooling production to *thermal loading* of the cogenerator (Section 4.5.2) and thus on the fuel network [ChM08c] (a GDS is considered here). Finally, when adopting an EDC or a GARG, that is, a separate cooling generation scheme with DFCs (Section 4.5.1), only the GDS is loaded.

In this context, it is possible to summarize the network loading framework as in Table 8.2, where also the *business-as-usual* case (electricity bought from the EDS, heat produced in a CHG, and cooling power produced in a CERG) is shown. In particular, from Table 8.2 it emerges how, in general, all the different trigeneration solutions tend to completely shift, or at least decrease, the network loading due to cooling demand from the electrical network to the gas network.

As discussed in Chapter 4, the cooling load transformations due to different generation solutions may lead to different selection and sizing of the prime mover when adopting different types of chillers. Apart from the equipment investment cost, it is interesting to point here how this could in turn impact consistently on the possible CHP operation strategies and energy flows exchanged with the energy networks. In particular, the role played by gas and electricity market prices could change depending on the (equivalent) loading levels for the various energy vectors throughout the year and on the selected equipment solution.

Table 8.2. Energy network loading for different multi-generation solutions

generation solution	*energy vector*		
	W	*Q*	*R*
business-as-usual scenario	EDS	GDS	EDS
CHP + CERG	EDS+GDS	GDS	EDS+GDS
CHP + DFC	EDS+GDS	GDS	GDS
CHP + IFAC	EDS+GDS	GDS	GDS

8.2.2. Evaluation of multi-generation alternatives with regard to network connection

In order to highlight some numerical aspects related to the lambda analysis and the relevant evaluation criteria that can be adopted for DMG systems, let us consider the threefold energy demand scenarios shown in Table 8.3. For each energy vector, the loads are reported in *per unit* of the base electrical load (200 kW), and for the sake of simplicity are assumed constant for a given characteristic day. Seasonal variations for winter (with duration of 40% of the year), midseason and summer (each lasting 30% of the year) are then considered. Practical applications could be relevant to typical conditions for aggregated district users in Mediterranean areas.

Table 8.3. Energy demand scenarios

demand scenario	*energy vector demand [pu]*		
	W	*Q*	*R*
winter	1	4	0
midseason	1	1	1
summer	1	1	3

On the generation side, let us consider three alternative CCHP solutions, with equipment and performance characteristics given in Table 8.4. For the CGP, a double-effect GARG is considered as DFC and a single-effect absorption chiller as IFAC. In addition, four CHP alternatives are analysed, namely, a cluster of one, two, three, or four MTs of 100 kW_e capacity each. The relevant CHP sizes correspond to 0.5, 1, 1.5, or 2 *per unit* (pu), respectively. CHG and chillers are sized on the basis of the relevant thermal and cooling load.

Table 8.4. Generation alternatives and equipment characteristics

generation alternative	*CHP (sizes of 0.5, 1, 1.5 and 2 [pu])*		*CHG*	*chiller*
	η_W	η_Q	η_t	*COP*
business-as-usual scenario	---	---	0.9	3
CHP + CERG	0.3	0.5	0.9	3
CHP + GARG	0.3	0.5	0.9	1.2
CHP + WARG	0.3	0.5	0.9	0.7

8.2.2.1. Energy network loading analysis

Figure 8.7 shows the loads seen from the energy network interconnection to the MG plant and the boiler loads for the base case.

As far as the CCHP alternatives are concerned, two control strategies are considered for the different CHP solutions (sizes), namely, *always-on* control strategy (in which the prime mover runs continuously at rated power), and *thermal load-following* control strategy (Section 2.2.2).

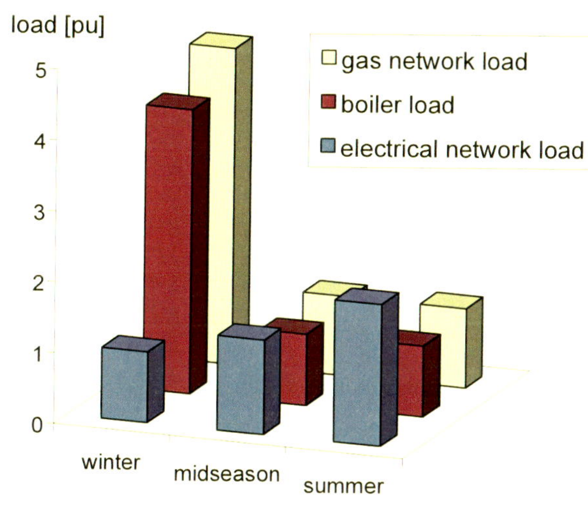

Figure 8.7. Energy network and boiler loads for the base (business-as-usual) scenario.

In Figure 8.8, the left-hand graphs show the relevant energy network and boiler loads in the *always-on* case. Negative boiler loads are used here to indicate energy wasted by the CHP group. Positive and negative electrical loads correspond to energy bought from or sold to the EDS, respectively. While increasing the size of the prime mover, the EDS load passes from positive to negative. At the same time, the GDS load increases, and so does the heat wasted when adopting the CERG or the GARG. The WARG leads to lower heat waste, but also higher fuel loads.

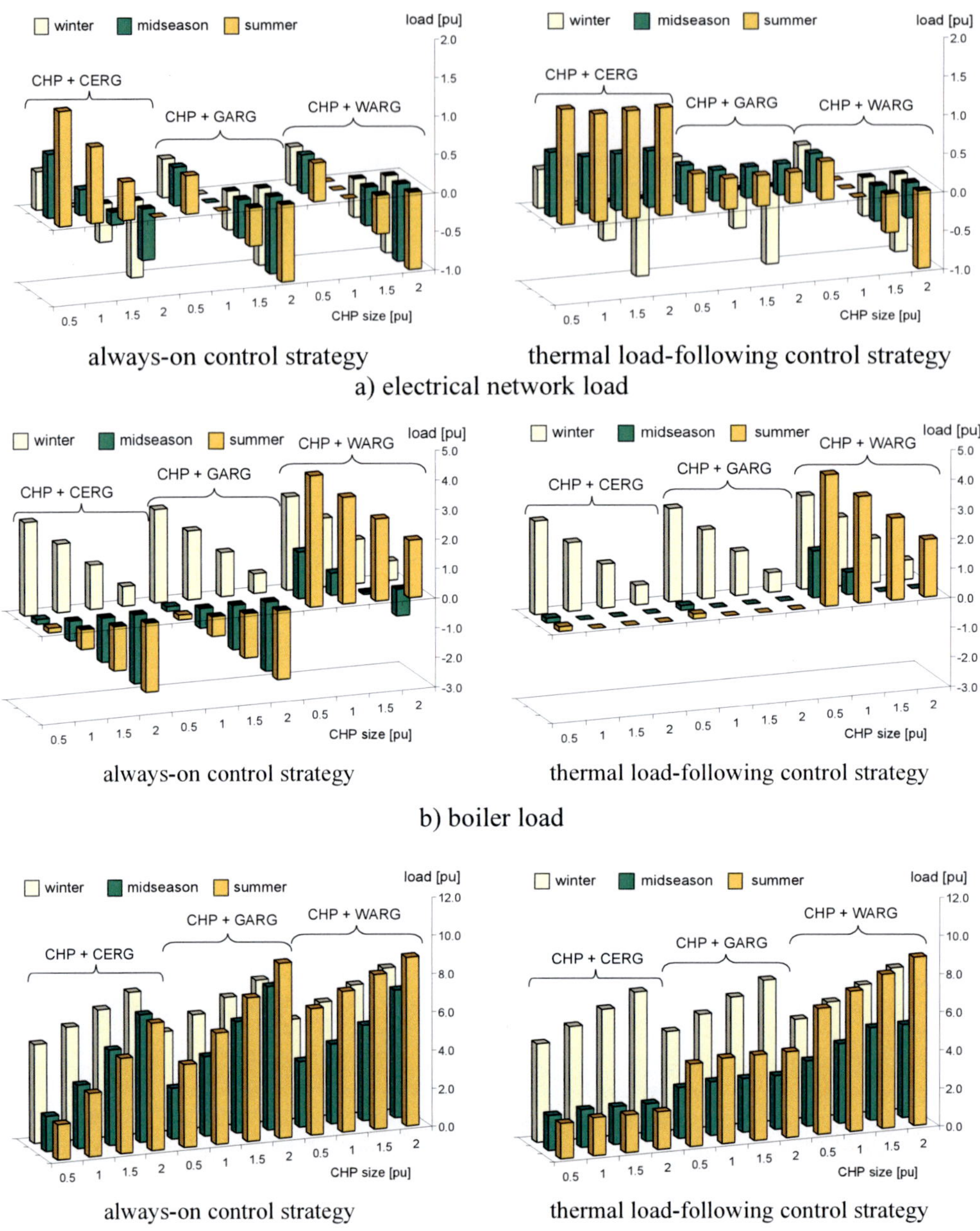

Figure 8.8. Energy network and boiler loads for different trigeneration scenarios.

In order not to waste heat, let us now consider that the prime mover is partialised if the heat load is lower than the cogenerated one. The results of the *thermal load-following* mode are shown in the right-hand graphs of Figure 8.8. The results for the CHP size equal to 0.5 pu are coincident with the ones from always-on control. As expected, no heat is now wasted, and the cases that previously corresponded to heat waste now yield lower GDS loads, higher levels of electricity bought and lower levels of electricity sold.

In comparison with the business-as-usual scenario, in general utilization of the WARG (and of an IFAC more in general) or the GARG (and of a DFC more in general) can provide the EDS with effective under-loading in the summertime. Thus, the distribution system operator could greatly benefit of such electrical load relief, particularly in the presence of severe grid congestions or overloading conditions, although consistent reverse flows might occur too. At the same time, as highlighted earlier the GDS load flows are always higher than in the base case, regardless of the CCHP solution. This might raise future gas network operational constraints with the increasing deployment of DG sources.

8.2.2.2. Primary energy saving analysis

The implications of adopting different equipment solutions and operation strategies on the primary energy saving are shown in Figure 8.9, where the *TPES* values are calculated with respect to SP references equal to $\eta_e^{SP} = 0.4$, $\eta_t^{SP} = 0.9$ and $COP^{SP} = 3$. The same SP figures are also used for the business-as-usual scenario.

In the case with smaller CHP size (when the two control strategies are coincident), the WARG brings *negative* energy saving (*i.e.*, the conventional SP is more efficient). Indeed, not enough heat is cogenerated to supply the thermal load and/or the heat-transformed cooling load. Thereby, the boiler is called in operation to shave the peaks for a considerable amount of time and energy, thus losing part of the cogeneration benefits. The same applies to the case with prime mover size equal to 1 pu, which brings smaller energy saving when adopting the WARG than when adopting the other two chiller technologies. On the other hand, when considering the two bigger prime mover sizes, adoption of the WARG allows getting higher *TPES* values for the always-on control strategy. In fact, in this case more cogenerated heat is usefully exploited for cooling production, whereas part of this heat is wasted when operating the other two chillers. Furthermore, when adopting the thermal load-following control strategy, which optimally exploits the two bigger CHP units partializing their production, a *TPES* even higher than when adopting the WARG can be reached.

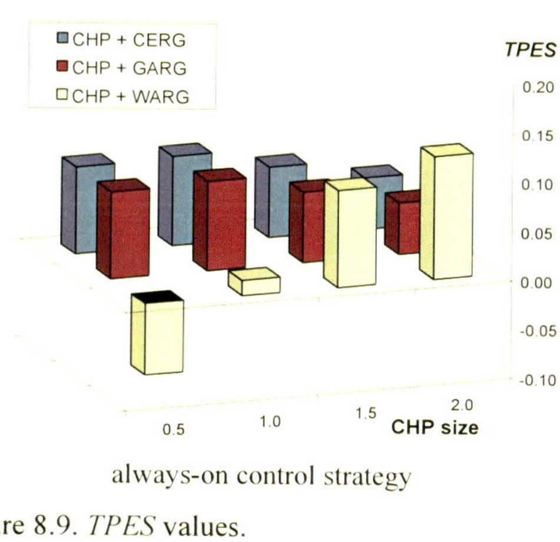

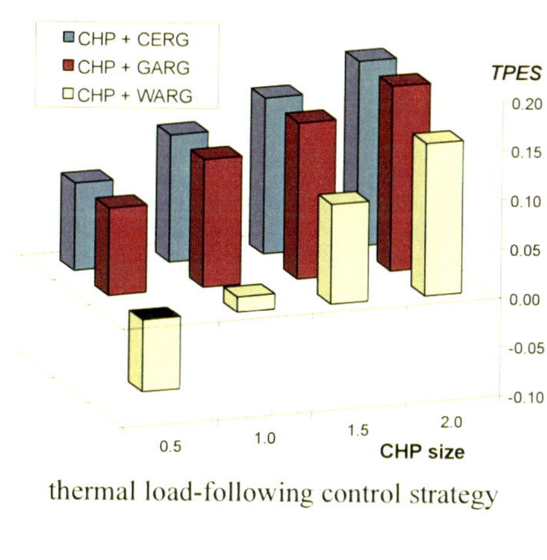

Figure 8.9. *TPES* values.

As a general consideration, this simple analysis highlights the key role played by the selection of a certain CGP option and of the CHP control strategy, as well as by the relative demand levels of the various energy vectors, in determining the overall DMG plant performance.

8.2.2.3. Economic analysis

The economic analysis regarding the *annual operational costs* due to purchase of electricity and gas, net of the profits in case earned from electricity sale, is summarized in

Figure 8.10, Figure 8.11 and Figure 8.12. More specifically, three different electricity-to-gas price ratio scenarios (considering 1 monetary unit – mu – as base value for the gas rate, fixed for all the cases) are considered. For the sake of simplicity, selling and buying prices for electricity are the same, and all prices are constant throughout the year for a given price scenario (set for instance within bilateral contracts).

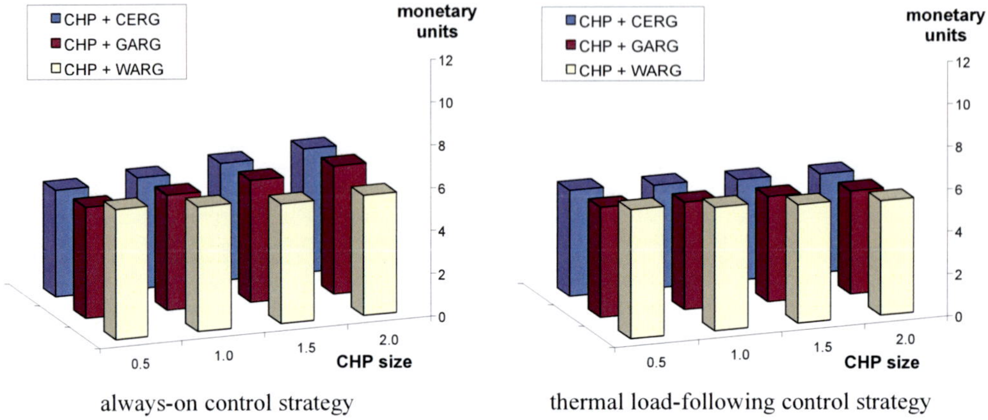

Figure 8.10. Operational costs in mu (electricity to fuel price ratio: 2 to 1).

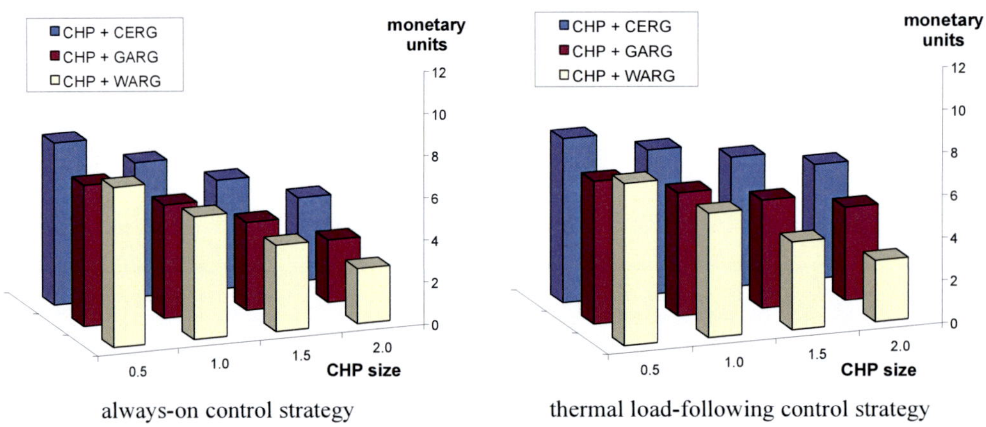

Figure 8.11. Operational costs in mu (electricity to fuel price ratio: 5 to 1).

For the (smaller) 2-to-1 price ratio (Figure 8.10), all the operational costs are of the same order of magnitude. Clearly, the solution with smaller CHP size should then be preferred, owing to the smaller investment costs (although not explicitly considered in this analysis). For the 5-to-1 price ratio (Figure 8.11), increasing the CHP size brings higher economic benefits, in particular with the WARG. Indeed, in this case cooling power is optimally generated from wasted heat, while exceeding electricity is profitably sold back to the EDS, thus scoring high profits for higher electrical rates. This is confirmed by the results in Figure 8.12, in which case further increase of the electricity rate makes it even more profitable to generate and sell electricity rather than buying it. Hence, even *negative costs* (incoming cash-flows) can be obtained for the bigger CHP size and when adopting GARG or WARG chillers (that allow higher amount of electricity sale). Conversely, adopting a CERG is now highly unprofitable, above all if the thermal load-following strategy is adopted. In fact, in this case part of the cogenerated electricity that could be profitably sold is used to supply the chiller, and furthermore the CHP group is often partialised, thus limiting the amount of electricity sold back to the EDS. The always-on control strategy is now more economic, being the CHP group exploited at its full potential to produce high-value electricity, although wasting heat.

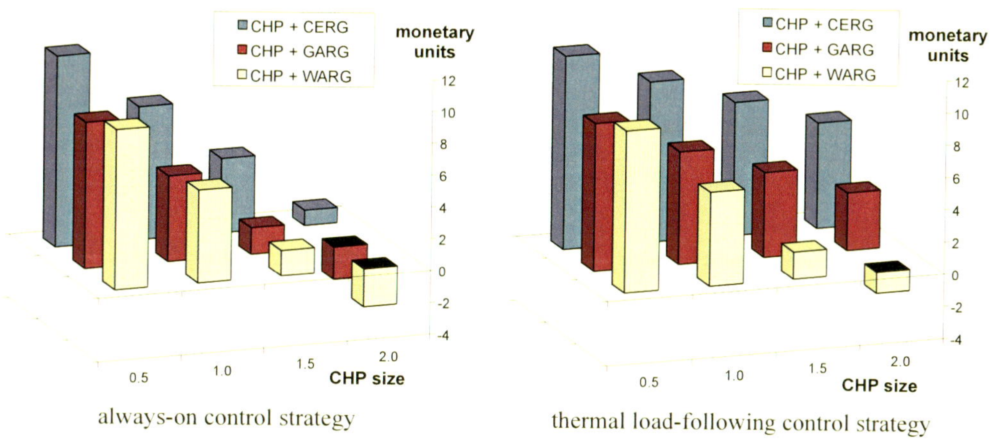

Figure 8.12. Operational costs in mu (electricity to fuel price ratio: 10 to 1).

As a general comparison term, the operational costs in the 2-to-1, 5-to-1 and 10-to-1 price scenarios for the business-as-usual solution are 5.2, 9.5 and 16.4 mu, respectively.

Comparing these base-case cost figures with the results in Figure 8.10, Figure 8.11 and Figure 8.12, significant savings could be obtained for the 5-to-1 and above all 10-to-1 price ratios. However, the operational costs give only first hints on the overall plant economic analysis, which should also consider equipment investment costs in order to select the best plant option. For instance, a Simple Pay-Back Time analysis could be effectively used to get preliminary indications for selecting the most profitable MG equipment mix within uncertain energy price scenarios [ChM06].

8.3. Economic Potential of Distributed Multi-Generation Solutions within Energy-Related Market Frameworks

8.3.1. Energy-related markets

It has been illustrated throughout this book that various multi-generation solutions already available exhibit significant potential for enhancing the sustainable development of the energy sector, by leading towards the development of energy alternatives capable to limit primary energy consumption together with CO_2 emissions from burning fossil source fuels. However, although technical and energy performance can be even outstanding, the success of promoting the adoption of such high-efficiency and low-emission plant schemes strongly depends on the associated economics. In this respect, within the present structure of the energy sector, MG technologies could be profitably exploited in competitive electricity markets, also in the light that several regulatory actions and incentives have been introduced worldwide with the aim of promoting efficient technologies and a more rational energy use.

Incentives of various forms represent a widespread tool to boost the market uptake of new technologies, above all when investment costs may be relatively high. On the other hand, market-based tools are often recognized to be soundly structured for promoting the most efficient generation options in a cost-effective way. From this outlook, different types of markets other than electricity markets, here called *energy-related markets*, have already arisen or could potentially arise in several countries. In these energy-related markets, the energy system operators are given the possibility of trading specific commodities such as greenhouse gas *emission allowances*, *white* certificates related to energy efficiency, and *green certificates* related to electricity production from RES. The rationale lying behind the operation of these markets is indeed to boost energy generation and end-use efficiency and to reduce GHG emissions, so as to improve the overall energy sector sustainability.

In general, in spite of the increasing trends towards harmonization of the different markets at the European level, differences do still persist in each country. In addition, the relevant regulation is under continuous evolution, so that changes might already be taking place with respect to the time of writing. However, some of the energy-related markets that can be encountered in a number of European countries are:

- *Emission markets*, based on the "emission trading" European Directive 2003/87/CE [EU03] and its successive modifications [EU04], which assign to the operators involved (relatively large energy systems, for the time being) a certain specific number of *allowances* or *certificates* to emit CO_2. Thereby, on the basis of the actual emissions generated, the positive or negative allowance spread can be traded on the relevant market. The certificate unitary price is typically measured in mu/tonCO_2eq (where mu stands for monetary units, as already used in the previous section).
- *Energy efficiency markets*, in which it is possible to trade the so-called *white certificates* (see for instance [AEEG] and [GMEww] for Italy), corresponding to acknowledged primary energy saving due to specific actions achieving reduction in electricity and/or gas consumption. The white certificate unitary price is typically measured in mu/tep.
- *RES markets*, in which the so-called *green certificates* can be traded. The green certicficates correspond to a specified amount of electricity produced from renewable sources [GMEww]. The green certificate unitary price is measured in mu/kWh_e.

As a follow-up of the primary objective of this work, namely, to address the energy and then environmental characteristics and benefits of DMG systems at the planning stage, this section presents a synthetic and general technical-economic model for indicative appraisal of the feasibility of small-scale CHP and CCHP plants within an energy-related market framework. More specifically, the simple model illustrated enables us to get a general picture of how much economically-efficient an MG system could be in dependence of a number of market variables. In particular, the model is constructed in order to allow for performing sound evaluations in the perspective of possible future and uncertain market configurations. The illustration addresses cogeneration as a representative case, in order to provide simple indications on how to formulate the problem and take into account the corresponding variables. Then, the analysis is extended to the evaluation of a typical CHP-WARG trigeneration scheme. Besides the "classical" variables such as fuel and electricity prices, the formulation specifically includes the potential role played by pricing models of CO_2 emissions and primary energy saving. In particular, since the pricing of the possible energy-related commodities might be different from country to country, and since the regulatory framework is continuously on the move, the analysis carried out here is based on the energy models presented in the previous chapters, and may go beyond actual operating schemes existing in specific markets. Hence, the main goal of this section is not to run actual economic analyses, but rather to study how the feasibility of MG systems could change if their superior energy performance characteristics (if any) were to be acknowledged within realistic market schemes. The considerations developed are exemplified through numerical applications with small-scale available technologies such as MTs and ICEs.

8.3.2. Profitability of cogeneration systems in energy-related markets

8.3.2.1. Cost components

The economic model presented in this section to assess the economic feasibility of DMG (and in particular CHP) plants is based on the *average cost of electricity production*. The starting point is represented by the *annual cost of production*, which is broken down into fixed (*in primis*, capital investment) and variable (mainly fuel, as well as operation and maintenance) cost components [Hor97][WiS00][BoK01]. Furthermore, as the specific objective is to address potential additional competitiveness for DMG systems owing to energy-related market participation, relevant additional cost/saving components are considered. More specifically, as in [ChM07b] we include in the analysis two *energy-related commodities*, namely, CO_2 allowances and efficiency certificates. On these bases, the annual cost of the electricity produced C_e [mu/year] in a CHP plant with energy-related market trading can be expressed as

$$C_e = \beta \cdot C_I + C_F + C_M + C_{CO_2} + C_y \tag{8.2}$$

where:

- C_I is the capital cost of the plant, in mu;
- β is the capital charge factor, related to the discount rate on capital and the plant useful life (according to the *present worth analysis* [WiS00];
- $\beta{\cdot}C_I$ is the annualised investment cost, in mu/year [Hor97];
- C_F is the annual fuel cost, in mu/year;
- C_M is the annual cost of operation and maintenance, in mu/year.

Regarding the entries in (8.2) relevant to energy-related markets, C_{CO_2} is the cost/saving due to GHG emission allowance trading, in mu/year, whereas the amount of cost/saving upon trading of white certificates is indicated as C_y [mu/year] (in particular, the subscript y points out that the potential energy saving, as well as emission reduction, is obtained owing to cogeneration production with respect to given SP references).

8.3.2.2. Performance assessment and determination of the cost components

The economic model associated to the energy-related market entries is based for illustrative purposes on primary energy saving according to the models discussed in this work. Relevant CO_2 emission reduction can be assessed in a same fashion, as also discussed in Section 8.4. In practice, specific application schemes might differ from the models hypothesized here, but the general considerations would still hold.

For the purpose of trading the relevant energy-related commodities (white certificates) in the efficiency market, we refer to the *absolute primary energy saving* ΔF defined in (5.27). Once evaluated ΔF, it is therefore possible to express C_y as

$$C_y = \rho_y \cdot \Delta F \tag{8.3}$$

where ρ_y is the unitary market price of white certificates, and ΔF is measured in the unit relevant to the market scheme (tep, for instance).

In a complementary fashion, given SP benchmark references, trading of relevant energy-related commodities (emission allowances) in the emission market can be modelled through an *absolute CO_2 emission reduction* Δm_{CO_2} [ChM07b]:

$$\Delta m_{CO_2} = \left(m_{CO_2}\right)^{\mathrm{SP}} - \left(m_{CO_2}\right)_y \tag{8.4}$$

in which $\left(m_{CO_2}\right)^{\mathrm{SP}}$ is the mass of CO_2 emitted in conventional separate production to produce W_y and Q_y, and $\left(m_{CO_2}\right)_y$ is the mass of CO_2 emitted in cogeneration. Hints on how to determine an equivalent model for environmental assessment starting from energy models are discussed in Section 8.4.

Once assessed the emission reduction (8.4) owing to cogeneration, it is possible to evaluate the entry C_{CO_2} as

$$C_{CO_2} = \rho_{CO_2} \cdot \Delta m_{CO_2} \tag{8.5}$$

where ρ_{CO_2} is the unitary market price of emission allowances, with Δm_{CO_2} measured in the unit relevant to the market scheme (tonCO_2eq, for instance).

8.3.2.3. Incremental evaluation of the fuel cost for electricity production

A possible approach to electricity production cost assessment in CHP systems takes into account that the plant is producing *also* heat besides electricity. Hence, somehow not all the fuel is actually used to produce electricity, in line with the incremental approach discussed in Section 5.4.4. Therefore, C_F could be evaluated on the basis of the *actual* fuel consumption for electricity production [Hor97][Man06][ChM06c]. Following the aforementioned incremental approach, the annual economic fuel cost balance for a cogeneration plant (assuming gas as fuel) can then be expressed as

$$\chi_F \cdot W_y = \rho_g \left(F_y - \frac{Q_y}{\eta_t^{\mathrm{SP}}} \right) \quad \Rightarrow \quad \chi_F = \rho_g \cdot EIHR \tag{8.6}$$

In (8.6), χ_F is the *average fuel cost for electricity production* [mu/kWh$_e$], ρ_g is the average gas price [mu/kWh$_t$], and *EIHR* is the Electrical Incremental Heat Rate (5.19), calculated on the basis of annual entries.

8.3.2.4. The economic model

In the light of the above considerations, from the annual production cost C_e, once normalised with respect to the electricity produced, it is possible to derive the *average production cost of electricity* (*"unitised" production cost*) χ_e [mu/kWh$_e$]:

$$\chi_e = \rho_g \cdot EIHR + \frac{\beta \cdot C_I + C_M + C_{CO_2} + C_y}{W_y} \tag{8.7}$$

The analysis could also be extended to account for the general situation in which the produced electricity W_y could fail or exceed the user's demand W_d, thus involving economic transactions with the conventional electricity markets or tariff schemes. Let us consider the annual electricity W_i purchased from the grid (*"in"*) and the annual electricity W_o sold to the grid (*"out"*) (all the entries are in [kWh$_e$/year]). In this case, the economic analysis should address the annual energy balance $W_y = W_d + W_o - W_i$. The complete economic flow model is shown in Figure 8.13, where ρ_{ei} and ρ_{eo} are the average prices of the bought-in and sold-out electricity (for instance tariff, market, or bilateral contract prices), respectively, and χ_e^* is the cost of production of the demand electricity W_d [Hor97][Man06].

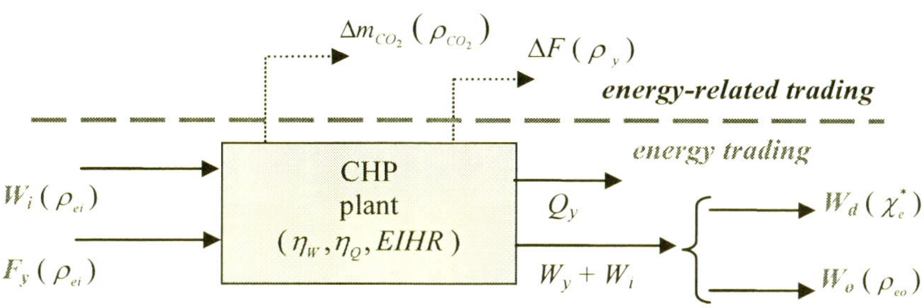

Figure 8.13. Economic model with energy and energy-related flows.

In order to focus on energy-related market issues, the details referred to electricity market transactions are not detailed here. An extension of the analysis in this direction can be readily carried out following the general model discussed in [Hor97]. However, a suitable economic formulation based on the *EITHR* indicator (5.47) for application to trigeneration systems, which also represents a formal generalization of the formulation discussed here taking into account electricity market transactions as well, is illustrated in Section 8.3.4.

On these premises, the model (8.7) is used to highlight the potential exploitation of cogeneration systems in trading commodities different from (although together with) electricity, by hypothesizing that all the cogenerated electricity W_y is used to supply the electrical demand W_d. Under this assumption, the expression (8.7) can be rewritten as

$$\chi_e = \rho_g \cdot EIHR + \frac{k + \rho_{CO_2} \Delta m_{CO_2} + \rho_y \Delta F}{W_y} \tag{8.8}$$

with $k = \beta C_I + C_M$.

Once evaluated all the entries in (8.8), assessing whether producing electricity (or building the plant) is convenient or not leads to set up the *feasibility condition*

$$\chi_e < \rho_{ei} \tag{8.9}$$

Comparative economic analyses among different CHP alternatives can then be run accounting for variable average prices for gas [ChM06][Man06], emission allowances, and white certificates. In particular, (8.8) and (8.9) can be effectively adopted for preliminary planning *multi-scenario analyses* aimed at estimating the feasibility conditions of exploiting a CHP system under price uncertainty conditions [Man06][ChM06][ChM06c].

8.3.2.5. Case study application

As an application example, let us consider two technological alternatives (an ICE and an MT), both with electrical capacity equal to 100 kW_e. In both cases, the equipment operates at full capacity for 5000 hours a year (only in wintertime and part of spring/autumn, when there is sufficient thermal load). All the cogenerated electricity and heat are consumed locally. In particular, no electricity is sold back to the EDS, in line with the assumptions made in the model (8.7). The main equipment information, together with a synthesis of the energy and environmental results, is shown in Table 8.5. Energy saving and CO_2 emission reduction are evaluated with respect to Italian SP average values. More specifically, for electricity the reference efficiency η_e^{SP} is equal to 0.4, and the average CO_2 emissions from the bulk thermal electricity generation are equal to 700 g/kWh_e (see also Section 8.4); for heat, the reference efficiency η_t^{SP} is equal to 0.8, *i.e.*, an average value for thermal residential boilers, while the average CO_2 emissions from an equivalent fuel mix are equal to 300 g/kWh_t. For the sake of simplicity, at first approximation the term *k* in (8.8) is assumed constant.

Table 8.5. Relevant data for the 100-kW_e MT and ICE in the case study

	investment costs	*operation and maintenance costs*	*electrical efficiency* η_W	*overall efficiency* $\eta_W + \eta_Q$	*EIHR*	*heat to power ratio*	*CO_2 emission reduction* Δm_{CO_2}	*energy saving* ΔF
	[mu/kW_e]	[cents/kWh_e]	[%]	[%]	[kWh_t/kWh_e]	---	[ton/year]	[tep/year]
MT	1750	1.5	29	75	1.47	1.6	240	44
ICE	1350	1.8	33	78	1.33	1.4	248	50

Figure 8.14 shows the results of a multi-scenario analysis for the MT and the ICE, with variable prices for CO_2 emission allowances and white certificates.

An average gas price equal to ρ_g = 0.013 mu/kWh_t is considered, taken from general average Italian transactions [ChM07b]. The selected ranges for emission and white certificate prices are also based on general average values for European transactions [EUCww] [GMEww], and envisaging variations from zero to about two times the current average market prices. In particular, a price equal to zero models the case in which the energy system does not participate in the relevant market.

As a general comment on the results, under the realistic prices and conditions considered here the electricity production cost could drop significantly if energy-related products were traded. In addition, from (8.8) the sensitivities of the electricity cost to the energy-related market prices are

$$\frac{\partial \chi_e}{\partial \rho_y} = \frac{\Delta F}{W_y} \qquad (8.10)$$

$$\frac{\partial \chi_e}{\partial \rho_{CO_2}} = \frac{\Delta m_{CO_2}}{W_y} \qquad (8.11)$$

Therefore, based on the data in Table 8.5, the electricity cost *variations* with respect to emission allowance and white certificate prices are slightly higher for the ICE than for the MT (owing to the higher energy saving and emission reduction), as confirmed by the graphical trends in Figure 8.14. In fact, the gap between electricity production costs for ICE and MT increases by increasing the prices of the energy-related commodities.

Deeper insights can be gained through further sensitivity analyses. In particular, it is useful to assess the sensitivity of the electricity production cost to gas price variations, as fuel prices may be subject to large variations over the plant lifetime. In this respect, from (8.8) it is possible to write

$$\frac{\partial \chi_e}{\partial \rho_g} = EIHR \tag{8.12}$$

Thus, the considered sensitivity is equal to *EIHR*, that is, it is inversely correlated to the "actual" efficiency for electricity production in cogeneration (see relation (5.20)). Therefore, in the specific case it is expected to obtain higher sensitivity for the MT, owing to an *EIHR* value about 10% higher than for the ICE. The relevant trends are pointed out in the multi-scenario analyses shown in Figure 8.14b (gas price increment of 50% related to the base gas price) and Figure 8.14c (gas price increment of 100%).

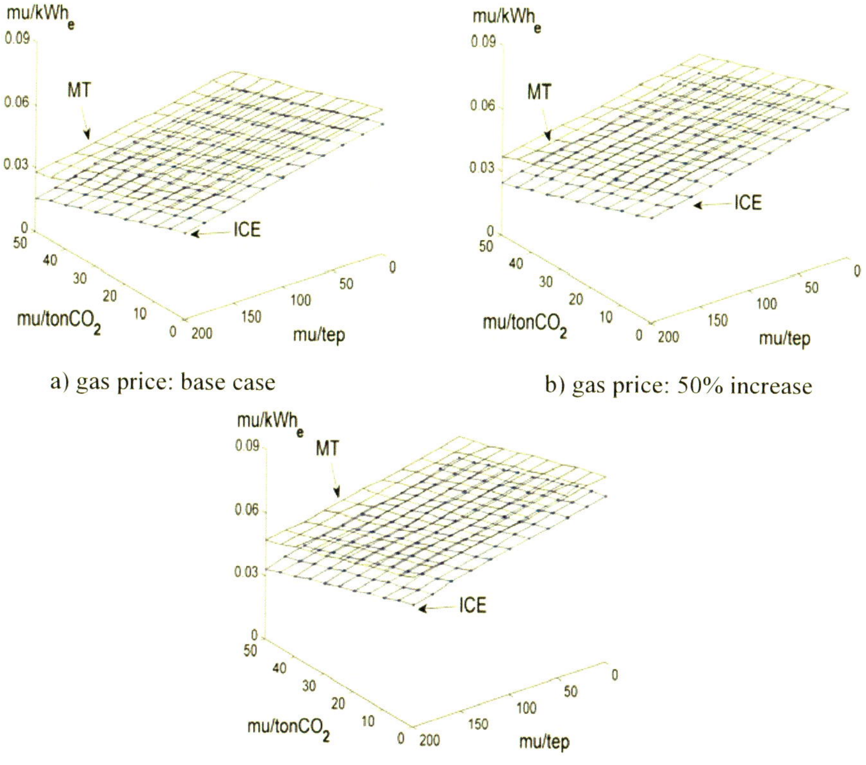

a) gas price: base case

b) gas price: 50% increase

c) gas price: 100% increase

Figure 8.14. Electricity production cost in the case study for the 100-kW$_e$ MT and the 100-kW$_e$ ICE.

Another type of sensitivity analysis can be run with respect to the separate production reference values. In fact, considering SP average values is only one of the possible approaches that might be envisaged within regulatory frameworks (Section 6.2.1 and Section 8.3.2, for instance). For comparison purposes, let us consider best available technology references for energy efficiency, namely, η_e^{SP} = 0.52 (CCGT, with efficiency penalty to include transmission and distribution losses) and η_t^{SP} = 0.95. Similarly, the CO_2 emission reduction is evaluated assuming natural gas as fuel input to the same SP systems considered for the energy saving analysis, with corresponding specific emissions equal to 390 g/kWh$_e$ and 215 g/kWh$_t$ for electricity and heat generation, respectively. In this case, the results for the relevant indicators change with respect to the ones in Table 8.5, becoming $EIHR$ = 1.78, ΔF = 6 tep/year, and Δm_{CO_2} = 15 ton/year for the MT, and $EIHR$ = 1.60, ΔF = 14 tep/year, and Δm_{CO_2} = 33 ton/year for the ICE. The new SP references lead to a dramatic decrease in the energy and environmental CHP performance, above all for the MT which is consistently penalised because of the relatively lower electrical efficiency.

As a consequence, also the plant economic performance changes consistently, as shown in Figure 8.15 (new reference values and base gas price), to be compared with Figure 8.14a. In practice, energy saving and emission reduction are now so limited that there is basically nothing to trade, so that the electricity production cost is nearly independent of the energy-related market prices. This applies primarily to the MT, for which the sensitivities (8.10) and (8.11) are negligible, as it emerges evidently from Figure 8.15. In addition, apart from the energy-related trading, it is interesting to notice that also the base electricity production cost (ρ_y = 0 and ρ_{CO_2} = 0) has slightly increased, due to the higher $EIHR$. In turn, this would also bring about a higher electricity cost sensitivity (8.12) to the gas price.

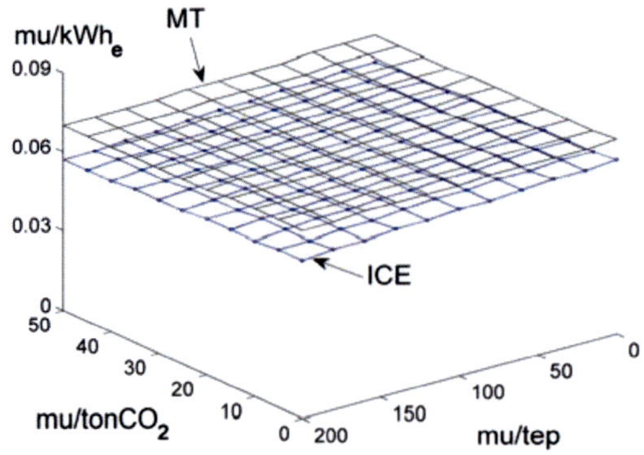

Figure 8.15. Electricity production cost in the case study for the 100-kW$_e$ MT and the 100-kW$_e$ ICE (state-of-the-art reference values).

The results from the simple analysis considered here highlight how the selection of the SP reference values is relevant not only to energy and environmental assessment, but also, subsequently, to potential economic feasibility of MG systems. In particular, too high reference values might hurdle the deployment of such systems, whereas a number of further considerations should be taken into account. For instance, small-scale systems cannot exhibit electrical efficiencies as high as CCGT, because of economy-of-scale reasons. However, distributed systems can guarantee high overall efficiency owing to the possibility of exploiting *in loco* the cogenerated heat, besides providing a degree of operational flexibility higher than in large centralized plants. In this respect, although not a crucial issue for the time being, it is expected that in the future, with increasing penetration of intermittent renewable sources, the value of flexibility that can be provided to the system by local and relatively fast-responding generators will increase substantially.

8.3.3. Extension to a trigeneration case and further comments

In the analyses shown above, it has been assumed that the cogeneration MT and ICE operate for 5000 hours owing to the presence of adequate thermal load to supply. Generally speaking, the energy system profitability increases with the operational time, as the investment costs can be recovered more quickly if more energy is generated, and the specific cost of energy production (8.7) decreases. As discussed broadly in this book, coupling a CHP system to an absorption or adsorption system for cooling generation leads indeed to increase the CHP operational time, by exploiting cogenerated heat for *seasonal trigeneration*.

As a practical application of this concept within the energy-related market framework considered in this section, let us assume that a certain cooling load is required by the user, and that in the base summertime case chilled water is produced through a conventional electric chiller, with the CHP system switched off due to lack of sufficient thermal load (only domestic hot water is needed). Instead, now let us suppose that a single-effect WARG is coupled to the CHP system to supply the cooling load. In addition, we assume that the "pure" electrical demand and the *cooling-transformed* electrical demand from adoption of the CERG are such that the cooling load does not affect consistently the overall electrical demand. On the other hand, we assume that the *cooling–transformed* thermal-demand from adoption of the WARG leads to generate, with the same cooling load, an equivalent thermal load that added to the "base" domestic hot water load is sufficient to make the CHP system run. In particular, owing to the "new" equivalent thermal load, the CHP group can now run for 6000 hours, thus increasing its operational time by 1000 hours with respect to the previous case. With these assumptions, we can still assume that all the cogenerated electricity and heat (with a major part of the latter now going to feed the absorption chiller) are consumed locally. In addition, the costs (for investment and maintenance) associated to the WARG are assumed at first approximation equal to the ones of the CERG, so that the relevant net equipment costs still refer to the CHP equipment only, as from Table 8.5.

On these bases, the model (8.8) still applies, and it is possible to address the feasibility of the trigeneration system starting from the cogeneration case and through the cogeneration electricity production cost model. In addition, the primary energy saving and emission reduction models used in the analysis refer to cogeneration production, including the heat for thermal purposes and cooling generation. This model is consistent with some European regulatory frameworks [EU04b][CaP05] that consider CHP-WARG trigeneration as a sub-case of cogeneration. Apparently, there are a number of assumptions in this reasoning. However, the final goal is to estimate how a trigeneration system can enhance the profitability of a "pure" cogeneration system for energy-related market applications, so that the preliminary (although approximated) indications drawn from the analysis can be useful before running further detailed analyses.

The economic results from the trigeneration case are reported in Figure 8.16, referring to the base gas price case. It can be appreciated how the increase in the number of operating hours is accompanied by a reduction in the electricity production cost, more pronounced especially when the price of the emission allowances increases. This is a consequence of the increased number of operating hours, which justifies the profitability and practical success of seasonal trigeneration schemes. In addition, as a consequence of the higher values of energy saving and emission reduction, there is also a better potential of exploiting the energy system within energy-related markets.

Summarizing the results presented, the electricity production cost for cogeneration and trigeneration systems (and, more extensively, any kind of MG system) could drop significantly in correspondence of the average prices currently occurring within energy-related markets. Hence, allowing such systems to participate in these schemes could consistently enhance economic viability and boost the implementation of DMG systems on a wider basis. This would have a positive knock-on effect on the energy and environmental performance of the overall energy sector. On the other hand, the economic profitability of DMG systems can be dramatically affected by the price variation width in the considered scenarios. This calls for setting up suitable strategies for maximizing profits under uncertainty and at the same time for hedging against the risk of price variations [DaL03]. Further remarks in this direction are provided in the final section.

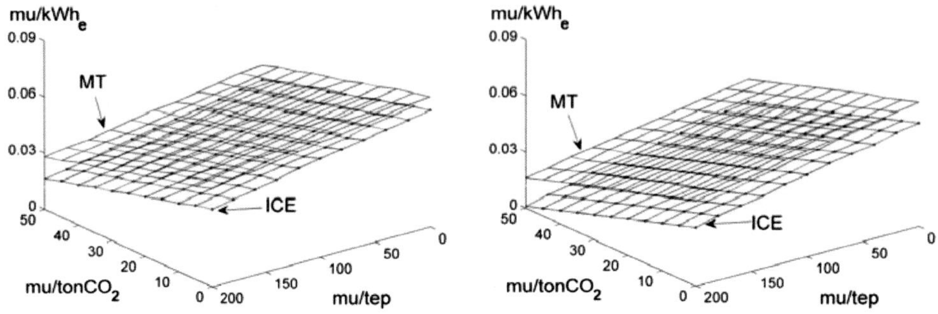

a) base case, 5000 hours b) trigeneration case, 6000 hours

Figure 8.16. Electricity production cost in the case study for the 100-kW$_e$ MT and the 100-kW$_e$ ICE. Comparison between the base cogeneration case and the trigeneration case.

8.3.4. Adoption of incremental indicators for trigeneration system economic comparison

The previous section has illustrated some potential economic benefits from the adoption of MG systems within energy-related markets. The economic model developed has been based on the cogeneration incremental indicator *EIHR*.

As a further exercise, this section shows the adoption of the trigeneration incremental indicator *EITHR* defined in (5.47) for economic assessment of CCHP systems [Man06][ChM06b]. In this case, no energy-related markets such as for emission or energy certificate trading are considered, but electricity purchases and sales with the electricity market are explicitly modelled, thus extending the model provided in the previous section. However, as far as the incremental energy performance model is concerned, its application for economic analysis is only approximated, as explained below. However, the approach illustrated exemplifies potential implications of energy saving also in economic terms.

8.3.4.1. Economic model for incremental trigeneration assessment

The starting point is the annual cost balance as in (8.2), but with no energy-related entries:

$$C_e = \beta \cdot C_I + C_F + C_M \qquad (8.13)$$

As in the cogeneration case, in this kind of approach only the investment capital is formally discounted to refer all the figures to annual values. However, in principle all the variables (in particular, cost and energy balances) should be updated year by year to their actual values.

As in Section 8.3.2.4, the annual production cost C_e, once referred to the energy unit, becomes the average production cost of electricity ("unitised" production cost) χ_e [mu/kWh$_e$] and has to be compared to the available (tariff, contract or market) average electricity price ρ_e, in order to assess the convenience of building the power plant.

However, a CCHP plant is doing more than just producing electricity, that is, it produces simultaneously also thermal and cooling energy, yielding a subsequent net fuel saving. Under this point of view, the relation (8.13) holds in general true also for the trigenerated electricity. Considering that C_F could take into account the fuel saving due to the other trigenerated vectors, in case of electrically "matched" plant (where the production equals the need), the annual economic fuel cost balance can be expressed through:

$$\chi_F \cdot W_z = \rho_g \left(F_z - \frac{Q_z}{\eta_t^{\mathrm{SP}}} - \frac{R_z}{COP^{\mathrm{SP}} \cdot \eta_e^{\mathrm{SP}}} \right) \tag{8.14}$$

that is,

$$\chi_F = \rho_g \cdot EITHR \tag{8.15}$$

where all the trigeneration entries are now highlighted through the subscript z. In particular, the trigeneration energy entries F_z, W_z, Q_z, and R_z are all evaluated on an annual basis [kWh/year] and represent the net energy flows at the CCHP black-box boundaries, as widely discussed in Section 5.5. *EITHR* is then calculated on the basis of annual entries and by averagely evaluating the benchmark SP references η_t^{SP}, η_e^{SP} and COP^{SP}. As in (8.6), χ_F [mu/kWh$_e$] is the average fuel cost for electricity production and ρ_g [mu/kWh$_t$] is the average gas price. The same price is assumed for the gas bought for boilers and prime movers in the specific case (in general, a more detailed formulation could be considered).

It can be noticed that (8.14) represents an approximation of cost balances carried out through energy balances. In fact, whereas the *EITHR* indicator can somehow address energy production effectiveness, as discussed above, the expression (8.14) would be fully correct if all the separate production energy vectors were actually all produced through gas input at the price ρ_g. However, this is not the case for cooling generation, which in the model is assumed to occur in electric chillers. Hence, rather than an economic assessment tool, the analyses developed in this section should be regarded as an estimate exercise of how energy balances might correspond to economic balances taking into account actual costs of electricity and gas. In particular, although approximate, still suitable comparative analyses can be run for different cooling generation alternatives to estimate the impact of more or less CCHP efficient solutions on economic assessment. Somehow, such an approach can be compared to the utilization of the value-weighted *EUF* (5.15), which, more than for a sheer economic assessment, can be used as a further performance indicator by adequately weighting the relevant energy balances through economic parameters. In addition, as detailed below, the formulation thus developed represents a formal extension of the model (8.8) to take into account electricity market transactions.

Using (8.13) together with (8.15) yields the "unitised" production cost of electricity χ_e for an electrically-"matched" trigeneration plant:

$$\chi_e = \rho_g \cdot EITHR + \frac{\beta \cdot C_I + C_M}{W_z} \tag{8.16}$$

which has the same form as (8.7) (with no entries relevant to energy-related markets), since it has been derived on the basis of the same incremental approach.

However, when the plant is not "matched" and the annual electricity produced W_z exceeds or fails the annual user's need W_d, the annual balance will yield:

$$W_z = W_d + W_o - W_i \tag{8.17}$$

where W_i is the annual electricity purchased from the grid ("in") and W_o is the annual electricity sold to the grid ("out"). All the entries are in [kWh$_e$/year]. In this case, in the economic balances also the cost of the electricity bought-in and the profit from the electricity sold-out need to be taken into account. In addition, the most effective economic metric to assess the cost of the different plant alternatives to be analysed is the cost for producing the given electrical output W_d, while simultaneously producing also the needed Q_d and R_d. Accordingly, the unitised "trigeneration" electricity cost χ_e^* [mu/kWh$_e$] to satisfy the annual electrical demand W_d in the trigeneration system, is

$$\chi_e^* = \frac{\chi_e \cdot W_z + \rho_{ei} \cdot W_i - \rho_{eo} \cdot W_o}{W_d} \tag{8.18}$$

where ρ_{ei} [mu/kWh$_e$] is the average price of the electricity purchased, and ρ_{eo} [mu/kWh$_e$] is the average price of the electricity sold. In general, both ρ_{ei} and ρ_{eo} can refer to EDS tariff prices, as well as bilateral contracts or market prices. The relation (8.18) represents a straightforward generalization of (8.7).

Considering (8.16), (8.17) and (8.18) together with the feasibility condition (8.19) that determines the economic convenience threshold for building the plant

$$\chi_e^* < \rho_{ei} \tag{8.19}$$

yields the alternative feasibility condition (8.20)

$$\rho_g \le \rho_g^* \tag{8.20}$$

where ρ_g^* can be seen as an *"equivalent" gas price*, namely:

$$\rho_g^* = \frac{\rho_{ei}(W_z - W_o) + \rho_{eo} W_o}{EITHR \cdot W_z} - \frac{\beta \cdot C_I + C_M}{EITHR \cdot W_z} \tag{8.21}$$

Equal values of the electricity prices can be considered for the sake of simplicity (as might occur for instance in bilateral contracts):

$$\rho_{ei} = \rho_{eo} = \rho_e \tag{8.22}$$

In this case, the relation (8.21) can be rewritten in a more compact and elegant form:

$$\rho_g^* = \frac{\rho_{ei}}{EITHR} - \frac{\beta \cdot C_I + C_M}{EITHR \cdot W_z} \tag{8.23}$$

After running suitable plant simulations over a one-year time span, yielding in output W_z, W_o and *EITHR*, the expression (8.23), although based on an approximate approach, allows a comparative evaluation of CCHP alternatives also on an economic basis, providing further insights on the plant performance.

8.3.4.2. Case study application

As a case study application, let us consider the same load patterns and the *five* CGP cases already discussed in Section 4.7. In addition, for each of the five CGP configurations, let us consider as prime mover alternatively a gas-fed 330-kW$_e$ ICE (as already done in Section 4.7) or a gas-fed 525-kW$_e$ ICE. For each case, hourly time-domain simulations over a one-year time span have been run, considering the same three *control strategies* of Section 7.6, namely:

- *A*: *electrical load-following*;
- *B*: *thermal load-following*;
- *C*: *engine always-on at full load.*

Applying the formulation shown above (without present worth analysis, that is, setting $\beta = 1$ for the sake of simplicity and since the focus here is on the methodology rather than on the numerical results) for an expected plant useful life of 15 years, the simulations yield the equivalent gas prices. In particular, the expression (8.23) is mapped against average electricity prices for all the cases when adopting the 330-kW$_e$ ICE (Figure 8.17) and the 525-kW$_e$ ICE (Figure 8.18). For the *EITHR*, SP references equal to $\eta_t^{SP} = 0.9$, $\eta_e^{SP} = 0.4$, and $COP^{SP} = 4$ have been considered. An average gas price (aligned with typical European values) is also shown as a reference.

The maps in Figure 8.17 and Figure 8.18 provide an interesting tool for comparing the CCHP alternatives with variable average electricity and gas prices. In fact, the graphical results can be interpreted as follows: for every average electricity price, viable solutions are represented by all cases with equivalent gas prices higher than the given average gas price. In this way, it is possible to consider (*a posteriori* with respect to the simulations) the performance of every solution for every possible gas and electricity average prices.

As general comments, the plant of *Case 5* (CERG covering the cooling base load and WARG modulating the peaks), with the 330-kW$_e$ ICE run under always-on control strategy as a prime mover, can be identified as the most effective solution, with the highest equivalent gas price for every electricity price. In addition, the 330-kW$_e$ ICE performs generally better than the 525-kW$_e$ ICE, as it matches better the load and therefore is better exploited. This kind of result could be anticipated beforehand, at first approximation, by means of the trigeneration load duration curves of Figure 4.22. A crucial indication, strengthening the considerations carried out in Section 4.7 and in Section 7.6, is that the benefits obtainable from the CCHP alternatives strongly depend on the control strategies. In particular, the *Strategy C* can be appreciated as the most effective one. This result is not surprising, because the always-on control strategy implies continuous production of electricity (apart from the scheduled stops for maintenance and possible outages), leading to a higher value of W_z, which contributes, in turn, to increase ρ_g^*, as from (8.23). These outcomes are consistent with the ones obtained in Section 7.6, in which case the always-on control mode resulted the most efficient in terms of energy saving. However, Figure 8.17 and Figure 8.18 also show that the comparative economic outcomes generally depend on the specific average electricity price, highlighting how complementary analyses are needed for assessing both energy and economic performance.

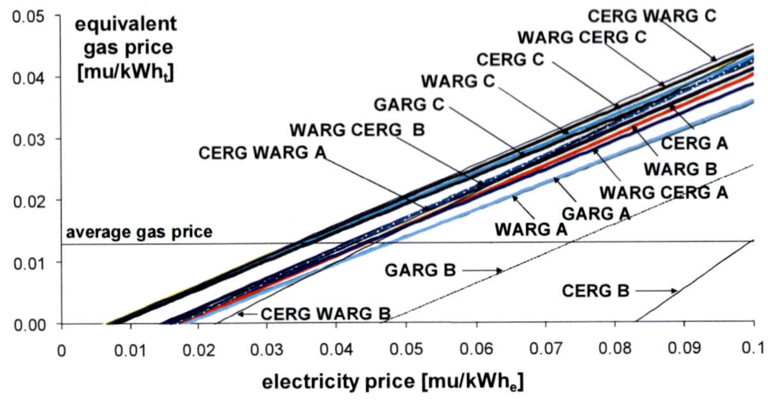

Figure 8.17. Equivalent gas price map for the 330-kW$_e$ ICE.

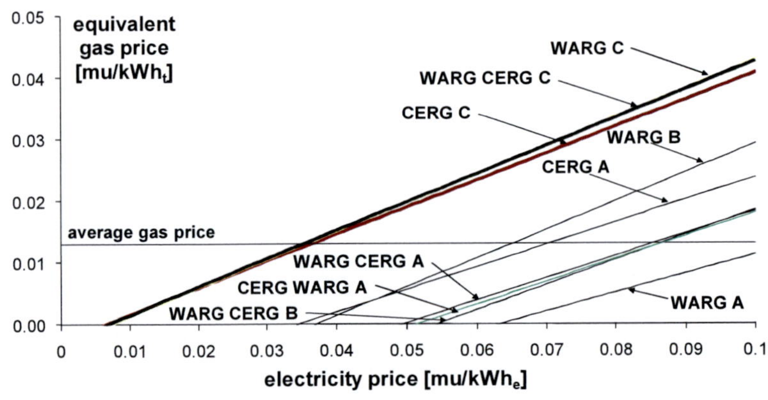

Figure 8.18. Equivalent gas price map for the 525-kW$_e$ ICE.

8.4. Environmental Impact Aspects

The models developed to address efficiency evaluation of combined energy production can be extended to encompass CO_2 emission assessment under a unified framework based on the black-box rationale. Hence, the various studies dedicated to energy performance assessment become relevant also in terms of environmental impact assessment of DMG systems.

Although a full analysis of these aspects is outside the scope of this work, a preliminary illustration is provided below to highlight the theoretical framework. Details can be found in [ChM08][ChM08b][MaC08][MaC08b][MaC09].

8.4.1. The emission factor model for CO_2 emission assessment

A suitable approach to model and characterize emissions from a generic combustion device is represented by the evaluation of the output-related *emission factor* [EDU01] [CaB02][Meu02][MiL03B][EPAww]. Although relevant emission factors can be defined with reference to any kind of pollutant (for cogeneration, see for instance the models illustrated in [MaC09]), in this work we specifically address the CO_2 case, given its GHG effect and the increasing importance of global warming issues within the agendas of most governments. Furthermore, the specific characteristics of carbon emission generation in combustion devices lead to formulate the unified structure for energy and CO_2 emission assessment.

Let us consider the mass $m_{CO_2}^X$, typically expressed in [g], of CO_2 emitted to produce the useful energy output X. For instance, with reference to trigeneration applications, the useful output X can be electrical energy W [kWh$_e$], heat Q [kWh$_t$], or cooling energy R [kWh$_c$]. In order to address the most general multi-generation case, as in Section 5.6.2 a set $\boldsymbol{D}$ can be defined, containing the types of useful net energy outputs (demand energy vectors) from the MG system. We recall that the set $\boldsymbol{D}$ is composed of pairs (X,x), each of which is formed by an energy vector X and the corresponding type of energy x characterizing the useful output (demand). Thereby, the mass $m_{CO_2}^X$ can be calculated by using the following model:

$$m_{CO_2}^X = \mu_{CO_2}^X \cdot X \tag{8.24}$$

where $\mu_{CO_2}^X$ represents the *emission factor* or *specific mass emissions* of CO_2 per unity of X, expressed in [g/kWh].

As a special case of the expression (8.24), it is possible to substitute X with the primary energy F released when burning a given fuel, which yields the definition of emission factor $\mu_{CO_2}^F$ referred to the fuel thermal energy content [EDU01][CaB02][MiL03b]. In this respect, the mass of CO_2 emitted when burning a given type of fuel can be estimated with good approximation on the basis of the nominal characteristics of the relevant chemical reactions, as well as the properties of the specific fuel (carbon content and *LHV*) [EDU01][CaB02]. At first approximation, the fuel-related emission factor $\mu_{CO_2}^F$ is a function of the type of fuel only, and can be considered constant under various operating conditions [EDU01][ChM08].

A generic energy output-related emission factor model can then be estimated by taking into account that the fuel input and the generic useful energy output are linked through the relevant generation efficiency η_x. For instance, the electrical efficiency η_e takes into account the generation of useful electricity W in a power plant from the fuel input F. Thus, equivalent CO_2 emission factors for $(X,x) \in \boldsymbol{D}$ can be derived as:

$$m_{CO_2} = \mu_{CO_2}^F \cdot F = \mu_{CO_2}^X \cdot X \tag{8.25}$$

from which the relation among output-related and fuel-related emission factors emerges to be

$$\mu_{CO_2}^X = \frac{\mu_{CO_2}^F}{\eta_x} \tag{8.26}$$

The expression (8.26) is the key formula to run unified energy and CO_2 emission analyses. In fact, provided that $\mu_{CO_2}^{F}$ is assumed constant for a given fuel, it highlights the inverse proportionality of energy efficiency and CO_2 emissions, or, equivalently, the direct proportionality of relevant primary energy rates (see for instance Section 5.2.4 for SP cases) and CO_2 emissions. This feature allows the generalization to environmental modelling of the energy chain and black-box concepts introduced in Chapter 5 for energy modelling, and, accordingly, the generalization of the energy performance indicators to environmental performance indicators [ChM08][ChM08b][MaC08][MaC08b][MaC09].

8.4.2. Emission reduction in multi-generation systems

Evaluation of the emission reduction from cogeneration and seasonal trigeneration systems (Section 7.2) has been discussed in [Meu02]. The key variables involved in the energy and environmental assessment of trigeneration systems have been highlighted in [MiL03b][ChM08][MaC08]. A generalisation of these models to multi-generation assessment, consistent with the concepts illustrated in this work, has been provided in [ChM08b]. Such a generalisation leads to a compact formulation of emission reduction indicators. The whole approach is constructed in analogy with the deduction of the *PPES* (5.50) carried out in Section 5.6.2 to provide a unified framework for energy and CO_2 emission impact assessment. In particular, again the term poly-generation is taken as synonym of the term multi-generation, and the superscript p is used to represent poly-generation entries.

According to the general model presented in [ChM08b], the Poly-generation CO_2 Emission Reduction (*PCO2ER*) indicator is defined as the relative saving in the emitted mass of CO_2 obtained in the multi-generation system when compared to the separate production. The *PCO2ER* is expressed as:

$$PCO2ER = \frac{\left(m_{CO_2}^{F}\right)^{SP} - \left(m_{CO_2}^{F}\right)^{p}}{\left(m_{CO_2}^{F}\right)^{SP}} = 1 - \frac{\left(\mu_{CO_2}^{F}\right)^{p} \cdot F^{p}}{\sum\limits_{(X,x)\in \mathbf{D}} \left(\mu_{CO_2}^{X}\right)^{SP} \cdot X^{p}} \tag{8.27}$$

The expression (8.27) refers to the production of the energy vector X^{p} in the multi-generation system requiring the fuel input F^{p}. The term $\left(\mu_{CO_2}^{F}\right)^{p}$ is the emission factor related to the production of X^{p}. The term $\left(m_{CO_2}^{F}\right)^{p}$ represents the CO_2 mass [g] emitted by the multi-generation system to cover the demand of the energy vectors included in $\mathbf{D}$; this mass can be estimated by using the equivalent model (8.25). To obtain the same energy vector X^{p} in separate production, it would be necessary to emit a CO_2 mass $\left(m_{CO_2}^{F}\right)^{SP}$, to be estimated by using the specific *reference* emission factors $\left(\mu_{CO_2}^{X}\right)^{SP}$ relevant to separate production. As an example, for a cogeneration system it would be necessary to use in (8.27) the emission factor $\left(\mu_{CO_2}^{F}\right)_{y}$ [g/kWh$_t$] related to the fuel input to the CHP system, as well as the emission factors $\left(\mu_{CO_2}^{W}\right)^{SP}$ [g/kWh$_e$] and $\left(\mu_{CO_2}^{Q}\right)^{SP}$ [g/kWh$_t$] representing the conventional separate production of electricity and heat, respectively.

Extension of the model (8.27) to include also other GHG pollutants can be readily carried out on the basis of the indicators introduced in [ChM08] for trigeneration systems.

8.4.3. Unified structure for energy and environmental indicators

The *PCO2ER* indicator (8.27) has the same structure as the *PPES* indicator (5.50). In particular, the "weights" of the black-box energy outputs in (5.50) are represented by the reciprocals of the SP reference efficiencies, and in (8.27) by the reference emission factors. However, the expression (8.27) contains the relevant emission factor weight also for the fuel energy F^p. This breaking of the symmetry between the two structures is intrinsic in the concepts of energy and emissions. Indeed, when assessing energy transformations starting from given outputs, what actually matters are only the efficiencies and not the *type of fuel* used in the various systems. In this respect, the specific technologies are addressed only for assessing the relevant efficiencies, and a "blind" black-box model can be formulated. On the other hand, when evaluating CO_2 emissions not only the technologies (that is, efficiencies) are relevant, but also the *fuel inputs* to the considered systems or equivalent black-boxes [ChM08]. In fact, different fuels exhibit different carbon chemical contents and therefore would generate different CO_2 emissions. This can be readily appreciated from the model (8.26), in which, given the same energy conversion efficiency, different fuel inputs (*i.e.*, different fuel-related emission factors) yield different output-related emission factors.

On these bases, it is possible to highlight a key theoretical result [ChM08][MaC08] of the unified structure proposed for the *PPES* and the *PCO2ER*. More specifically, if we hypothesize that all the systems involved in the analysis (multi-generation and separate generation) are supplied by the same type of fuel, the formulation of the *PCO2ER* and *PPES* indicators would *coincide*, as it is straightforward to notice by substituting (8.26) in (8.27). In particular, under this assumption the two indicators would yield the *same numerical values*. Hence, the CO_2 emission reduction would depend only upon the relevant energy efficiencies considered in the analysis, since the energy saving would correspond to the CO_2 emissions reduction brought by the combined MG process.

The comprehensive energy efficiency and environmental impact framework presented here enables us to use the general models and trends illustrated in the previous chapters also for environmental assessment. In addition, this framework could be adopted by specific regulations aimed at assessing the MG system performance, to extend the regulation referred to cogeneration systems, already in force [EU04b][CaP05]. Further policy development, aimed at promoting a larger diffusion of high-efficiency and low-emission MG systems, can be envisioned by applying the indicators described in Section 5.6.2 and in this section. In this respect, properly addressing energy and environmental performance of energy systems is also becoming more relevant in the light of the possibility of enhancing the MG system economic profitability. More specifically, this could occur either by granting MG systems adequate incentives on the basis of proven energy and emission saving as for cogeneration [CaP05], or by enabling their participation to different types of *energy-related* markets [ChM07b] at national and transnational levels, for instance dedicated to emission trading [EU03] and to white and green certificates [GMEww], as discussed in Section 8.3.

For all these issues, in analogy to what widely discussed with respect to energy analyses, again an appropriate selection of the reference models assumes a fundamental role.

8.4.4. Local emissions

Besides the aspects concerning CO_2 and GHG emissions (that we can consider *global pollutants* due to the scale of their global warming effects), the diffusion of DMG technologies closer to the consumers is raising the issue of taking into account the effects of a number of hazardous pollutants, emitted by the generators distributed over the territory, on the human health. Considering the environmental impact of such pollutants (that can be dubbed *local pollutants*, again due to their more or less limited impact radius [MaC09]) is essential when dealing with consumer sites located sufficiently close to the sources. The studies on local air quality are challenging, due to the need for taking into account, for each pollutant emitted from the local sources (also at partial-load conditions [MaC09]), the characteristics (location, height) of the point of emission of the pollutants, the time-dependent exposure of the site to atmospheric agents to determine the pollutant dispersion [Ary99], the potential hazard to the human health, and so forth.

The environmental impact from multi-generation systems can be addressed on the local and global levels by using the *emission balance approach* [EDU01][CaC08][MaC09], according to which the mass of pollutant emitted is compared to the corresponding mass emitted in separate production to satisfy the same demand of the various energy vectors. Suitable assessment models, also considering the spatial effects of each relevant pollutant, can be elaborated within the emission balance approach starting from the emission factors described in Section 8.4.1 [MaC09]. In particular, the methodology illustrated above for CO_2 emission assessment, leading to the definition of the *PCO2ER* indicator (8.27), is a practical example of emission balance applied to global pollutants. A similar approach can be applied to other GHGs by defining equivalent emission factors, depending on the *GWP* of the respective GHG [ChM08].

The environmental impact also depends on the type of fuel used for the local generators. Direct comparisons are possible among technologies adopting the same fuel (*e.g.*, natural gas). If the analyses entail the presence of different fuels (such as gas, diesel, bio-masses), and more extensively of different primary energy sources (*e.g.*, sun or wind) supplying the equipment included in the multi-generation system, more detailed evaluations are possible within the energy and environmental LCA framework [Bou96][MaH96][Peh01][Reb04] [PeP04][ExE05][ChM05][CaC08]. In this case, the analyses are carried out by taking into account the upstream and downstream impacts related to fuel extraction, processing, transportation, and distribution, following a cradle-to-grave approach [GoS03][BaR04] [WiH06][FiC07].

8.5. Final Remarks on Distributed Multi-Generation Perspectives

The issues presented in this book have been selected to provide an introductory view with some details on models and system-based analysis techniques concerning the development of DMG solutions. The effective implementation and exploitation of these solutions requires further studies and the deployment of a significant amount of new resources from the technical, economic and scientific points of view. This concluding section summarizes some perspectives for the application of the DMG paradigm in the comprehensive "triple E" scenario indicated in Section 1.5, entailing Energy, Environment and Economics.

Besides the specific characteristics of the various technologies, the major upsides of DMG systems come from the overall flexibility due to the potential interactions among interconnected multi-generation sites. Managing these interactions requires significant efforts towards the formulation of enhanced planning and operation procedures. In particular, suitable tools are needed to deal with the high level of complexity of DMG energy planning and management, requiring different types of analysis techniques. The major challenges refer to the formulation and solution of comprehensive optimisation problems encompassing the variety of issues involved in the study. In this respect, DMG assessment problems can be schematically grouped on the basis of the relevant time frame. According to typical classifications, plant design-orientated *planning* problems are formulated in the *long-term* time frame, over the plant useful life (*e.g.*, 10÷20 years). On the other hand, system *operation* is addressed in the *short-term* time frame, over an assigned period (*e.g.*, one year). In general, the specific features of the different time frames call for different types of modelling and analyses [ChM09]. A number of aspects related to DMG assessment for both planning and operational purposes are included within the more general discussions presented below. Integrated operational planning tools could also be developed, as shown for instance in [ArF07] and [PiC07].

8.5.1. DMG planning perspectives

From the planning point of view, acquisition of new resources and substitution of existing ones need to be rethought in the light of foreseeing the trends of technological development, availability of energy sources, and evolution of energy prices. In this respect, a key role is played by the many uncertainties involved in the study. To cope with these uncertainties, it is possible to carry out sensitivity analyses considering price scenarios associated to probabilistic weights. The objective functions of these analyses can be extended beyond the classical ones based on investment and operation and management costs. Further aspects to be incorporated include the representation of the environmental impact of DMG solutions and the related economics. More extensively, possible benefits may come from the DMG interactions with various types of energy-related markets, as shown in Section 8.3.2 for what concerns assessment of the sensitivity of the electricity production cost to variations in the prices of white certificates and emission allowances.

In the overall development of composite energy systems, environmental issues are becoming more and more important also in the light of assessing the *external costs* of energy. The notion of external costs encompasses the costs not directly included in the cash flows related to the construction, operation and dismantling of a plant. In this respect, damages caused to human health, monuments, ecosystems, and so forth, from emissions referred to energy production represent a classical example of external costs on the environment and on the society [Gul06b]. External cost assessment can be carried out according to detailed, comprehensive, and often cumbersome procedures. For instance, the ExternE project [ExE05] introduced a *bottom-up* approach based on the concept of tracking back the various emissions to which the receptors are exposed to their initial causes, determining the so-called *impact pathways*. These determinations should lead to an economic quantification of the potential damages to the receptors. The use of the emission factors described in Section 8.4 can assist the process of determining the characteristics of the emissions from the generation sources, to be used as inputs to external cost assessment models. It is worth mentioning that in general full-load emissions are insufficient to provide a clear picture of the emissions of hazardous pollutants during the energy system operation. More detailed evaluations, supported by experimental analyses, have to be carried out to represent the partial-load behaviour of the emissions from the local generators. Once assessed the external costs, it is possible to devise specific procedures to internalise the external costs [HoA07], in order to obtain overall information to be used in economic evaluations.

Further elements can be added to the approaches and tools illustrated in this book, in line with the innovative views on the evaluation of energy systems and markets. The current trend is to construct suitable tools for the appraisal of different types of *risks* referred to the acquisition and exploitation of new technological solutions, including the effects of the various uncertainties in the evolution of demand and energy prices, as well as those referred to financial aspects and possible changes in the environmental regulations. On these bases, suitable strategies can be formulated for *hedging* the risks depending on the various forms of uncertainty [DaL03]. In a DMG-based context, development of sound hedging mechanisms needs to take into account the different nature of the various energy vectors and sources. In particular, electricity storage capabilities are very limited with respect to other commodities. However, in a comprehensive DMG system, electricity storage limitations could be mitigated by the possibility of producing electricity from other forms of stored energy (e.g., heat or hydrogen) through suitable energy vector transformations. In the planning horizon, possible significant impact on the electrical side may come from the development of hydrogen-based systems [ArS00][Dun02][GrD07], storage solutions [ClI94][KoH06][KoK08], and electric and hybrid vehicles [Cha02] with potential system-wide consequences of deploying plug-in options [GoL08]. Exploiting storage solutions, particularly on the thermal side, could lead to significant changes in both plant schemes and operational strategies of DMG systems.

The perspectives of flexible management of multiple energy vectors are also linked to an effective integration of the different energy networks. Besides the natural link through the electrical network, that guarantees the delivery of large amounts of energy in a very efficient way, additional interactions with district heating or district cooling networks can be taken into account to condition the thermal energy usage. The energy delivery through networks could be extended to other forms of energy (such as hydrogen), provided that an effective way to transport these forms of energy is found [GeK07]. The objective of making different energy vectors easily available to the user on a wide scale encompasses the rationale of existing developments such as Micro-grids and Virtual Power Plants (VPPs), moving towards effective and ambitious horizons of managing in a combined way multi-product systems coming from multiple sources [HeZ07].

The planning methodologies can be formulated in such a way to address *multi-objective* problems, in which technical, economic, and environmental objectives could be conflicting with each other. More generally, the decision-making process may require comparing a specific set of alternative solutions. In the presence of different scenarios of evolution of the energy systems, with many variables subject to high uncertainty, a deterministic or even a pure probabilistic approach would be insufficient to set up sound techniques of analysis. This situation can be addressed under a *multi-criteria* approach [And90][DeP03][AlK07], searching for the most suitable trade-off solution within a set of weighted scenarios identified by the decision-maker.

8.5.2. Perspectives for DMG system operation

In the presence of manifold energy demand, multi-generation plants with various types of equipment can provide significant opportunities to diversify the operational strategies. In this respect, it is important to consider technological, modelling, and computational aspects. The following remarks complement the indications on the perspectives concerning the interaction of multi-generation systems with the energy networks provided in Section 1.5, where a number of aspects related to the need for coordinated control and management, as well as to the role of power electronics and ICT in the development of advanced solutions, have already been remarked.

On the technological side, the information on the rated values of the equipment energy characteristics and emission factors is generally not sufficient to deal with the complexity of the operational mechanisms of a DMG system. Indeed, the expected benefits from exploiting DMG solutions highly depend on efficiently managing the partial-load operation of the interacting plant components. Yet, in partial-load conditions there may occur various drawbacks, ranging from possible efficiency reductions to increased operation and maintenance costs (especially in case of multiple switch-on/off operations, or more generally of dynamic operation), as well as to increased emissions of local pollutants [CaC08][MaC09]. Adequate information should then be obtained on the specific short-term operation of the various plant components. In many cases, the experimental set-up needed to gather accurate information is rather expensive, and manufacturers' data are often not disclosed. Thus, availability of a sufficient amount of data to be introduced into the specific models could be strongly limited.

On the modelling side, when considering a certain multi-generation plant scheme, besides the specific data of the equipment it is essential to represent the interactions among the plant components in an effective way. In this respect, *dispatch factors* can play the essential role of control variables [ChM09b]. Suitable set-up and adjustment of the dispatch factors allow the implementation of different control strategies for the MG system operation. This opens the possibility of optimally managing different types of equipment, for instance for cooling production, enabling the energy operator to score additional economic benefits with respect to classical strategies such as electrical and thermal load-following. Other dispatch factor adjustments could come from exploiting the interactions between combined multi-generation and RES. However, the determination of the most convenient control strategies is a challenging task, since in general economic benefits may not coincide with energy saving benefits or with reduction of the environmental impact, depending on the loading levels for the various energy vectors, on the market prices, and on the relevant network interactions. Thus, enhanced tools are required to schedule and coordinate the operation of DMG systems under comprehensive multi-objective or multi-criteria optimisations. More generally, the level of modelling details (*e.g.*, with possible incorporation of component and system dynamics) has to be decided depending on the specific type of calculation or simulation addressed. In many cases the system operation is considered as a succession of steady-state conditions, for instance using a sufficiently long time step, so as to neglect the faster effects of electro-mechanical dynamics on the equipment operation. More detailed modelling for time-domain simulations could be needed for investigating the interactions with networks, for instance to check the specific operation of protection devices whose tripping could put one or more plant components off-line. In a DMG scenario, the computational models need to encompass the effects of applying energy-shifting strategies and managing storage capabilities [RoL08]. This leads to strengthen coupling-in-time aspects in the formulation of the operational strategies in time-domain simulation problems.

On the computational point of view, significant issues may arise for formulating and solving specific optimisation problems according to given technical, economic, environmental, or combined objective functions. Challenges like possible non-convexity of the objective function domain [RoL05b], existence of multiple local optima, and the presence of time-dependent operational constraints (such as limits to the duration of on-off operation, ramp-rate constraints, or storage capabilities) are similar to those already encountered in specific problems in various fields. However, in a DMG system the individual challenges add up and can become critical in terms of problem dimensionality and interacting constraints.

The coexisting issues belonging to different fields call for reshaping the formulation of the problems in a more comprehensive way, for creating upgraded conceptual frameworks and tools, and for interpreting the solutions under a truly interdisciplinary view. Providing adequate responses to these issues in the DMG context in order to develop more sustainable energy systems is thus the major challenge of present and forthcoming research.

REFERENCES

[AbA06] Abu-Sharkh S, Arnold RJ, Kohler J, Li R, Markvart T, Ross JN, Steemers K, Wilson P, Yao L. Can microgrids make a major contribution to UK energy supply? Renewable and Sustainable Energy Reviews, 2006: 10 (2): 78-127.

[AcA01] Ackermann T, Andersson G, Söder L. Distributed generation: a definition. Electric Power Systems Research, 2001; 57 (3): 195–204.

[Ack07] Ackermann T. Distributed resources and re-regulated electricity markets. Electric Power Systems Research, 2007; 77 (9): 1148-1159.

[AEEG] Italian Electricity and Gas Authority (Autorità per l'energia elettrica e il gas). Available (September 2008): www.autorita.energia.it.

[Afo06] Afonso CFA. Recent advances in building air conditioning systems. Applied Thermal Engineering, 2006; 26 (16): 1961-1971.

[AkO05] Aki H, Oyama T, Tsuji K. Analysis of energy pricing in urban energy service systems considering a multiobjective problem of environmental and economic impact. IEEE Transactions on Power Systems, 2005; 18 (4): 1275-1282.

[AkO06] Aki H, Oyama T, Tsuji K. Analysis of energy service systems in urban areas and their CO_2 mitigations and economic impacts. Applied Energy, 2006; 83 (10): 1076-1088.

[AkY06] Aki H, Yamamoto S, Ishikawa Y, Kondoh J, Maeda T, Yamaguchi H, Murata A, Ishii I. Operational strategies of networked fuel cells in residential homes. IEEE Transactions on Power Systems, 2006; 21 (3): 1405-1414.

[AlF97] Alfano G, Filippi M, Sacchi E. Impianti di climatizzazione per l'edilizia (Air conditioning plants for building applications, in Italian). Masson, Milano, Italy, 1997.

[AlK07] Al Mansour F, Kožuh M. Risk analysis for CHP decision making within the conditions of an open electricity market. Energy, 2007; 32: 1905-1916.

[AlL02] Allison J, Lents J. Encouraging distributed generation of power that improves air quality: can we have our cake and eat it too? Energy Policy, 2002; 30: 737-752.

[AlP07] Alfonso D, Pérez-Navarro A, Encinas N, Àlvarez C, Rodríguez J, Alcázar M. Methodology for ranking customer segments by their suitability for distributed energy resources applications. Energy Conversion and Management, 2007; 48 (5): 1615-1623.

[AlS06] Alanne K, Saari A. Distributed energy generation and sustainable development. Renewable & Sustainable Energy Reviews, 2006; 10: 539-558.

[AmH04] Ameri M, Hejazi SH. The study of capacity enhancement of the Chabahar gas turbine installation using an absorption chiller. Applied Thermal Engineering, 2004; 24 (1): 59-68.

[AmW05]Amin SM, Wollenberg BF. Toward a smart grid: power delivery for the 21st century. IEEE Power and Energy Magazine, 2005; 3 (5): 34-41.

[And90] Anders GJ. Probability concepts in electric power systems. Wiley, New York, 1990.

[Ang98] Anglesio P. Elementi di Impianti Termotecnici (Elements of thermotechnical plants, in Italian). Pitagora Editrice, Bologna, Italy, 1998.

[AnL07] Andersen AN, Lund H. New CHP partnerships offering balancing of fluctuating renewable electricity productions. Journal of Cleaner Production, 2007; 15 (3): 288-293.

[ArF07] Arcuri P, Florio G, Fragiacomo P. A mixed integer programming model for optimal design of trigeneration in a hospital complex. Energy, 2007; 32: 1430-1447.

[ArS00] Árnason B, Sigfússon TI. Iceland — a future hydrogen economy. International Journal of Hydrogen Energy, 2000; 25 (5): 389-394.

[Ary99] Arya SP. Air pollution meteorology and dispersion. Oxford University Press, New York, 1999.

[AsB03] Ashok S. Baneryee R. Optimal cool storage capacity for load management. Energy, 2003; 28: 115-126.

[ASH00] American Society of Heating, Refrigerating and Air-conditioning Engineers, ASHRAE HVAC Systems and Equipment Handbook, SI Edition, ASHRAE, 2000.

[AsM07] Asif M, Muneer T. Energy supply, its demand and security issues for developed and emerging economies. Renewable and Sustainable Energy Reviews, 2007; 11 (7): 1388-1413.

[AwP97] Awerbuch S, Preston A. The Virtual Utility: Accounting, Technology & Competitive Aspects of the Emerging Industry. Kluwer Academic Publishers, Massachusetts, USA, 1997.

[BaC99] Barsali S, Ceraolo M, Giglioli R, Pelacchi P. Microturbines for dispersed generation, Proceedings of CIRED 1999, Nice, France, 1-4 June 1999.

[BaC06] Batrinu F, Chicco G, Fazari L, Spertino F. Grid-connected PV installations in social houses: experimental evaluation and diffusion perspectives. Proceedings 21st European Photovoltaic Solar Energy Conference and Exhibition, Dresden, Germany, 4-8 September 2006, 2686-2690.

[BaE07] Baran ME, El-Markabi IM. A Multiagent-Based Dispatching Scheme for Distributed Generators for Voltage Support on Distribution Feeders. IEEE Transactions on Power Systems, 2007; 22 (1): 52-59.

[BaJ07] Bahaj AS, James PAB. Urban energy generation: The added value of photovoltaics in social housing. Renewable and Sustainable Energy Reviews, 2007; 11 (9): 2121-2136.

[BaK02] Bassols J, Kuckelkorn B, Langreck J, Schneider R, Veelken H. Trigeneration in the food industry. Applied Thermal Engineering, 2002; 22 (6): 595-602.

[BaL01] Bazilian MD, Leenders F, Van Der Ree BGC, Prasad D. Photovoltaic cogeneration in the built environment. Solar Energy, 2001; 71: 57-69.

[BaR04] Bargigli S, Raugei M, Ulgiati S. Comparison of thermodynamic and environmental indexes of natural gas, syngas and hydrogen production processes. Energy, 2004; 29: 2145-2159.

[BeD06] Bernal-Augustín JL, Dufo-López R, Rivas-Ascaso DM. Design of isolated hybrid systems minimizing costs and pollutant emissions. Renewable Energy, 2006; 31: 2227-2244.

[BeH06] Bertoldi P, Huld T. Tradable certificates for renewable electricity and energy savings. Energy Policy, 2006; 34 (2): 212-222.

[Bej97] Bejan A. Advanced engineering thermodynamics. 1997; John Wiley & Sons, New York.

[BeK03] Bejan A, Kraus AD. Heat transfer handbook. John Wiley & Sons, Hoboken, New Jersey, USA, 2003.

[Ber02] Berntsson T. Heat sources – technology, economy and environment. International Journal of Refrigeration, 2002; 25: 428-438.

[BeT96] Bejan A, Tsatsaronis G, Moran M. Thermal design and optimization, John Wiley and Sons, New York, USA, 1996.

[BiM04] Bizzarri G, Morini GL. Greenhouse gas reduction and primary energy savings via adoption of a fuel cell hybrid plant in a hospital. Applied Thermal Engineering, 2004; 24: 383-400.

[BiS06] Biezma MV, San Cristóbal JR. Investment criteria for the selection of cogeneration plants—a state of the art review. Applied Thermal Engineering, 2006; 26 (5-6): 583-588.

[BoB07] Bolund B, Bernhoff H, Leijon M. Flywheel energy and power storage systems. Renewable and Sustainable Energy Reviews, 2007; 11 (2): 235-258.

[Bod06] Bode S. Multi-period emissions trading in the electricity sector—winners and losers. Energy Policy, 2006; 34 (6): 680-691.

[Boe97] Boehm RF (editor). Developments in the design of thermal systems. 1997; Cambridge University Press, Cambridge, UK.

[BoK01] Borbely AM, Kreider JF (ed.). Distributed generation – The power paradigm of the new millennium, CRC Press, Boca Raton, FL, 2001.

[Bou96] Boustead I. LCA - How it Came about, The Beginning in the U.K. The International Journal of Life Cycle Assessment, 1996; 1: 147-150.

[BrV05] Bruno JC, Valero A, Coronas A. Performance analysis of combined microgas turbines and gas fired water/LiBr absorption chillers with post-combustion. Applied Thermal Engineering, 2005; 25: 87-99.

[BuC03] Bua F, Calderazzi L. Design and building of an industrial cogeneration plant (in Italian: Progetto e realizzazione di un impianto di cogenerazione industriale). Power Technology, 2003; 2 (4).

[BuT03] Burer M, Tanaka K, Favrat D, Yamada K. Multi-criteria optimization of a district cogeneration plant integrating a solid oxide fuel cell-gas turbine combined cycle, heat pumps and chillers. Energy, 2003; 28 (6): 497-518.

[Bys07] Bysveen M. Engine characteristics of emissions and performance using mixtures of natural gas and hydrogen. Energy, 2007; 32: 482-489.

[CaB02] Cârdu M, Baica M. Regarding the greenhouse gas emissions of thermopower plants. Energy Conversion and Management, 2002; 43: 2135-2144.

[CaB04] Calì M, Borchiellini R. District heating networks calculation and optimization. In Frangopoulos CA (editor) Exergy, Energy System Analysis, and Optimization. In Encyclopedia of Life Support Systems (EOLSS), Eolss Publishers, Oxford, UK, 2004.

[CaC08] Canova A, Chicco G, Genon G, Mancarella P. Emission characterization and evaluation of natural gas-fueled cogeneration microturbines and internal combustion engines. Energy Conversion and Management, 2008; 49 (10): 2900-2909.

[CaF06] Carrasco JM, Franquelo LG, Bialasiewicz JT, Galvan E, Portillo Guisado RC, Prats MAM, Leon JI, Moreno-Alfonso N. Power-Electronic Systems for the Grid Integration of Renewable Energy Sources: A Survey. IEEE Transactions on Industrial Electronics, 2006; 53 (4): 1002–1016.

[CaG96] Calì M, Gregorio P. Thermodynamics (in Italian: Termodinamica). Progetto Leonardo, Bologna, Italy, 1996.

[Cam00] Campanari S. Full-load and part-load performance prediction for integrated SOFC and microturbine systems. Journal of Engineering for Gas Turbines and Power, 2000; 122 (2): 239-246.

[CaM09] Campanari S, Manzolini G, de la Iglesia FG. Energy analysis of electric vehicles using batteries or fuel cells through well-to-wheel driving cycle simulations. Journal of Power Sources, 2009; 186: 464-477.

[CaP03] Cardona E, Piacentino A. A method for sizing a trigeneration plant in Mediterranean areas, Applied Thermal Engineering, 2003; 23: 1665-1680.

[CaP05] Cardona E, Piacentino A. Cogeneration: a regulatory framework toward growth, Energy Policy, 2005; 33: 2100-2111.

[CaP06] Cardona E, Piacentino A, Cardona F. Matching economical, energetic and environmental benefits: an analysis for hybrid CHCP-heat pump systems, Energy Conversion and Management, 2006; 47: 3530-3542.

[CaP06b] Cardona E, Piacentino A, Cardona F. Energy saving in airports by trigeneration. Part I: Assessing economic and technical potential. Applied Thermal Engineering, 2006; 26 (14-15): 1427-1436.

[CaP06c] Cardona E, Piacentino A. A new approach to exergoeconomic analysis and design of variable demand energy systems. Energy, 2006; 31 (4): 490-515.

[CaP07] Cardona E, Piacentino A. Optimal design of CHCP plants in the civil sector by thermoeconomics. Applied energy, 2007; 84: 729-748.

[CaR02] Cattaneo F, Rigoni L. Heat pumps for industrial use. Tests on a prototype working with R-134a, and first economic, energetic and environmental evaluations (in Italian: Pompe di calore per uso industriale. Sperimentazione su di una macchina prototipale funzionante con fluido refrigerante R-134° e prime valutazioni economiche, energetiche e ambientali). CESI report, 2002. Available (September 2008): www.ricercadisistema.it.

[CaS06] Cardona E, Sannino P, Piacentino A, Cardona F. Energy saving in airports by trigeneration. Part II: Short and long term planning for the Malpensa 2000 CHCP plant. Applied Thermal Engineering, 2006; 26 (14-15): 1437-1447.

[Cat01] Cattaneo F. Applications of the heat pump technology in the main industrial sectors. (in Italian: Applicazioni della tecnologia a pompa di calore (PDC) nei principali settori industriali). CESI report, 2001. Available (September 2008): www.ricercadisistema.it.

[Cat01b] Cattaneo F. Adjustment, installation and test of a heat pump experimental platform for industrial use (in Italian: Messa a punto, installazione e collaudo di una piattaforma sperimentale a pompa di calore (PDC) per uso industriale). CESI report, 2001. Available (September 2008): www.ricercadisistema.it.

[CaT05] Calva E. Teopa, Núñez M. Picón, Toral M.A. Rodríguez. Thermal integration of trigeneration systems. Applied Thermal Engineering, 2005; 25 (7): 973-984.

[CaT07] Carbon Trust. Micro-CHP accelerator, Interim Report, UK, November 2007. Available (September 2008): www.carbontrust.co.uk.

[Cha02] Chan C-C. The state of the art of electric and hybrid vehicles. Proceedings of the IEEE, 2002; 90 (2): 247–275.

[ChB05] Chen Z, Blaabjerg F, Pedersen JK. Hybrid compensation arrangement in dispersed generation systems. IEEE Transactions on Power Delivery, 2005; 20 (2): 1719-1727.

[ChM05] Chevalier C, Meunier F. Environmental assessment of biogas co- or tri-generation units by life cycle analysis methodology. Applied Thermal Engineering, 2005; 25 (17-18): 3025-3041.

[ChM06] Chicco G, Mancarella P. From cogeneration to trigeneration: profitable alternatives in a competitive market, IEEE Transactions on Energy Conversion, 2006; 21 (1): 265-272.

[ChM06b] Chicco G, Mancarella P. Planning Evaluation and Economic Assessment of the Electricity Production from Small-scale Trigeneration Plants, WSEAS Transactions on Power Systems, 2006; 1 (2): 393-400.

[ChM06c]Chicco G, Mancarella P. Incremental indicators for assessing the performance of cogeneration systems with heat pumps. WSEAS Transactions on Power Systems, 2006; 1 (8): 1491-1498.

[ChM07a] Chicco G, Mancarella P. Trigeneration Primary Energy Saving Evaluation for Energy Planning and Policy Development. Energy Policy, 2007; 35 (12): 6132-6144.

[ChM07b]Chicco G, Mancarella P. Exploiting small-scale cogeneration in energy-related markets. Proc. IEEE Power Tech 2007 Conference. Lausanne, Switzerland; 1-5 July 2007.

[ChM07c]Chicco G, Mancarella P. Enhanced Energy Saving Performance in Composite Trigeneration Systems. Proc. IEEE Powertech 2007 Conference. Lausanne, Switzerland;1-5 July 2007.

[ChM07d]Chicco G, Mancarella P, Spertino F. Primary energy saving from microturbine-photovoltaic combined energy systems. Proc. 22nd European Photovoltaic Solar Energy Conference. Milano, Italy; 3-7 September 2007.

[ChM08] Chicco G, Mancarella P. Assessment of the greenhouse gas emissions from cogeneration and trigeneration systems. Part I: Models and indicators. Energy, 2008; 33 (3): 410-417.

[ChM08b]Chicco G, Mancarella P. A unified model for energy and environmental performance assessment of natural gas-fueled poly-generation systems. Energy Conversion and Management, 2008; 49 (8): 2069-2077.

[ChM08c]Chicco G, Mancarella P. Evaluation of multi-generation alternatives: an approach based on load transformations. Proceedings IEEE PES General Meeting 2008, Pittsburgh, PA, 20-24 July 2008, paper 08GM1283.

[ChM09] Chicco G, Mancarella P. Distributed Multi-Generation: A Comprehensive View, Renewable and Sustainable Energy Reviews, 2009; 13 (3): 535-551.

[ChM09b]Chicco G, Mancarella P. Matrix modelling of small-scale trigeneration systems and application to operational optimization. Energy, 2009; 34 (3): 261-273.

[ChN03] Chicco G, Napoli R, Postolache P, Scutariu M, Toader C. Customer Characterization Options for Improving the Tariff Offer. IEEE Transactions on Power Systems, 2003; 18 (1): 381-387.

[ChN05] Chicco G, Napoli R, Piglione F, Scutariu M, Postolache P, Toader C. Emergent Electricity Customer Classification. IEE Proceedings Generation Transmission and Distribution, 2005; 152 (2): 164-172.

[ChS09] Chicco G, Schlabbach J, Spertino F. Experimental assessment of the waveform distortion in grid-connected photovoltaic systems. Solar Energy, 2009, doi:10.1016/j.solener.2009.01.005.

[ClI04] Clark W, Isherwood W. Distributed generation: remote power systems with advanced storage technologies. Energy Policy, 2004; 32, (14): 1573-1589.

[CoA07] Colombo PML, Armanasco F, Perego O. Experimentation on a cogenerative system based on a microturbine. Applied Thermal Engineering, 2007; 27: 705-711.

[CoG03] Colonna P, Gabrielli S. Industrial trigeneration using ammonia-water absorption refrigeration systems (AAR). Applied Thermal Engineering, 2003; 23 (4): 381-396.

[COG05] COGEN EUROPE. Position statement on the definition of harmonised efficiency reference values; February 2005. Available (September 2008): http://www.cogen.org/Downloadables/Publications/Position_Paper_CHP_Reference_Values.pdf.

[CoG07] Costantini V, Gracceva F, Markandya A, Vicini G. Security of energy supply: Comparing scenarios from a European perspective. Energy Policy, 2007; 35 (1): 210-226.

[CoP07] Costa A, Paris J, Towers M, Browne T. Economics of trigeneration in a kraft pulp mill for enhanced energy efficiency and reduced GHG emissions. Energy, 2007; 32 (4): 474-481.

[CoP07b] Coll-Mayor D, Paget M, Lightner E. Future intelligent power grids: analysis of the vision in the European Union and in the United States. Energy Policy, 2007; 35: 2453-2465.

[CrT06] Crawford RH, Treloar GJ, Fuller RJ, Bazilian M. Life-cycle energy analysis of building integrated photovoltaic systems (BiPVs) with heat recovery unit. Renewable and Sustainable Energy Reviews, 2006; 10 (6): 559-575.

[CTwww] Cogeneration Technologies, Trigeneration website, Available (September 2008): www.trigeneration.com.

[CuV02a]Curti V, von Spakovsky MR, Favrat D. An environomic approach for the modeling and optimization of a district heating network based on centralized and decentralized heat pumps, cogeneration and/or gas furnace. Part I – Methodology. International Journal of Thermal Sciences, 2002; 39 (7): 721-730.

[CuV02b]Curti V, von Spakovsky MR, Favrat D. An environomic approach for the modeling and optimization of a district heating network based on centralized and decentralized heat pumps, cogeneration and/or gas furnace. Part II – Application. International Journal of Thermal Sciences, 2002; 39 (7): 731-741.

[DaG99] Davis MW, Gifford AH, Krupa TJ. Microturbines – An economic and reliability evaluation for commercial, residential, and remote load applications. IEEE Transactions on Power Systems, 1999; 14: 1556-1562.

[DaL03] Dahlgren R, Liu CC, Lawarrée J. Risk assessment in energy trading. IEEE Transactions on Power Systems, 2003; 18 (2) 503-511.

[Dan06] Danny Harvey LD. A handbook on low-energy buildings and district energy systems: fundamentals, techniques, and examples, 2006, James and James, UK.

[DaW06] Daou K, Wang RZ, Xia ZZ. Desiccant cooling air conditioning: a review. Renewable and Sustainable Energy Reviews, 2006; 10: 55-77.

[DeD07] De Ruyck J, Delattin F, Bram S. Co-utilization of biomass and natural gas in combined cycles through primary steam reforming of the natural gas. Energy, 2007; 32 (4): 371-377.

[DeL07] Delarue E, Lamberts H, D'haeseleer W. Simulating greenhouse gas (GHG) allowance cost and GHG emission reduction in Western Europe. Energy, 2007; 32 (8): 1299-1309.

[DeM90] De Vault RC, Marsala J. Ammonia-water triple effect absorption cycle, ASHRAE Transactions, 1990; part 1: 676-682.

[DeP03] Denton M, Palmer A, Masiello R, Skantze P. Managing Market Risk in Energy. IEEE Transactions on Power Systems, 2003; 18 (2): 494-502.

[DeW08] Deng J, Wang R, Wu J, Han G, Wu D, Li S. Exergy cost analysis of a micro-trigeneration system based on the structural theory of thermoeconomics. Energy, 2008; 33 (9): 1417-1426.

[DiA08] Dikici A, Akbulut A. Exergetic performance evaluation of heat pump systems having various heat sources. International Journal of Energy Research, 2008, 32: 1279-1296.

[DiH05] Dimeas AL, Hatziargyriou ND. Operation of a Multiagent System for Microgrid Control. IEEE Transactions on Power Systems, 2005; 20 (3): 1447-1455.

[Din99] Dincer I. Environmental impacts of energy. Energy Policy, 1999; 27: 845-854.

[DiR07] Dincer I, Rosen MA. Exergy. Energy, Environment and Sustainable Development, 2007, Elsevier, Oxford, UK.

[DOE04] U.S. Department of Energy, Review of Thermally activated technologies, Distributed energy program report, July 2004. Available (September 2008): http://www.eere.energy.gov/de/publications.html.

[DOEww] U.S. Department of Energy, Distributed Energy Program. Web site (available September 2008) http://www.eere.energy.gov/de/.

[DoW05] Dorer V, Weber R, Weber A. Performance assessment of fuel cell micro-cogeneration systems for residential buildings. Energy and Buildings, 2005; 37: 1132-1146.

[DuK06] Dugan RC, Key TS, Ball GJ. Distributed resources standards. IEEE Industry Applications Magazine, 2006; 12 (1): 27-34.

[Dun02] Dunn S. Hydrogen futures: toward a sustainable energy system. International Journal of Hydrogen Energy, 2002; 27: 235-264.

[DuT84] Dugan RC, Thomas SA, Rizy DT. Integrating dispersed storage and generation (DSG) with an automated distribution system, IEEE Transactions on Power Apparatus and Systems, 1984, 103, (6): 1142-1146.

[ECD01] European Commission, Directorate-general for energy. Energy savings by CHCP plants in the hotel sector. Report from European Commission, Save II programme, May 2001.

[EDU01] EDUCOGEN. The European Educational Tool on Cogeneration. December 2001. Available (September 2008): www.cogen.org/projects/educogen.htm.

[ElC03] El-Fadel M, Chedid R, Zeinati M, Hmaidan W. Mitigating energy-related GHG emissions through renewable energy. Renewable Energy, 2003; 28 (8): 1257-1276.

[ElH02] Elhadidy MA. Performance evaluation of hybrid (wind/solar/diesel) power systems. Renewable Energy, 2002; 26 (3): 401-413.

[ElS07] El-Sharkh MY, Sisworahardjo NS, Uzunoglu M, Onar O, Alam MS, Dynamic behavior of PEM fuel cell and microturbine power plants, Journal of Power Sources, 2007; 164 (1): 315-321.

[EnA07] Encinas N, Alfonso D, Àlvarez C, Perez-Navarro A, Garcia-Franco F. Energy market segmentation for distributed energy resources implementation purposes. IET Generation, Transmission & Distribution, 2007; 1 (2): 324–330.

[EPAww] US Environmental Protection Agency. Catalogue of CHP Technologies. Web site (available September 2008): www.epa.gov.

[Ert07] Ertesvag IS. Exergetic comparison of efficiency indicators for combined heat and power (CHP). Energy, 2007; 32: 2038-2050.

[ErV06] Eriksson M, Vamling L. Heat pumps and tradable emission permits: On the carbon dioxide emissions of technologies that cross a tradable emission market boundary. Energy Conversion and Management, 2006; 47 (20): 3510-3518.

[EU03] Directive 2003/87/EC of the European Parliament and of the Council of 13 October 2003, Establishing a scheme for greenhouse gas emission allowance trading within the Community and amending Council Directive 96/61/EC. Official Journal of the European Union; 25.10.2003, L 275, 32-46. http://europa.eu.int/eur-lex/pri/en/oj/dat/2003/l_275/l_27520031025en00320046.pdf (Available: September 2008).

[EU04] Directive 2004/101/EC of the European Parliament and of the Council of 27 October 2004, Amending Directive 2003/87/EC establishing a scheme for greenhouse gas emission allowance trading within the Community, in respect of the Kyoto Protocol's project mechanisms, Official Journal of the European Union; 13.11.2004, L 338, 18-23.

[EU04b] Directive 2004/8/EC of the European Parliament and of the Council of 11 February 2004 on the promotion of cogeneration based on a useful heat demand in the internal energy market and amending Directive 92/42/EEC. Official Journal of the European Union, 21.02.2004, L 52, 50–60. http://europa.eu.int/eur-lex/pri/en/oj/dat/2004/l_052/l_05220040221en00500060.pdf (Available September 2008).

[EU06] European Union, Directorate-General for Research, Sustainable Energy Systems European SmartGrids Technology Platform. Vision and Strategy for Europe's Electricity Network of the Future (ISBN 92-79-01414-5). European Communities, 2006.

[EUCww] CO_2 Prices.eu. Analysis of the EU CO2 Market. Available (September 2008): http://www.co2prices.eu.

[ExE05] Bickel P, Friedrich R (ed.). European Commission, ExternE: Externalities of Energy, Methodology, 2005 update, Available (September 2008): www.externe.info/brussels/methup05.pdf.

[FeC98] Feng X, Cai YN, Qian LL. A new performance criterion for cogeneration system. Energy Conversion and Management, 1998; 39 (15): 1607-1609.

[FiC07] Fiaschi D, Carta R. CO2 abatement by co-firing of natural gas and biomass-derived gas in a gas turbine. Energy, 2007; 32 (4): 549-567.

[FlT02] Florides GA, Tassou SA, Kalogirou SA, Wrobel LC. Review of solar and low energy cooling technologies for buildings. Renewable and Sustainable Energy Reviews, 2002; 6: 557-572.

[FoD91] Fortuin JB, Darmoutomo A, Corneth EJP. Absorption cooling and cogeneration, Klima Behersing, 1991; 20 (11): 339-344.

[FoD00] Foley G, De Vault R, Sweetser R. Future of Absorption Technology in America: A critical look at the impact of BCHP and innovation, Oak Ridge National Laboratory, June 2000. Available (September 2008): http://www.eere.energy.gov/de/publications.html.

[FtA06] Fthenakis V, Alsema E. Photovoltaics energy payback times, greenhouse gas emissions and external costs: 2004-early 2005 status. Progress in photovoltaics, research and applications, 2006; 14 (3): 275-280.

[GaJ04] Gao L, Jin H, Liu Z, Zheng D. Exergy analysis of coal-based polygeneration system for power and chemical production. Energy, 2004; 29 (12-15): 2359-2371.

[Gan96] Ganesan V. Internal Combustion Engines. McGraw-Hill, New York, 1996.

[GaN99] Gama Cerqueira, da SAA, Nebra SA. Cost attribution methodologies in cogeneration systems. Energy Conversion and Management, 1999; 40: 1587-1597.

[GaR04] Gareta R, Romeo LM, Gil A. Methodology for the economic evaluation of gas turbine air cooling systems in combined cycle applications. Energy, 2004; 29: 1805-1818.

[GaW08] Gao L, Wu H, Jin H, Yiang M. System study of combined cooling, heating and power system for eco-industrial parks. International Journal of Energy Research, 2008, 32: 1107-1118.

[GeA07] Geidl M, Andersson G. Optimal power flow of multiple energy carriers. IEEE Transactions on Power Systems, 2007; 22: 145-155.

[Gei07] Geidl M. Integrated modelling and optimization of multi-carrier energy systems. Doctoral thesis, ETH Zürich, Switzerland, 2007.

[GeK07] Geidl M, Koeppel G, Favre-Perrod P, Klöckl B, Andersson G, Fröhlich K. Energy Hubs for the Future. IEEE Power and Energy Magazine, 2007; 1: 25-30.

[GiJ06] Gil HA, Joos G. On the Quantification of the Network Capacity Deferral Value of Distributed Generation. IEEE Transactions on Power Systems, 2006; 21 (4): 1592-1599.

[GMEww]Italian Market Operator (Gestore del Mercato Elettrico - GME). Web site http://www.mercatoelettrico.org.

[GoB07] Godefroy J, Boukhanouf R, Riffat S. Design, testing and mathematical modelling of a small-scale CHP and cooling system (small CHP-ejector trigeneration). Applied Thermal Engineering, 2007; 27 (1): 68-77.

[Gog05] Gogus YA. Thermoeconomic optimization. International Journal of Energy Research, 2005; 29: 559-580.

[GoG07] Gommed K, Grossman G. Experimental investigation of a liquid desiccant system for solar cooling and dehumidification. Solar Energy, 2007; 81: 131-138.

[GoH08] Gomri R, Hakimi R. Second law analysis of double effect vapour absorption cooler system. Energy Conversion and Management, 2008; 49: 3343-3348.

[GoL08] Gong Q, Li Y, Peng Z-R. Trip-Based Optimal Power Management of Plug-in Hybrid Electric Vehicles. IEEE Transactions on Vehicular Technology, 2008; 57 (6): 3393:3401.

[GoS03] González A, Sala JM, Flores I, López LM. Application of thermoeconomics to the allocation of environmental loads in the life cycle assessment of cogeneration plants. Energy, 2003; 28: 557-574.

[GoV08] Gómez VH, Vidal A, Best R, García-Valladares O, Velázquez N. Theoretical and experimental evaluation of an indirect-fired GAX cycle cooling system. Applied Thermal Engineering, 2008; 28: 975-987.

[GrD07] Granovskii M, Dincer I, Rosen MA. Greenhouse gas emissions reduction by use of wind and solar energies for hydrogen and electricity production: Economic factors. International Journal of Hydrogen Energy, 2007; 32, (8): 927-931.

[GrF07] Grekas DN, Frangopoulos CA. Automatic synthesis of mathematical models using graph theory for optimisation of thermal energy systems. Energy Conversion and Management, 2007; 48 (11): 2818-2826.

[GrH07] Gross R, Heptonstall P, Leach M, Anderson D, Greem T, Skea J. Renewables and the grid: understanding intermittency. Proceedings of the Institution of Civil Engineers, Energy, 2007; 160 (issue EN1): 31-41.

[GrK07] Greiner CJ, Korpås M, Holen AT. A Norwegian case study on the production of hydrogen from wind power. International Journal of Hydrogen Energy, 2007; 32: 1500-1507.

[GrM03] Grillo O, Magistri L, Massardo AF. Hybrid systems for distributed power generation based on pressurization and heat recovery of an existing 100 kW molten carbonate fuel cell. Journal of Power Sources, 2003; 115 (2): 252-267.

[GrW04] GridWise Alliance. Rethinking energy from generation to consumption. Web site www.gridwise.org.

[Gul06] Gulli F. Small distributed generation versus centralised supply: a social cost–benefit analysis in the residential and service sectors. Energy Policy, 2006; 34: 804-832.

[Gul06b] Gulli F. Social choice, uncertainty about external costs and trade-off between intergenerational environmental impacts: The emblematic case of gas-based energy supply decentralization. Ecological Economics, 2006; 57 (2): 282-305.

[HaA07] Hatziargyriou N, Asano H, Iravani R, Marnay C. Microgrids. An overview of ongoing research, development, and demonstration projects. IEEE Power and Energy, 2007; 5 (4); 78-94.

[HaL07] Hawkes AD, Leach MA. Cost-effective operating strategy for residential micro-combined heat and power. Energy, 2007; 32: 711-723.

[Han04] Hanssen JH, Riikonen A, Noren C, Karlsson G, Malmrup L, Ernebrandt S, Stokholm R, Veland BJ, Fock , Mosbech H, de Wit J, Pedersen AH. Operating experiences from 18 microturbine applications for CHP and industrial purposes. Danish Gas Technology Centre, Report C0404. Available (September 2008): http://www.dgc.dk/publikationer/konference/jdw_igrc04.pdf.

[Has98] Hasnain SM. Review on sustainable thermal energy storage technologies, Part I: heat storage materials and techniques. Energy Conversion and Management, 1998; 39: 1127-1138.

[Has98b] Hasnain SM. Review on sustainable thermal energy storage technologies, Part II: cool thermal storage. Energy Conversion and Management, 1998; 39 (11): 1139-1153.

[Hav99] Havelsky V. Energetic efficiency of cogeneration systems for combined heat, cold and power production, International Journal of Refrigeration, 1999; 22: 479-485.

[HaV03] Haibo Y, Vittal V, Zhong Y. Self-healing in power systems: an approach using islanding and rate of frequency decline-based load shedding. IEEE Transactions on Power Systems, 2003; 18 (1): 174-181.

[HeB02] Heteu PMT, Bolle L. Economie d'énergie en trigénération. International Journal of Thermal Sciences, 2002; 41 1151-1159.

[HeE09] Hepbasli A, Erbay Z, Icier F, Colak N, Hancioglu E. A review of gas engine-driven heat pumps (GEHPs) for residential and industrial applications. Renewable and Sustainable Energy Reviews, 2009; 13: 85-99.

[HeG05] Hernandez-Aramburo CA, Green TC, Mugniot N. Fuel consumption minimization of a microgrid, IEEE Transactions on Industry Applications, 2005, 41 (3): 673-681.

[Hen07] Henning HM. Solar assisted air conditioning of buildings – an overview. Applied Thermal Engineering, 2007; 27: 1734-1749.

[Hep05] Hepbasli A. Thermodynamic analysis of a ground-source heat pump system for district heating. International Journal of Energy Research, 2005, 29, 671-687.

[HeP07] Henning H.-M, Pagano T, Mola S, Wiemken E. Micro tri-generation system for indoor air conditioning in the Mediterranean climate. Applied Thermal Engineering, 2007; 27 (13): 2188-2194.

[HeS03] Hernandez-Santoyo J, Sanchez-Cifuentes A. Trigeneration: an alternative for energy savings. Applied Energy, 2003, 76 (1-3): 219-227.

[HeZ07] Hemmes K, Zachariah-Wolff JL, Geidl M, Andersson G. Towards multi-source multi-product energy systems. International Journal of Hydrogen Energy, 2007; 32: 1332-1338.

[HiS07] Hiremath RB, Shikha S, Ravindranath NH, Decentralized energy planning; modeling and application – a review. Renewable and Sustainable Energy Reviews, 2007; 11: 729-752.

[HoA07] Holmgren K, Amiri S. Internalising external costs of electricity and heat production in a municipal energy system. Energy Policy, 2007; 35 (10): 5242-5253.

[HoA08] Houwing M, Ajah AN, Heijnen PW, Bouwmans I, Herder PM. Uncertainties in the design and operation of distributed energy resources: The case of micro-CHP systems. Energy, 2008; 33: 1518-1536.

[HoC04] Ho JC, Chua KJ, Chou SK. Performance study of a microturbine system for cogeneration application, Renewable Energy, 2004; 29: 1121-1133.

[Hon05] Hondo H. Life cycle GHG emission analysis of power generation systems: Japanese case. Energy 2005; 30: 2042-2056.

[Hor97] Horlock JH. Cogeneration-Combined Heat and Power (CHP). Krieger, Malabar, FL, 1997.

[Hor07] Horlock JH. Advanced Gas Turbine Cycles. Krieger, Malabar, FL, 2007.

[Huf92] Hufford E. Absorption chillers improve cogeneration energy efficiency. ASHRAE Journal 1992: 46-53.

[HuW07] Huangfu Y, Wu JY, Wang RZ, Xia ZZ. Experimental investigation of adsorption chiller for micro-scale BCHP system application. Energy and Building 2007; 39: 120-127.

[HuW07b]Huangfu Y, Wu JY, Wang RZ, Kong XQ, Wei BH. Evaluation and analysis of novel micro-scale combined cooling, heating and power (MCCHP) system. Energy Conversion and Management, 2007; 48: 1703-1709.

[Hwa04] Hwang Y. Potential energy benefits of integrated refrigeration system with microturbine and absorption chiller. International Journal of Refrigeration, 2004; 27 (8): 816-829.

[IsK05] Ise T, Kita M, Taguchi A. A hybrid energy storage with a SMES and secondary battery. IEEE Transactions on Applied Superconductivity, 2005; 15 (2): 1915-1918.

[InI05] Invernizzi C, Iora P. Heat recovery from a micro-gas turbine by vapour jet refrigeration systems. Applied Thermal Engineering, 2005; 25 (8-9): 1233-1246.

[Jäg07] Jäger-Waldau A. Photovoltaics and renewable energies in Europe. Renewable and Sustainable Energy Reviews, 2007; 11: 1414-1437.

[JeA00] Jenkins N, Allan R, Crossley P, Kirschen D, Strbac G. Embedded generation. IEE Power and Energy Series 31, The IEE, London, UK, 2000.

[JeI06] Jebaraj S, Iniyan S. A review of energy models. Renewable and Sustainable Energy Reviews, 2006; 10: 281-311.

[INTww] Intergovernmental Panel on Climate Change, Working Group III. Data on carbon dioxide intensities. Available (September 2008): http://arch.rivm.nl/env/int/ipcc/pages_media/SROC-final/Tables/t0305.pdf.

[IPCww] Intergovernmental Panel on Climate Change (IPCC), Working Group III, Data on CO2 intensities. Available (September 2008): http://arch.rivm.nl/env/int/ipcc/pages_media/SROC-final/Tables/t0305.pdf.

[KaG03] Karlsson Å, Gustavsson L. External costs and taxes in heat supply systems. Energy Policy, 2003; 31 (14): 1541-1560.

[KaI06] Katiraei F, Iravani MR. Power Management Strategies for a Microgrid With Multiple Distributed Generation Units, IEEE Transactions on Power Systems, 2006; 21 (4): 1821-1831.

[KaK07] Kaynakli O, Kilic M. Theoretical study on the effect of operating conditions on performance of absorption refrigeration system. Energy Conversion and Management, 2007; 48: 599-607.

[KeP08] Keil C, Plura S, Radspieler M, Schweigler C. Application of customized absorption heat pumps for utilization of low-grade heat sources. Applied Thermal Engineering, 2008; 28 (16): 2070-2076.

[KhH05] Khan FI, Hawboldt K, Iqbal MT. Life Cycle Analysis of wind–fuel cell integrated system. Renewable Energy, 2005; 30 (2): 157-177.

[KhR04] Khan KH, Rasul MG, Khan MMK. Energy conservation in buildings: cogeneration and cogeneration coupled with thermal-energy storage. Applied Energy 2004; 77: 15-34.

[KiI08] Kim DS, Infante Ferreira CA. Analytic modelling of steady state single-effect absorption cycles. International Journal of Refrigeration, 2008; 31: 1012-1020.

[KiK07] Kilic M, Kaynakli O. Second law-based thermodynamic analysis of water-lithium bromide absorption refrigeration system. Energy, 2007; 32: 1505-1512.

[Kil99] Kilkis IB. Utilization of wind energy in space heating and cooling with hybrid HVAC systems and heat pumps. Energy and Buildings, 1999; 30: 147-153.

[KiN81] Kirkham H, Nightingale D, Koerner T. Energy management system design with dispersed storage and generation, IEEE Transactions on Power Apparatus and Systems, 1981; 100, (7): 3432-3440.

[KiR00] Kim TS, Ro ST. Power augmentation of combined cycle power plants using cold energy of liquefied natural gas. Energy, 2000; 25: 841-856.

[Kir03] Kirschen DS. Demand-side view of electricity markets. IEEE Transactions on Power Systems, 2003; 18 (2): 520-527.

[KiZ02] Kim JS, Ziegler F, Lee H. Simulation of th compressor-assisted triple-effect H2O/LiBr absorption cooling cycles. Applied Thermal Engineering, 2002; 22: 295-308.

[KoH06] Korpås M, Holen AT. Operation planning of hydrogen storage connected to wind power operating in a power market. IEEE Transactions on Energy Conversion, 2006; 21 (3): 742-749.

[KoK07] Kondili EM, Kaldellis JK. Biofuel implementation in East Europe: Current status and future prospects. Renewable and Sustainable Energy Reviews, 2007; 11 (9): 2137-2151.

[KoK08] Koeppel G, Korpås M. Improving the network infeed accuracy of non-dispatchable generators with energy storage devices. Electric Power Systems Research, 2008; 78 (12): 2024-2036.

[KuK03] Kueck JD, Kirby BJ. The Distribution System of the Future. The Electricity Journal, 2003; 16 (5): 78-87.

[KoW04] Kong XQ, Wang RZ, Huang XH. Energy efficiency and economic feasibility of CCHP driven by Stirling engine. Energy Conversion and Management, 2004; 45: 1433-1442.

[KoW05] Kong X.Q, Wang R.Z, Wu J.Y, Huang X.H, Huangfu Y, Wu D.W, Xu Y.X. Experimental investigation of a micro-combined cooling, heating and power system driven by a gas engine. International Journal of Refrigeration, 2005; 28 (7): 977-987.

[KoW07] Kong XQ, Wang RZ, Huang XH. Energy optimization model for a CCHP system with available gas turbines. Applied Thermal Engineering, 2007; 25: 377-391.

[Kre01] Kreider JF (editor). Handbook of Heating, Ventilation and Air Conditioning, CRC Press, Boca Raton, FL, 2001.

[KrG07] Kreith F, Goswami DY (editors). Handbook of energy efficiency and renewable energy. CRC Press, Boca Raton, USA, 2007.

[KuK08] Kuhn V, Klemes J, Bulatov I. MicroCHP: Overview of selected technologies, products and field test results. Applied Thermal Engineering, 2008; 28: 2039-2048.

[LaA02] Lasseter RH, Akhil A, Marnay C, Stephens J, Dagle J, Guttromson R, Meliopoulous A, Ginger R, Eto J. The CERTS Microgrid Concept. White paper on Integration of Distributed Energy Resources, prepared for Transmission Reliability Program, Office of Power Technologies, U.S. Department of Energy, April 2002. Available (September 2008): http://eetd.lbl.gov/CERTS/pdf/50829-app.pdf.

[LaC07] Lazzarin R, Castellotti F. A new heat pump desiccant dehumidifier for supermarket application. Energy and Buildings, 2007; 39: 59-65.

[LaN06] Lazzarin R, Noro M. Local or district heating by natural gas: Which is better from energetic, environmental and economic point of views? Applied Thermal Engineering, 2006; 26: 244-250.

[LaN06b] Lazzarin R, Noro M. District heating and gas engine heat pump: Economic analysis based on a case study. Applied Thermal Engineering, 2006; 26: 193-199.

[Lev06] Levy R. A Vision of Demand Response – 2016. The Electricity Journal, 2006; 19, (8): 12-23.

[LiC01] Limmeechokchai B, Chungpaibulpatana S. Application of cool storage air-conditioning in the commercial sector: an integrated resource planning approach for power capacity expansion planning and emission reduction. Applied Energy 2001; 68: 289-300.

[LiF06] Li H, Fu L, Geng K, Jiang Y. Energy utilization evaluation of CCHP systems. Energy and buildings, 2006; 38: 253-257.

[LiG04] Liu XH, Geng KC, Lin BR, Jiang Y. Combined cogeneration and liquid-desiccant system applied in a demonstration building. Energy and Buildings, 2004; 36: 945-953.

[LiN06] Li H, Nalim R, Haldi PA. Thermal-economic optimization of a distributed multi-generation energy system – A case study of Beijing. Applied Thermal Engineering, 2006; 26 (7): 709-719.

[LiR07] Liao X, Radermacher R. Absorption chiller crystallization control strategies for integrated cooling heating and power systems. International Journal of Refrigeration, 2007; 30 (5): 904-911.

[LiW01] Lin F, Yi J, Weixing Y, Xuzhong Q. Influence of supply and return water temperatures on the energy consumption of a district cooling system. Applied Thermal Engineering, 2001; 21 (4): 511-521.

[LiW06] Lien D, Wang Y. Cross-hedging with futures and options: the effects of disappointment aversion. Journal of Multinational Financial Management, 2006; 16: 16–26.

[LiW07] Lin L, Wang Y, Al-Shemmeri T, Ruxton T, Turner S, Zeng S, Huang J, He Y, Huang X. An experimental investigation of a household size trigeneration. Applied Thermal Engineering, 2007; 27: 576-585.

[LiZ05] Li S, Zhang W, Zhang R, Lv D, Huang Z. Cascade fuzzy control for gas engine driven heat pump, Energy Conversion and Management, 2005; 46: 1757-1766.

[LiZ07] Lior N, Zhang N. Energy, exergy, and second law performance criteria. Energy, 2007; 32: 281-296.

[LoC02] Lozza G, Campanari S. Strategie di utilizzo delle microturbine a gas, con eventuale impiego di pompe di calore, per la cogenerazione nel settore civile (Strategies of utilization of gas microturbines, with possibile use of heat pumps, for cogeneration applications for buildings, in Italian). Giornata di studio sulle fonti rinnovabili e la cogenerazione nel futuro mercato dell'energia, paper C5, 22 January 2002, Roma, Italy.

[LoG05] Longo GA, Gasparella A, Claudio Z. Analysis of absorption machine driven by the heat recovery of an I.C. reciprocating engine. International Journal of Energy Research, 2005; 29: 711-722.

[Loz00] Lozza G. Turbine a gas e cicli combinati (Gas turbines and combined cycles, in Italian). Progetto Leonardo, Esculapio, Bologna, Italy, 2000.

[LuB03] Lunghi P, Bove R, Desideri U. Analysis and optimization of hybrid MCFC gas turbine plants. Journal of Power Sources, 2003; 118: 108-117.

[Luc00] Lucas K. On the thermodynamics of cogeneration. International Journal of Thermal Sciences, 2000; 39: 1039-1046.

[Luc01] Lucas K. Efficient energy systems on the basis of cogeneration and heat pump technology. International Journal of Thermal Sciences, 2001; 40: 338-343.

[LuM06] Lund H, Munster E. Integrated energy systems and local energy markets. Energy Policy, 2006; 34: 1152-1160.

[Lun05] Lund H. Large-scale integration of wind power into different energy systems. Energy, 2005; 30: 2402-2412.

[Lun08] Lund P. Analysis of advanced energy chain typologies. International Journal of Energy Research, 2008; 32: 144-153.

[LuU01] Lunghi P, Ubertini S, Desideri U. Highly efficient electricity generation through a hybrid molten carbonate fuel cell-closed loop gas turbine plant. Energy Conversion and Management, 2001; 42: 1657-1672.

[MaB02] Mavroudaki P, Beggs CB, Sleigh PA, Halliday SP. The potential for solar powered single-stage desiccant cooling in southern Europe. Applied Thermal Engineering, 2002; 22: 1129-1140.

[MaC01] Macchi E, Campanari S. Potential development for gas microturbines. Hybrid cycles and trigeneration, Cogeneration and On-Site Power Production, 2001; 2 (2): 41-49.

[MaC02] Macchi E, Chiesa P, Scenario studies on electricity generation in Italy in the next 20 years, CESI Report n. 71/00298, 31/03/2002 (in Italian: Studi di scenario sulla generazione elettrica in Italia nei prossimi 20 anni). Available (September 2008): www.ricercadisistema.it.

[MaC03] Markvart T, Castaner L, Practical handbook of photovoltaics: fundamentals and applications. Elsevier Science, Oxford, UK, 2003.

[MaC08] Mancarella P, Chicco G. Assessment of the greenhouse gas emissions from cogeneration and trigeneration systems. Part II: Analysis techniques and application cases. Energy, 2008; 33 (3): 418-430.

[MaC08b]Mancarella P, Chicco G. Energy and CO_2 emission assessment of cooling generation alternatives: A comprehensive approach based on black-box models. WSEAS Transactions on Power Systems, 2008; 3 (4): 151-161.

[MaC09] Mancarella P, Chicco G. Global and local emission impact assessment of distributed cogeneration systems with partial-load models. Applied Energy, 2009; 86 (10): 2096-2106.

[MaD02] Masoum MAS, Dehbonei H, Fuchs EF. Theoretical and experimental analyses of photovoltaic systems with voltage- and current-based maximum power-point tracking. IEEE Transactions on Energy Conversion, 2002; 17 (4): 514-522.

[MaF07] Maribu KM, Firestone R, Marnay C, Siddiqui AS. Distributed energy resources market diffusion model. Energy Policy, 2007; 35 (9): 4471-4484.

[MaH96] Matsuhashi R, Hikita K, Ishitani H. Model analyses for sustainable energy supply taking resource and environmental constraints into considerations. Energy Conversion and Management, 1996; 37: 1253-1258.

[Man06] Mancarella P. From cogeneration to trigeneration: energy planning and evaluation in a competitive market framework. Doctoral thesis, Politecnico di Torino, Torino, Italy, April 2006.

[Man09] Mancarella P. Cogeneration systems with electric heat pumps: energy-shifting properties and equivalent plant modelling. Energy Conversion and Management 2009, in press, doi:10.1016/j.enconman.2009.04.010.

[MaP00] Maidment GG, Prosser G. The use of CHP and absorption cooling in cold storage. Applied Thermal Engineering, 2000; 20: 1059-1073.

[Mar98] Martens A. The energetic feasibility of CHP compared to the separate production of heat and power, Applied Thermal Engineering, 1998; 18: 935-946.

[Mar04] Marcic M. Long-term performance of central heat pumps in Slovenian homes. Energy and Buildings, 2004; 36: 185-193.

[MaR05] Makarov YV, Reshetov VI, Stroev VA, Voropai NI. Blackout prevention in the United States, Europe and Russia. Proceedings of the IEEE, 2005; 93 (11): 1942-1955.

[MaS07] Mäenpää I, Siikavirta H. Greenhouse gases embodied in the international trade and final consumption of Finland: An input–output analysis. Energy Policy, 2007; 35 (1): 128-143.

[MaT02] Maidment GG, Tozer RM. Combined cooling heat and power in supermarkets, Applied Thermal Engineering, 2002; 22: 653-665.

[MaV04] Macken KJP, Vanthournout K, Van den Keybus J, Deconinck G, Belmans RJM. Distributed control of renewable generation units with integrated active filter. IEEE Transactions on Power Electronics, 2004; 19 (5): 1353-1360.

[McD00] McDonald CF. Low cost recuperator concept for microturbine applications, ASME paper 2000-GT-166.

[McE06] McDowall W, Eames M. Forecasts, scenarios, visions, backcasts and roadmaps to the hydrogen economy: A review of the hydrogen futures literature. Energy Policy, 2006; 34: 1236-1250.

[MeB08] Medrano M, Brouwer J, McDonell V, Mauzey J, Samuelsen S. Integration of distributed generation systems into generic types of commercial buildings in California. Energy & Buildings, 2008; 40 (4): 537-548.

[MeC08] Medrano M, Castell A, Fontanals G, Castellón C, Cabeza L.F. Economics and climate change emissions analysis of a bioclimatic institutional building with trigeneration and solar support. Applied Thermal Engineering, 2008; 28 (17-18): 2227-2235.

[Meu02] Meunier F, Co- and tri-generation contribution to climate change control, Applied Thermal Engineering, 2002; 22: 703-718.

[MiK07] Mittelman G, Kribus A, Dayan A. Solar cooling with concentrating photovoltaic/thermal (CPVT) systems. Energy Conversion and Management, 2007; 48 (9): 2481-2490.

[MiL03] Minciuc E, Le Corre O, Athanasovici V, Tazerout M, Bitir I. Thermodynamic analysis of trigeneration with absorbtion chilling machine. Applied Thermal Engineering, 2003; 23 (11): 1391-1405.

[MiL03b] Minciuc E, Le Corre O, Athanasovici V, Tazerout M. Fuel savings and CO_2 emissions for tri-generation systems, Applied Thermal Engineering, 2003; 23: 1333-1346.

[MiM04] Míguez J.L, Murillo S, Porteiro J, López L.M. Feasibility of a new domestic CHP trigeneration with heat pump: I. Design and development. Applied Thermal Engineering, 2004; 24 (10): 1409-1419.

[MoA98] M.Mostafavi, A.Alaktiwi, and B.Agnew, Thermodynamic analysis of combined opencycle-twin-shaft gas turbine (Brayton cycle) and exhaust gas operated absorption refrigeration unit, Applied Thermal Engineering, 1998; 18: 847-856.

[MoC01] Moné CD, Chau DS, Phelan PE. Economic feasibility of combined heat and power and absorption refrigeration with commercially available gas turbines. Energy Conversion and Management, 2001; 42: 1558-1573.

[MoM07] Morais MS, Marangon Lima JW. Combined natural gas and electricity network pricing. Electric Power Systems Research, 2007; 77 (5-6): 712-719.

[MoR07] Moreira CL, Resende FO, Peças Lopes JA. Using low voltage Microgrids for service restoration. IEEE Transactions on Power Systems, 2007; 22 (2): 395-403.

[MoR74] Molina MJ, Rowland FS. Stratospheric sink of chlorofluoromethanes: chlorine atoms catalyzed destruction of ozone. Nature, 1974; 249: 810-812.

[MTJ06] Report on the experimental performance of a 60 kWe microturbine available on the Japanese market. Private communication, January 2006.

[MuO07] Muñoz M, Oschmann V, David Tàbara J. Harmonization of renewable electricity feed-in laws in the European Union. Energy Policy, 2007; 35 (5): 3104-3114.

[NaS07] Narayan PK, Smyth R, Prasad A. Electricity consumption in G7 countries: A panel cointegration analysis of residential demand elasticities. Energy Policy, 2007; 35 (9): 4485-4494.

[NeD06] Nesreddine H, Dostie M, Millette J, Poulin A, Handfield L. Economic assessment of microturbine implementation in commercial and institutional buildings. WSEAS Transactions on Power Systems, 2006; 1: 1563-1572.

[Nor03] Noro M. Celle a combustibile. Tecnologia e possibilità applicative (Fuel cells. Technology and application opportunities, in Italian). Flaccovio Editore, Palermo, Italy, 2003.

[Oba06] Obara S. Load response of a fuel cell micro-grid with control of number of units. International Journal of Hydrogen Energy, 2006; 31: 1819-1830.

[OhL07] Oh SD, Lee HJ, Jung JY, Kwak HY. Optimal planning and economic evaluation of cogeneration system. Energy, 2007; 32: 760-771.

[Ome08] Omer AM. Ground-source heat pumps systems and applications. Renewable and Sustainable Energy Reviews, 2008; 12: 344–371.

[OnU06] Onovwiona HI, Ugursal VI. Residential cogeneration systems: review of the current technology. Renewable and Sustainable Energy Reviews, 2006; 10 (5): 389-431.

[OsV97] Ossebaard, ME; Van Wijk, AJM; Van Wees, MT. Heat supply in the Netherlands: a systems analysis of costs, exergy efficiency, CO2 and NOx emissions. Energy, 1997; 22: 1087-1098.

[PaK08] Park CW, Koo J, Kang YT. Performance analysis of ammonia absorption GAX cycle for combined cooling and hot water supply modes, International Journal of Refrigeration, 2008; 31: 727-733.

[Pal04] Palmer J. Rating scales for seasonal efficiency of boilers and chillers, Partners in innovation Report 39-3-720, Deliverable 7, Faber Maunsell, 28th July 2004. Available (September 2008): www.fabermaunsell.com.

[PeD05] Pepermans G, Driesen J, Haeseldonckx D, Belmans R, D'haeseleer W. Distributed generation: definition, benefits and issues. Energy Policy, 2007; 33 (6): 787-798.

[Peh01] Pehnt M. Life-cycle assessment of fuel cell stacks. International Journal of Hydrogen Energy, 2001; 26: 91-101.

[Peh06] Pehnt M. Dynamic life cycle assessment (LCA) of renewable energy technologies. Renewable Energy, 2006; 31 (1): 55-71.

[Peh06b] Pehnt M, Cames M, Fischer C, Praetorius B, Schneider L, Schmacher K, Voß J-P. Micro Cogeneration. Towards decentralized energy systems. Springer-Verlag, Berlin Heidelberg, 2006.

[PeH07] Peças Lopes JA, Hatziargyriou N, Mutale J, Djapic P, Jenkins, N. Integrating distributed generation into electric power systems: A review of drivers, challenges and opportunities. Electric Power Systems Research, 2007; 77 (9): 1189-1203.

[PeL04] Perego O, Liviero DU, Arosio S, Colombo L. Experimental characterization of the CESI microturbine cogeneration plant; operational simulation of a similar plant for residential users; economic evaluations (in Italian: Caratterizzazione sperimentale dell'impianto cogenerativo a microturbina situato al CESI; simulazione d'esercizio di un impianto analogo per utenze residenziali; valutazioni economiche). CESI Report, 2004. Available (September 2008): www.ricercadisistema.it.

[PeL07] Perpiñan O, Lorenzo E, Castro A. On the calculation of energy produced by a PV grid-connected system, Progress in Photovoltaics: Research and Applications, 2007; 15: 265-274.

[Pen06] Penner SS. Steps toward the hydrogen economy. Energy, 2006; 31: 33-43.

[PeP04] Pennington DW, Potting J, Finnveden G, Lindeijer E, Jolliet O, Rydberg T, Rebitzer G. Life cycle assessment. Part 2: Current impact assessment practice. Environment International, 2004; 30: 721-739.

[Per05] Perego O. Impianti di cogenerazione e trigenerazione con microturbine a gas (Cogeneration and trigeneration plants with gas microturbines, in Italian). Seminario AEIT su Generazione Distribuita ed Energie Rinnovabili, Torino, 5 April 2005.

[PiC07] Piacentino A, Cardona F. On thermoeconomics of energy systems at variable load conditions: Integrated optimization of plant design and operation. Energy Conversion and Management, 2007; 48 (8): 2341–2355.

[PiC08] Piacentino A, Cardona F. An original multi-objective criterion for the design of small-scale polygeneration systems based on realistic operating conditions. Applied Thermal Engineering, 2008; 28 (17-18): 2391-2404.

[Pil02] Pilavachi PA. Mini- and micro-gas turbines for combined heat and power. Applied Thermal Engineering, 2002; 22: 2003-2014.

[PiW05] Pippattanasomporn M, Willingham M, Rahman S. Implications of on-site distributed generation for commercial/industrial facilities. IEEE Transactions on Power Systems, 2005; 20 (1): 206-212.

[PoK06] Pourbeik P, Kundur PS, Taylor CW. The anatomy of a power grid blackout. IEEE Power & Energy Magazine, 2006; 4 (5): 22-29.

[PoM82] Porter RW, Mastanaiah K. Thermal-economic analysis of heat-matched industrial cogeneration systems. Energy, 1982; 7 (2): 171-187.

[PoM04] Porteiro J, Míguez J.L, Murillo S, López L.M. Feasibility of a new domestic CHP trigeneration with heat pump: II. Availability analysis. Applied Thermal Engineering, 2004; 24 (10): 1421-1429.

[PrD07] Prodanovic M, De Brabandere K, Van Den Keybus J, Green T, Driesen J. Harmonic and reactive power compensation as ancillary services in inverter-based distributed generation. IET Generation, Transmission & Distribution, 2007; 1 (3): 432-438.

[PuR07] Pudjianto D, Ramsay C, Strbac G. Virtual power plant and system integration of distributed energy resources. IET Renewable Power Generation, 2007; 1 (1): 10-16.

[RaR06] Raineri R, Ríos S, Schiele D. Technical and economic aspects of ancillary services markets in the electric power industry: an international comparison. Energy Policy, 2006; 34 (13): 1540-1555.

[RDC03] Resource Dynamics Corporation, Cooling, Heating, and Power for Industry: A Market Assessment, Final Report prepared for U.S. Department of Energy and Oak Ridge National Laboratory, Vienna, VA, August 2003. Available (September 2008): www.cogeneration.net.

[Reb04] Rebitzer G, Ekvall T, Frischknecht R, Hunkeler D, Norris G, Rydberg T, Schmidt W-P, Suh S, Weidema BP, Pennington DW. Life cycle assessment. Part 1: Framework, goal and scope definition, inventory analysis, and applications. Environment International, 2004; 30: 701-720.

[ReK07a] Rebours YG, Kirschen DS, Trotignon M, Rossignol S. A Survey of Frequency and Voltage Control Ancillary Services—Part I: Technical Features. IEEE Transactions on Power Systems, 2007; 22 (1): 350-357.

[ReK07b] Rebours YG, Kirschen DS, Trotignon M, Rossignol S. A Survey of Frequency and Voltage Control Ancillary Services—Part II: Economic Features. IEEE Transactions on Power Systems, 2007; 22 (1): 358-366.

[RiD06] Riva A, D'Angelosante S, Trebeschi A. Natural gas and the environment results of life cycle assessment. Energy, 2006; 31: 138-148.

[RiJ01] Ribeiro PF, Johnson BK, Crow ML, Arsoy A, Liu Y. Energy storage systems for advanced power applications. Proceedings of the IEEE, 2001; 89 (12): 1744-1756.

[RoL05] Rosen MA, Le MN, Dincer I. Efficiency analysis of a cogeneration and district energy system. Applied Thermal Engineering, 2005; 25: 147-159.

[RoL05b] Rong A, Lahdelma R. An efficient linear programming model and optimization algorithm for trigeneration. Applied Energy, 2005; 82: 40-63.

[RoL07] Rong A, Lahdelma R. CO2 emissions trading planning in combined heat and power production via multi-period stochastic optimization. European Journal of Operational Research, 2007; 176 (3): 1874-1895.

[RoL08] Rong A, Lahdelma R, Luh PB. Lagrangian relaxation based algorithm for trigeneration planning with storages. European Journal of Operational Research, 2008; 188 (1): 240-257.

[Ros99] Rosen MA. Second-law analyses: approaches and implications. International Journal of Energy Research, 1999; 23: 415-429.

[Ros08] Rosen MA. Allocating carbon dioxide emissions from cogeneration systems: descriptions of selected output-based methods. Journal of cleaner Production, 2008; 16: 171-177.

[SaH07] Sartori I, Hestnes AG. Energy use in the life cycle of conventional and low-energy buildings: A review article. Energy & Buildings, 2007; 39 (3): 249-257.

[Sar07] Sareygh MA. The solar contribution to air conditioning systems for residential buildings. Desalination 2007; 209: 171-176.

[SaS92] C.E.Salas and M.Salas, Guide to refrigeration CFCs, Fairmont Press, Liburn, GA, 1992.

[Sch07] Schaumann G. The efficiency of the rational use of energy. Applied Energy, 2007; 84 (7-8): 719-728.

[Sci05] Sciubba E. Exergo-economics: thermodynamic foundation for a more rational resurce use. International Journal of Energy Research, 2005; 29: 613-636.

[SeG07] Sezgen O, Goldman CA, Krishnarao P. Option value of electricity demand response. Energy, 2007; 32, (2): 108-119.

[SeP06] Selimovic A, Palsson J. Networked solid oxide fuel cell stacks combined with a gas turbine cycle. Journal of Power Sources, 2006; 106 (1-2): 76-82.

[ShF05] Shahidehpour M, Fu Y, Wiedman T. Impact of natural gas infrastructures on electric power systems. Proceedings of the IEEE, 2005; 93 (5): 1042-1056.

[ShN02] Shivakumar K, Narasimhan S. A robust and efficient NLP formulation using graph theoretic principles for synthesis of heat exchanger networks. Computers and Chemical Engineering, 2002; 26 (11): 1517–32.

[ShR05] Shrestha RM, Ram M, Marpaung COP, Charles OP. Supply- and demand-side effects of power sector planning with demand-side management options and SO2 emission constraints. Energy Policy, 2005; 33 (6): 815-825.

[SiE95] Sick F, Erge T. Photovoltaics in Buildings: a Design Handbook for Architects and Engineers. James & James, London, UK, 1995.

[SiL01] Silveira JL, Leal EM, Ragonha Jr LF. Analysis of a molten carbonate fuel cell: cogeneration to produce electricity and cold water. Energy, 2001; 26: 891-904.

[SmGww] ETP_SmartGrid. Technology Platform for the Electricity Networks of the Future. Web site (available September 2008) http://ec.europa.eu/research/energy/nn/nn_rt/nn_rt_dg/article_2262_en.htm.

[SöP06] Söderman J, Pettersson F. Structural and operational optimisation of distributed energy systems. Applied Thermal Engineering, 2006; 26: 1400-1408.

[Soz01] Sozen A. Effect of heat exchangers on performance of absorption refrigeration systems. Energy Conversion and Management, 2001; 42: 1699-1716.

[StF06] Strachan, N, Farrell, A. Emission from distributed vs. centralized generation: The importance of system performance. Energy Policy, 2006; 34: 2677-2689.

[SuG06] Sun ZG, Guo KH. Cooling performance and energy saving of a compression-absorption refrigeration systems driven by a gas engine. International Journal of Energy Research, 2006; 30: 1109-116.

[Sun07] Sun ZG. A combined heat and cold system driven by a gas industrial engine. Energy Conversion and Management, 2007; 48: 366-369.

[SuW04] Sun ZC, Wang RZ, Sun WZ. Energetic efficiency of a gas-engine-driven cooling and heating system. Applied Thermal Engineering, 2004; 24: 941-947.

[SwM06] Swisher JN, McAlpin MC. Environmental impact of electricity deregulation. Energy, 2006; 31 (6-7): 1067-1083.

[TaC07] Tassou S.A, Chaer I, Sugiartha N, Ge Y.-T, Marriott D. Application of tri-generation systems to the food retail industry. Energy Conversion and Management, 2007; 48 (11): 2988-2995.

[TeB04] Temir G, Bilge D. Thermoeconomic analysis of a trigeneration system. Applied Thermal Engineering, 2004; 24 (17-18): 2689-2699.

[Tho08] Thomas B, Benchmark testing of Micro-CHP units. Applied Thermal Engineering, 2008; 28: 2049-2054.

[ToJ98] Tozer R, James RW. Heat powered refrigeration cycles. Applied Thermal Engineering, 1998; 18: 731-743.

[ToS01] Tolmasquim MT, Szklo A, Soares JB. Economic potential of natural gas fired cogeneration plants at malls in Rio de Janeiro. Energy Conversion and Management, 2001; 42: 663-674.

[TRIww] Trigeneration technologies website: www.trigeneration.com.

[TrN02] Tripanagnostopoulos Y, Nousia T, Souliotis M, Yianoulis P. Hybrid photovoltaic/thermal solar systems. Solar Energy, 2002; 72: 217-234.

[TsH07] Tsikalakis AG, Hatziargyriou ND. Environmental benefits of distributed generation with and without emissions trading. Energy Policy, 2007; 35 (6): 3395-3409.

[UcS04] Uche J, Serra L, Sanz A. Integration of desalination with cold-heat-power production in the agro-food industry. Desalination, 2004; 166: 379-391.

[USC03] U.S.-Canada Power System Outage Task Force. Interim Report: Causes of the August 14th Blackout in the United States and Canada. November 2003. Available (September 2008): https://reports.energy.gov/814BlackoutReport.pdf.

[UzO07] Uzunoglu M, Onar O, El-Sharkh MY, Sisworahardjo NS, Rahman A, Alam, MS. Parallel operation characteristics of PEM fuel cell and microturbine power plants. Journal of Power Sources, 2007, 168 (2): 469-476.

[VaL97] Valero A, Lozano MA. An introduction to thermoeconomics. In [Boe97], Chapter 8.

[VaO07] Valero S, Ortiz M, Senabre C, Alvarez C, Franco FJG, Gabaldon A. Methods for customer and demand response policies selection in new electricity markets. IET Generation, Transmission & Distribution, 2007; 1, (1):104–110.

[VaP05] Valenciaga F, Puleston PF. Supervisor control for a stand-alone hybrid generation system using wind and photovoltaic energy. IEEE Transactions on Energy Conversion 2005; 20 (2): 398-405.

[VeB02] Velázquez N, Best R. Methodology for the energy analysis of an air cooled GAX absorption heat pump operated by natural gas and solar energy. Applied Thermal Engineering, 2002; 22: 1089-1103.

[VeC08] Verda V, Calì Quaglia M. Solid oxide fuel cell systems for distributed power generation and cogeneration. International Journal of Hydrogen Energy, 2008; 33: 2087-2096.

[Via00] Viadana R. Analysis of the electro-technologies for space heating and hot water service – Part I (in Italian: Esame delle elettrotecnologie per il riscaldamento e per la produzione di acqua calda sanitaria – Parte I). CESI report, 2000. Available (available September 2008): www.ricercadisistema.it.

[Via00b] Viadana R. Analysis of the electro-technologies for space heating and hot water service – Part II (in Italian: Esame delle elettrotecnologie per il riscaldamento e per la produzione di acqua calda sanitaria – Parte II). CESI report, 2000. Available (available September 2008): www.ricercadisistema.it.

[ViB07] Vidal A, Bruno JC, Best R, Coronas A. Performance characteristics and modeling of a micro gas turbine for their integration with thermally activated cooling technologies. International Journal of Energy Research, 2007; 31: 119-134.

[VoB00] Voorspools KR Brouwers EZ, D'haeseleer WD. Energy content and indirect greenhouse gas emissions embedded in 'emission-free' power plants: results for the Low Countries. Applied Energy, 2000; 67 (3): 307-330.

[VoC06] Vokas G, Christiandonis N, Skittides F. Hybrid photovoltaic-thermal systems for domestic heating and cooling – A theoretical approach. Solar Energy, 2006; 80: 607-615.

[VoD03] Voorspools KR, D'haeseleer WD. The impact of the implementation of cogeneration in a given energetic context. IEEE Transactions. on Energy Conversion, 2003; 18: 135-141.

[WaL07] Wang Y, Lior N. Fuel allocation in a combined steam-injected gas turbine and thermal seawater desalination system. Desalination, 2007; 214: 306-326.

[WaN07] Wang C, Nehrir MH. Fuel cells and load transients. IEEE Power & Energy Magazine, 2007; 5 (1): 58-63.

[Wan08] Wang R.Z. Efficient adsorption refrigerators integrated with heat pipes. Applied Thermal Engineering, 2008; 28 (4): 317-326.

[WaO08] Wang R.Z, Oliveira R.G. Adsorption refrigeration—An efficient way to make good use of waste heat and solar energy. Progress in Energy and Combustion Science, 2006; 32 (4): 424-458.

[Wei07] Weisser D. A guide to life-cycle greenhouse gas (GHG emissions from electric supply technologies. Energy, 2007; 32: 1543-1559.

[WeM06] Weber C, Maréchal F, Favrat D, Kraines S. Optimization of an SOFC-based decentralized polygeneration system for providing energy services in an office-building in Tōkyō. Applied Thermal Engineering, 2006; 26: (13): 1409-1419.

[WiH06] Wietschel M, Hasenauer U, de Groot A. Development of European hydrogen infrastructure scenarios — CO2 reduction potential and infrastructure investment. Energy Policy, 2006; 34: 1284-1298.

[WiS00] Willis HL, Scott WG. Distributed power generation: planning and evaluation, Dekker, New York, 2000.

[Win05] Winter CJ. Into the hydrogen economy-milestones. International Journal of Hydrogen Energy, 2005; 30: 681-685.

[Win05b] Winter CJ. Electricity, hydrogen- competitors, partners? International Journal of Hydrogen Energy, 2005; 30: 1371-1374.

[Wul99] Wulfinghoff DR. Energy Efficiency Manual. Energy Institute Press, Wheaton, MD, USA, 1999.

[WuW06] Wu DW, Wang RZ. Combined cooling, heating and power: a review. Progress in Energy and Combustion Science, 2006; 32: 459-495.

[YuC05] Yu FW, Chan KT. Experimental determination of the energy efficiency of an air-cooled chiller under part-load conditions. Energy, 2005; 30: 1747-1758.

[ZaP06] Zaltash A, Petrov AY, Rizy DT, Labinov SD, Vineyard EA, Linkous RL. Laboratory R&D on integrated energy systems (IES). Applied Thermal Engineering, 2006; 26: 18-35.

[ZhG07] Zhang HF, Ge XS, Ye H. Modeling of a space heating and cooling system with seasonal energy storage. Energy, 2007; 32 (1): 51-58.

[ZhL05] Zhang RR, Lu XS, Li SZ, Lin WS, Gu AZ. Analysis on the heating performance of a gas engine driven air to water heat pump based on a steady-state model, Energy Conversion and Management, 2005; 46: 1714-1730.

[ZhS03] Zhao Y, Shigang Z, Haibe Z. Optmization study of combined refrigeration cycles driven by an engine. Applied Energy, 2003; 76: 379-389.

[ZhT02] Zhu Y, Tomsovic K. Development of models for analyzing the load-following performance of microturbines and fuel cells. Electric Power System Research, 2002; 62: 1-11.

[Zie02] Ziegler F. State of the art in sorption heat pumping and cooling technologies. International Journal of Refrigeration, 2002; 25: 450-459.

[ZiP06] Ziher D, Poredos A. Economics of a trigeneration system in a hospital. Applied Thermal Engineering, 2006; 26 (7): 680-687.

[ZiR93] Ziegler F, Riesch P. Absorption cycles: a review with regard to energetic efficiency, Heat recovery systems & CHP, 1993; 13 (2): 147-159.

[ZoG06] Zoulias EI, Glockner R, Lymberopoulos N, Tsoutsos T, Vosseler I, Gavalda O, Mydske HJ, Taylor P. Integration of hydrogen energy technologies in stand-alone power systems analysis of the current potential for applications. Renewable and Sustainable Energy Reviews, 2006; 10 (5): 432-462.

[ZoS07] Zoka Y, Sugimoto A, Yorino N, Kawahara K, Kubokawa J. An economic evaluation for an autonomous independent network of distributed energy resources. Electric Power Systems Research, 2007; 77 (7): 831-838.

ABOUT THE AUTHORS

Pierluigi Mancarella graduated in Electrical Engineering (honors) at Politecnico di Torino (PdT), Torino, Italy, in 2002, and received the Ph.D. degree in Electrical Engineering from the PdT School in 2006, with a PhD thesis entitled "From cogeneration to trigeneration: energy planning and evaluation in a competitive market framework". In 2004 he was a visiting student at the Electric Power Engineering Department of the NTNU, Trondheim, Norway.

In 2006 and 2007 he worked as a Research Fellow in the Department of Electrical Engineering at PdT in the field of distributed multi-generation and energy externalities. Since 2007, he has presented energy efficiency and multi-generation lectures in various international courses, seminars and tutorials. In particular, in May 2007 he co-taught at the PdT the Doctoral course "Characterization and planning of small-scale multigeneration systems". In September 2007 he participated in the 2nd Manchester Seminar for Young Researchers in Power Systems. Since January 2008 he works as a Research Associate in the Department of Electrical and Electronic Engineering at Imperial College London, UK.

He is author or coauthor of more than 40 publications appeared in international journals and conference proceedings, and acts as a reviewer in several ISI international journals in the energy and power system fields. He is a member of IEEE, CIGRE NGN-UK, and AEIT. His research interests cover electrical distribution systems and distributed generation, distributed multi-generation systems and integrated energy networks, energy efficiency in power system applications, assessment of the environmental impact from energy generation, energy-related markets, and policy development.

E-mail: p.mancarella@imperial.ac.uk, pierluigi.mancarella@polito.it.

Gianfranco Chicco graduated in Electrotechnical Engineering (with honors and the "Giancarlo Vallauri" 1988 Award) at Politecnico di Torino (PdT), Torino, Italy, in 1987, and received the Ph.D. degree in Electrotechnical Engineering from the PdT School in 1992. In 1995 he joined PdT, where he is currently an Associate Professor of Electric Distribution Systems. In 2003 he was a member of the Administration Board of PdT. Since October 2007 he is the President of the University Academic Planning Council in Electrical Engineering at PdT. In 1999, he visited the Electrical and Computer Engineering Department of the University of Illinois at Urbana-Champaign, Urbana, IL.

He has been the scientific responsible of various projects in the distributed generation area, funded by public and private Institutions. He has been Chairman and Editor of the Proceedings of the VI World Energy System Conference (Torino, Italy, July 10-12, 2006), co-Chairman of the VII World Energy System Conference (Iaşi, Romania, June 30-July 2, 2008), and co-editor of the Proceedings of the International Conference Bulk Power System Dynamics and Control VI (Cortina d'Ampezzo, Italy, August 22-27, 2004). He has been reviewer and chairman of the scientific sessions in several international conferences. He is a reviewer of various ISI international journals in the power and energy fields.

He is author or coauthor of over 130 publications appeared in national and international journals and Conference Proceedings. He is Senior Member of the IEEE, Member of the AEIT and registered professional Engineer in the Province of Torino, Italy. His research activities refer to power system analysis, distribution system analysis, energy efficiency, load management, artificial intelligence applications to electrical systems, and power quality.

E-mail: gianfranco.chicco@polito.it.

INDEX